Biological Inorganic Chemistry

An Introduction

Biological Inorganic Chemistry
An Introduction

Robert R. Crichton

Unité de Biochimie
Université Catholique de Louvain
Louvain-La-Neuve
Belgium

With the collaboration of
Fréderic Lallemand, Ioanna S.M. Psalti and
Roberta J. Ward

ELSEVIER

Amsterdam • Boston • Heidelberg • London • New York • Oxford
Paris • San Diego • San Francisco • Singapore • Sydney • Tokyo

Elsevier
Radarweg 29, PO Box 211, 1000 AE Amsterdam, The Netherlands
The Boulevard, Langford Lane, Kidlington, Oxford OX5 1GB, UK

First edition 2008

Library of Congress Cataloging-in-Publication Data
A catalog record for this book is available from the Library of Congress

British Library Cataloguing in Publication Data
A catalogue record for this book is available from the British Library

ISBN: 978-0-444-52740-0

Cover illustration: © 2007, Ioanna SM PSALTI, DIME Creative Dimensions,
Oxford, OX4 4PE, UK. Reproduced by permission.

There are instances where we have been unable to trace or contact
the copyright holder. If notified the publisher will be pleased to rectify
any errors or omissions at the earliest opportunity.

For information on all Elsevier publications
visit our website at books.elsevier.com

Printed and bound in Italy

08 09 10 11 12   10 9 8 7 6 5 4 3 2 1

Preface

When one ponders on the question 'why did you decide to write this book?', there are at least two options. Firstly, that you felt that there was a need for a book of this type, and that, however pretentious it might sound, you were the right person to write it. Alternatively, you might argue that, having taught undergraduate and postgraduate courses on the subject, you could share your teaching with others who might appreciate the fruits of your experience in the area.

While there is an element of both of these in my decision to put 'pen to paper' (which really means word processor/cut and paste), I have had a third, and finally more powerful motivation, to write this introduction to biological inorganic chemistry. I have, since the beginning of my scientific career, been involved with metalloproteins. I cut my teeth on the haem-binding peptide of cytochrome c, which could be generated by Ag or Hb cleavage from the native protein (and which subsequently became famous/notorious as a 'miniperoxidase'). I then graduated to insect haemoglobins in the laboratory of Gerhard Braunitzer, at the Max-Planck-Institut für Biochemie in Munich, and upon my return to Glasgow, certainly influenced by the pioneering work of Hamish Munro on the regulation of the biosynthesis of the iron storage protein ferritin, to the field of iron metabolism. I have remained faithful to my favourite metal since then, as underlined by my organization of the Second International Meeting on Iron Metabolism in 1975. To quote my colleague Phil Aisen:

> 'The first conference (in July 1973 at University College Hospital Medical School, London) was sufficiently successful to provoke Bob Crichton to follow up with a second meeting two years later at Louvain-la-Neuve, where Bob was newly appointed as head of biochemistry. That meeting established the pattern for all its successors: meetings held on a biannual basis, organizers elected by conferees, partial funding sought and secured from outside agencies, a formal conference programme with informal discussions after each presentation, a conference banquet and suitable diversion to lighten the event'.

Over the years I have sought to continue the creation of relaxed atmospheres to facilitate scientific exchange. Examples are the seventeen advanced courses which Cees Veeger and I organized over the last twenty years, training more than 750 doctoral and postdoctoral students on the multidisciplinary approaches required to study metals in biology, and the recent COST[1] Chemistry Action D34 'Molecular Targeting and Drug Design in Neurological and Bacterial Diseases', of which I am chairman.

But enough of these reminiscences of the past. I owe an enormous debt of gratitude to my three collaborators without whom this book could not have been completed. Ioanna Psalti has not only rewritten my chapter on coordination chemistry, but also carried out two

---

[1] COST is one of the longest-running instruments supporting co-operation among scientists and researchers in 35 member countries across Europe and enables scientists to collaborate in a wide spectrum of activities in research and technology.

monumental tasks in proofreading the text for chemical incoherencies and in compiling the index, not to forget the absolutely brilliant cover which she has designed. Bobbie Ward has given me enormous help in the chapter on iron in brain as well as dealing with the problems of getting permission to reproduce the figures. Fréderic Lallemand has been there to sort out all of my computer crises (and there have been quite a lot), as well as drawing a lot of figures for the early chapters.

I also would like to thank a large number of colleagues, including Ernesto Carafoli, Bernard Mahieu, Brian Hoffmann, Peter Kroneck, Istvan Marko, Bill Rutherford and many others for their guidance in the scientific content of the text.

However, I remain responsible for errors or mistakes which have been perpetrated in what I hope will be the first of many editions of a book which is written with the clear and unequivocal objective to incite students coming from either a biology or chemistry background, not to forget those coming from medical or environmental formations, to develop their interest in the extremely important role that metals play in biology, in medicine, and in the environment.

Finally, I would like to dedicate this book to Antonio Xavier, not only in memory of his outstanding contributions to metals in biology, and to establishing the Society and the *Journal of Biological Inorganic Chemistry*, but for the outstanding personal qualities that made him a friend that one will not quickly forget.

Louvain-la-Neuve, 3rd October, 2007
Robert R. Crichton, FRSC

Contents

1   An Overview of Metals in Biology 1
Introduction 1
Why Do We Need Anything Other Than C, H, N and O (Together with Some P and S)? . . . 2
What are the Essential Metal Ions? 3
References 12

2   Basic Coordination Chemistry for Biologists 13
Introduction 13
    Ionic bonding 14
    Covalent bonding 14
Hard and Soft Ligands 15
    The chelate effect 16
Coordination Geometry 18
Crystal Field Theory and Ligand Field Theory 19
References 26

3   Biological Ligands for Metal Ions 27
Introduction 27
Protein Amino Acid Residues (and Derivatives) as Ligands 27
An Example of a Non-Protein Ligand: Carbonate and Phosphate 29
Engineering Metal Insertion into Organic Cofactors 30
Chelatase: Terminal Step in Tetrapyrrole Metallation 30
Iron–Sulfur Cluster Containing Proteins 32
Iron–Sulfur Cluster Formation 33
Copper Insertion into Superoxide Dismutase 35
More Complex Cofactors: MoCo, FeMoCo, P-Clusters, H-Clusters and CuZ 36
Siderophores 39
References 42

4   Structural and Molecular Biology for Chemists 43
Introduction 43
The Structural Building Blocks of Proteins 43
Primary, Secondary, Tertiary and Quaternary Structures of Proteins 47
    The structural building blocks of nucleic acids 55
Secondary and Tertiary Structures of Nucleic Acids 56
    Carbohydrates 59
    Lipids and biological membranes 64
    A brief overview of molecular biology 66
    Replication and transcription 67
    Translation 71
    Postscript 75
References 76

5 An Overview of Intermediary Metabolism and Bioenergetics 77
   Introduction 77
   Redox Reactions in Metabolism 78
   The Central Role of ATP in Metabolism 79
   The Types of Reaction Catalysed by Enzymes of Intermediary Metabolism 82
   An Overview of Intermediary Metabolism: Catabolism 86
   Selected Case Studies: Glycolysis and the Tricarboxylic Acid Cycle 88
   An Overview of Intermediary Metabolism: Anabolism 92
   Bioenergetics: Generation of Phosphoryl Transfer Potential at the Expense
      of Proton Gradients 97
   References 104

6 Methods to Study Metals in Biological Systems 105
   Introduction 105
   Magnetic Properties 107
   Electron Paramagnetic Resonance (EPR) Spectroscopy 108
   Mössbauer Spectroscopy 109
   NMR Spectroscopy 110
   Electronic and Vibrational Spectroscopies 112
   Circular Dichroism and Magnetic Circular Dichroism 113
   Resonance Raman Spectroscopy 114
   Extended X-Ray Absorption Fine Structure 115
   X-Ray Diffraction 115
   References 116

7 Metal Assimilation Pathways 117
   Introduction 117
   Metal Assimilation in Bacteria 117
      Iron 117
      Copper and zinc 120
   Metal Assimilation in Plants and Fungi 121
      Iron 121
      Copper and zinc 124
   Metal Assimilation in Mammals 126
      Iron 126
      Copper and zinc 127
   References 129

8 Transport, Storage and Homeostasis of Metal Ions 131
   Introduction 131
   Metal Storage and Homeostasis in Bacteria 131
      Iron 131
      Copper and zinc 135
   Metal Transport, Storage and Homeostasis in Plants and Fungi 136
      Iron storage and transport in fungi and plants 136
      Iron homeostasis in fungi and plants 137
      Copper transport and storage in fungi and plants 139

Copper homeostasis in fungi and plants. 142
Zinc transport and storage in fungi and plants 142
Zinc homeostasis in fungi and plants. 143
Metal Transport, Storage and Homeostasis in Mammals. 144
Iron transport and storage in mammals 144
Iron homeostasis in mammals 146
Copper and zinc transport and storage in mammals. 148
Copper and zinc homeostasis in mammals. 148
References 150

9   Sodium and Potassium—Channels and Pumps 151
Introduction: —Transport Across Membranes. 151
Sodium *Versus* Potassium 152
Potassium channels 153
Sodium Channels 155
The sodium-potassium ATPase 157
Active transport driven by Na^+ gradients. 158
Sodium/proton exchangers 159
Other roles of intracellular K^+. 161
References 163

10 Magnesium–Phosphate Metabolism and Photoreceptors 165
Introduction 165
Magnesium-Dependent Enzymes 166
Phosphoryl Group Transfer: Kinases. 167
Phosphoryl Group Transfer: Phosphatases 170
Stabilization of Enolate Anions: The Enolase Super Family 173
Enzymes of Nucleic Acid Metabolism 175
Magnesium and Photoreception 178
References 181

11 Calcium: Cellular Signalling 183
Introduction: —Comparison of Ca^{2+} and Mg^{2+} 183
The Discovery of a Role for Ca^{2+} Other Than as a Structural Component. 183
Plasma Membrane Uptake Pathways. 185
Calcium Export from Cells. 185
Ca^{2+} Transport Across Intracellular Membranes 188
Ca^{2+} and Cell Signalling. 191
References 195

12 Zinc: Lewis Acid and Gene Regulator. 197
Introduction 197
Mononuclear Zinc Enzymes 198
Carbonic Anhydrase 199
Carboxypeptidases and Thermolysins 200
Alcohol Dehydrogenases 202

Other Mononuclear Zinc Enzymes 203
Multinuclear and Cocatalytic Zinc Enzymes 205
Zinc Fingers – DNA- and RNA-Binding Motifs 208
References 210

13   Iron: Essential for Almost All Life 211
Introduction 211
Iron and Oxygen 212
The Biological Importance of Iron 214
Biological Functions of Iron-Containing Proteins 216
Haemoproteins 217
      Oxygen transport 217
      Activators of molecular oxygen 220
      Electron transport proteins 222
Iron–Sulfur Proteins 226
Other Iron-Containing Proteins 231
      Mononuclear non-haem iron enzymes 231
      Dinuclear non-haem iron enzymes 235
References 239

14   Copper: Coping with Dioxygen 241
Introduction 241
Blue Copper Proteins Involved in Electron Transport 242
Copper-Containing Enzymes in Oxygen Activation and Reduction 244
      Type 2 Copper Oxidases and Oxygenases 244
      Dinuclear Type 3 Copper Proteins 245
      Multi-Copper Oxidases 247
      Cytochrome c Oxidases 248
      Superoxide Dismutation in Health and Diseases 250
Copper Enzymes Involved with Other Low-Molecular Weight Substrates 251
Mars and Venus: The Role of Copper in Iron Metabolism 253
References 254

15   Nickel and Cobalt: Evolutionary Relics 257
Introduction: Comparison of Nickel and Cobalt 257
Nickel Enzymes 258
      Urease 258
      Ni–Fe–S Proteins 259
Methyl-Coenzyme M Reductase 263
Cobalamine and Cobalt Proteins 263
B_{12}-Dependent Isomerases 264
B_{12}-Dependent Methyltransferases 266
Non-Corrin Co-Containing Enzymes 268
References 269

16 **Manganese: Water Splitting, Oxygen Atom Donor** 271
    Introduction: Manganese Chemistry 271
    Mn^{2+} and Detoxification of Oxygen Free Radicals 272
    Non-Redox di-Mn Enzymes: Arginase 274
    Photosynthetic Oxidation of Water: Oxygen Evolution 276
    References 278

17 **Molybdenum, Tungsten, Vanadium and Chromium** 279
    Introduction 279
    Molybdenum and Tungsten 279
    Molybdenum Enzyme Families 282
    Tungsten Enzymes 285
    Nitrogenases 286
    Vanadium Biochemistry 291
    Vanadium Biology 292
    Chromium in Biology 294
    References 295

18 **Metals in Brain and Their Role in Various Neurodegenerative Diseases** 297
    Introduction: Metals in Brain 297
    Calcium 297
    Zinc 300
    Copper 301
    Disorders of Copper Metabolism: Wilson's and Menkes Diseases 301
    Aceruloplasminaemia 303
    Creutzfeldt–Jakob and Other Prion Diseases 303
    Iron 306
    Redox Metal Ions, Oxidative Stress and
        Neurodegenerative Diseases 308
        Parkinson's Disease, PD 311
        Alzheimer's Disease, AD 313
        Huntington's Disease 317
        Friedreich's Ataxia 319
    References 320

19 **Biomineralization** 321
    Introduction 321
    Iron Deposition in Ferritin 322
        Iron pathways into ferritin 323
        Iron oxidation at dinuclear centres 324
        Ferrihydrite nucleation sites 326
        Biomineralization 327
    Calcium-Based Biominerals: Calcium Carbonates
        in Ascidians and Molluscs 330

Biomineralization in Bone and Enamel Formation 333
The Organic Matrix, Mineral Phase and Bone Mineralization 334
References 336

20   Metals in Medicine and the Environment 339
Introduction 339
Metallotherapeutics with Lithium 340
Cisplatin: An Anti-Cancer Drug 341
Contrast Agents for Magnetic Resonance Imaging 344
Metals in the Environment 346
    Cadmium 346
    Aluminium 350
References 352

Index 353

– 1 –

An Overview of Metals in Biology

INTRODUCTION

The importance of metals in biology, the environment and medicine has become increasingly evident over the last 25 years. The movement of electrons in the electron-transfer pathways of photosynthetic organisms and in the respiratory chain of mitochondria, coupled to proton pumping to enable the synthesis of ATP, is carried out by iron- and copper-containing proteins (cytochromes, iron–sulfur proteins and plastocyanins). The water-splitting centre of green plants (photosystem II), which produces oxygen, is based on the sophisticated biological use of manganese chemistry. Metals such as cadmium, manganese and lead in our environment represent a serious health hazard. Cadmium is present in substantial amounts in tobacco leaves, so that cigarette smokers on a packet a day can easily double their cadmium intake. Yet, while many metals are toxic, many key drugs are metal based—examples are cisplatin and related anticancer drugs, and lithium carbonate, used in the treatment of manic depression. Paramagnetic metal complexes are widely used as contrast agents for magnetic resonance imaging (MRI). Numerous trace metals are also required to ensure human health; and while metal deficiencies are well known (for example inadequate dietary iron causes anaemia), it is evident that excessive levels of metals in the body can also be toxic.

It has been clear from the outset that the study of metals in biological systems can only be approached by a multidisciplinary approach, involving many branches of the physical and biological sciences. The study of the roles of metal ions in biological systems represents the exciting and rapidly growing interface between inorganic chemistry and the living world. It has been defined by chemists as bioinorganic chemistry, and by biochemists as inorganic biochemistry. From 1990 to 1997 the European Science Foundation funded a programme on the Chemistry of Metals in Biological Systems[1]. This resulted, in the course of what turned out to be monumentally important meeting held in the Tuscan town of San Miniato, in the

---

[1] The steering committee of this programme, which I joined in 1992, included Helmut Sigel (Basle, Switzerland) as chair, Ivano Bertini (Florence, Italy) who organized the San Miniato meeting; Sture Forsen (Lund, Sweden), Dave Garner (Manchester, UK), Carlos Gomez-Moreno (Zaragoza, Spain), Paco Gonzales-Vilchez (Seville, Spain), Imre Sovago (Debrecen, Hungary), Alfred Trautwein, Lübeck, Germany), Jens Ulstrup (Lyngby, Denmark), Cees Veeger (Wageningen, Holland), Raymond Weiss (Strasbourg, France) and Antonio Xavier (Oeiras, Portugal).

launching of important initiatives around the international consensus name 'Biological Inorganic Chemistry'. The outcome was the creation of the Society of Biological Inorganic Chemistry (SBIC) and the Journal of Biological Inorganic Chemistry (JBIC). These then joined the already existing International Congress of Biological Inorganic Chemistry (ICBIC) and European Congress of Biological Inorganic Chemistry (EUROBIC) to form a series of acronyms; all now use the stylized French word for a ballpoint pen 'bic' to designate the term biological inorganic chemistry. I use this definition in this book, but would like to indicate to the prospective reader that this text will deal to a much greater extent with the biochemical aspects of metals in living systems rather than with their inorganic chemistry.

WHY DO WE NEED ANYTHING OTHER THAN C, H, N AND O (TOGETHER WITH SOME P AND S)?

Organic is defined as 'designating the branch of chemistry dealing with carbon compounds', or 'designating any chemical compound containing carbon', although the interesting codicil is added, in the latter definition, that some of the simple compounds of carbon, such as carbon dioxide, are frequently classified as inorganic compounds. Of course, in the world of organic foodstuffs (grown with only animal or vegetable fertilizers) the word takes a broader connotation, signifying production from the detritus of living organisms. And, when we come to examine the biotope, we quickly perceive that carbon alone does not suffice for life. We also need oxygen, hydrogen, nitrogen, a non-negligible dose of phosphorus, as well as some sulfur.

But these elements alone do not enable life as we know it to exist, in its multiple and varied forms we need components of inorganic chemistry as well. If we were to ask for a definition of inorganic chemistry (previously defined in French as mineral chemistry), we would find ourselves confronted with a world that was not organic, nor of animal or vegetable origin— most inorganic compounds do not contain carbon, and are derived from mineral sources. Yet this inanimate chemistry, apparently with nothing to do with living systems, has a crucial role to play in our understanding of the biological world. So we can recognize that in the course of evolution, Nature has selected constituents not only from the organic world but also from the inorganic world to construct living organisms. Some of these inorganic elements, such as sodium and potassium, calcium and magnesium, are present in quite large concentrations, and tend to be known as 'bulk elements', on a scale with those cited in the first paragraph. Others, such as cobalt, copper, iron and zinc, are known as 'trace elements', with dietary requirements that are much lower than the bulk elements.

Indeed, the human body is made up of 99.9% of just 11 elements, 4 of which (hydrogen, oxygen, carbon and nitrogen) account for 99% of the total (62.8%, 25.4%, 9.4% and 1.4%, respectively). Why we require as many as 25 elements in total from the periodic table will become clearer as we advance in this chapter, but one thing shines out, namely that these elements have been selected on the basis of their suitability for the functions that they are called upon to play, in what is predominantly an aqueous environment[2].

---

[2] Another important distinction between organic chemistry and the chemistry of living organisms (biochemistry) is that the former is carried out almost entirely in non-aqueous media, whereas the latter occurs essentially in approximately 56 M H_2O.

Table 1.1

Correlations between ligand binding, mobility and function of some biologically relevant metal ions

Metal ion	Binding	Mobility	Function
Na^+, K^+	Weak	High	Charge carriers
Mg^{2+}, Ca^{2+}	Moderate	Semi-mobile	Triggers, transfers structural
Zn^{2+}	Moderate/strong	Intermediate	Lewis acid, structural
Fe, Cu, Mn, Mo[a]	Strong	Low	Redox catalysts, oxygen chemistry

[a]Charge not given, since this varies with oxidation state.

Na^+ and K^+ (together with H^+ and Cl^-), which bind weakly to organic ligands (Table 1.1), are ideally suited in generating ionic gradients across membranes and for the main-tenance of osmotic balance. In contrast, Mg^{2+} and Ca^{2+} with intermediate-binding strengths to organic ligands, can play important structural roles, and in the particular case of Ca^{2+}, serve as a charge carrier and a trigger for signal transmission. Zn^{2+} not only plays a structural role but can also fulfil a very important function as a Lewis acid. Redox metal ions, such as iron and copper, which bind tightly to organic ligands, participate in innu-merable redox reactions, besides playing an important role in oxygen transport. We now discuss the essential metal ions and thereafter briefly review their roles.

WHAT ARE THE ESSENTIAL METAL IONS?

If we look at the periodic table we can find around 25 elements that are required by most, if not all, biological systems. A somewhat idiosyncratic version of this is given in Figure 1.1 (an equally idiosyncratic version can be found in Levi, 1985).

Element number 1, hydrogen, is extremely important in biology. It can be incorporated into covalent bonds with many non-metals, such as carbon and nitrogen, notably by the action of light. It can be transferred in an important number of biological redox reactions involving one or two electron transfers, and it can participate in the generation of the pro-ton gradients across biological membranes, which are universally used for ATP synthesis.

Helium, like the other members of its family, is an inert gas often used in balloons on account of its low density, and when inhaled results in a comic transposition of the human voice to a significantly higher register (not a realistic way to mimic counter-tenors, but very effective in well-loved Walt Disney cartoon characters)!

Lithium, while not required for life, is used therapeutically in the form of lithium car-bonate for the treatment of manic depression; although its mechanism of action remains a mystery. Effective treatment requires attaining serum lithium concentrations of between 0.8 and 1.2 mmol/L.

Boron is an essential trace element for plants, and may well turn out to be essential for mammals as well. The boron-containing polyether–macrolide antibiotic, boromycin, was isolated as a potent anti-HIV agent.

The non-metals carbon, nitrogen and oxygen are all essential for man, as is element number 9, fluorine. Some of the biological effects of the important intracellular mes-senger, nitric oxide, NO, which is derived from the amino acid arginine, are illustrated in

Periodic table with annotations:

No.	Element	Status	Note
1	Hydrogen H	Essential	Hydrogen is incorporated into carbon compounds by light and can be transferred as H+, H- or H•
2	Helium He	Inert gas	Good for filling balloons and making people's voices high pitched
3	Lithium Li	Drugs	Li2CO3 Treatment of manic depression
4	Beryllium Be	Toxic	BeO, BeF2 Potent toxins
5	Boron B	Essential	Boromycin - antibiotic, anti-HIV drug
6	Carbon C	Essential	Carbohydrates
7	Nitrogen N	Essential	NO
8	Oxygen O	Essential	Redox chemistry
9	Fluorine F	Essential	5-fluorodeoxyuridylate - a suicide inhibitor of thymoxylate synthase
10	Neon Ne	Inert gas	In a vacuum discharge tube, neon glows reddish orange
11	Sodium Na	Essential	Chlorophylls
12	Magnesium Mg	Essential	Chlorophylls
13	Aluminium Al	Toxic	Acid rain
14	Silicon Si	Essential	Diatoms
15	Phosphorous P	Essential	ATP
16	Sulphur S	Essential	Fe-S cluster
17	Chlorine Cl	Essential	Chloride channels in cystic fibrosis
18	Argon Ar	Inert gas	Heavier than air - good in glove boxes
19	Potassium K	Essential	Potassium channel
20	Calcium Ca	Essential	Intracellular signalling - calmodulin
21	Scandium Sc	Non essential	Predicted to exist by Mendeleev. Discovered 1879 in Scandinavia
22	Titanium Ti	Non essential	Budotitane, an anti-tumour drug
23	Vanadium V	Essential	bromoperoxidase
24	Chromium Cr	Essential	Promotes glucose tolerance by an, as yet, unknown mechanism
25	Manganese Mn	Essential	Oxygen evolving complex e1 Photosystem II
26	Iron Fe	Essential	Haem
27	Cobalt Co	Essential	Vitamin B12
28	Nickel Ni	Essential	Active site of hydrogenase
29	Copper Cu	Essential	Japanese lacquered lunch boxes
30	Zinc Zn	Essential	Zinc finger in DNA binding proteins
31	Gallium Ga	Non essential	67 Ga (78% G emitter). Tumor imaging
32			
33	Arsenic As	Essential ?	Arsenite and arsenic trioxide (treatment of malignancies)
34	Selenium Se	Essential	Glutathione peroxidase
35	Bromine Br	Essential ?	Stratospheric ozone-depleting refrigerants and extinguishing agents
36			
42	Molybdenum Mo	Essential	FeMoCo in nitrogenase
48	Cadmium Cd	Non essential	Toxic ubiquitous environmental pollutant
50	Tin Sn	Essential ?	Triorganotin carboxylates. Bacteriocides and pesticides
53	Iodine I	Essential	Thyroid hormones, T3, T4
74	Tungsten W	Essential	Tungsten lamps
78	Platinum Pt	Non essential	Cis-platin anticancer drug
79	Gold Au	Non essential	Auranofin Therapy for rheumatoid arthritis
80	Mercury Hg	Non essential	The Mad Hatter
82	Lead Pb	Non essential	Pb binding site in phosphobilinogen synthase
64	Gadolinium Gd	Non essential	Gd-DTPA MRI contrast agent

Figure 1.1 Periodic table.

A

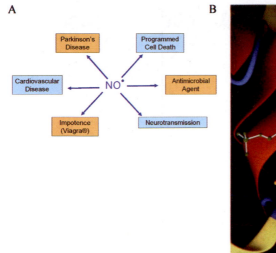

B

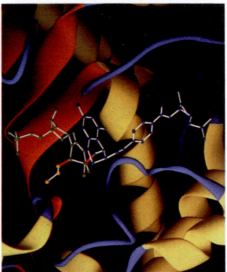

Figure 1.2 (a) Some biological effects of NO and (b) the structure of thymidylate synthase complexed with the suicide substrate 5-fluorodeoxyuridylate. (From Voet and Voet, 2004. Reproduced with permission from John Wiley & Sons., Inc.)

Figure 1.2. The addition of fluoride in drinking water to retard dental caries, particularly in children, has been criticized on the grounds of potential toxicity, but the concentrations used are many orders of magnitude below that which would be required to inhibit enzymes such as enolase in the glycolytic pathway. The key enzyme of DNA synthesis, thymidylate synthase, is inhibited by the anti-tumour drug 5-fluorodeoxythymidylate (Figure 1.2), a so-called 'suicide substrate', because it inhibits the enzyme only after undergoing part of its normal catalytic reaction. Neon[3], of course is an inert gas, but has the property of emitting light in a tube filled with the gas when an electric discharge is applied.

Sodium is involved in ionic gradients and in osmotic regulation, and, despite its much higher extracellular concentration, has to be kept out of many cells by the action of an energy-consuming Na^+/K^+ ATPase. The way in which biological systems manage to select the ions that are transported across membranes will be discussed in later chapters—Figure 1.3 illustrates the selective-binding sites for Na^+, K^+, Ca^{2+} and Cl^- in transport proteins.

Magnesium has its role intimately intertwined with phosphate: in many phosphoryl transfer reactions, as Mg-ATP in muscle contraction, in the stabilization of nucleic acid structures as well as in the catalytic activity of ribozymes (catalytic RNA molecules). It also serves as a structural component of enzymes, and is found as the metal centre in chlorophylls, which absorbs light energy in photosynthesis.

Aluminium, while extremely abundant in the earth's crust, is not used by living organisms: it is a notorious neurotoxin, but its involvement as a cause of Alzheimer's disease

---

[3] Neon is derived from the Greek 'neos', meaning 'new'.

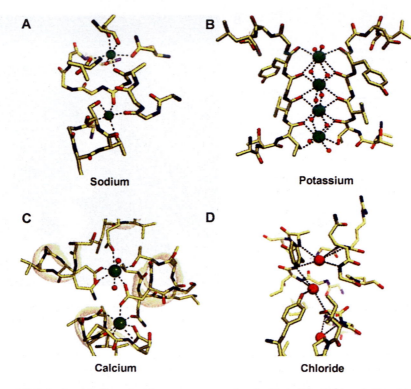

Figure 1.3 Selective-binding sites in transport proteins for Na^+, K^+, Ca^{2+} and Cl^-. (a) Two Na^+ binding sites in the LeuT Na^+-dependent pump. (b) Four K^+ binding sites in the KcsA K^+ channel. (c) Two Ca^{2+} binding sites in the Ca^{2+} ATPase pump. (d) Two central Cl^- binding sites in a mutant ClC Cl^-/H^+ exchanger. (From Gouax and MacKinnon, 2005. Copyright (2005) American Association for the Advancement of Science.)

seems less likely than was thought a few years ago. It is clear that acid rain, due to sulfur dioxide and nitrogen oxide emissions, increases the solubility and hence the bioavailability of aluminium. In the forests on the mountain slopes of Szklaska Poreba, on the Polish border with the Czech Republic, the pH values reached below 3, with disastrous effects on the tree population. Another effect of acid rain could have been to change the usual association of aluminium in the soil with silicate (predominant above pH 6.5) for phosphate, rendering aluminium more toxic.

This may be the reason why silicon is essential, namely that it keeps aluminium in a non-toxic form as aluminium silicate. While silicon is required as a trace element in most animals, in plants, particularly grasses, and in many unicellular organisms, such as diatoms[4], it is a major structural element. The importance of phosphorus and sulfur is obvious, the latter often associated with iron in an important family of proteins that contains iron–sulfur clusters.

---

[4] Diatoms are microscopic unicellular algae, with siliceous cell walls and the power of locomotion.

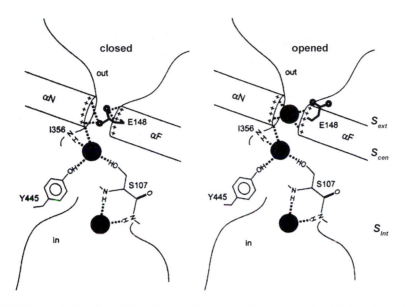

Figure 1.4 Schematic drawing of the closed and open conformations of a chloride channel. (From Dutzler et al., 2003. Copyright (2003) American Association for the Advancement of Science.)

Chlorine is another essential element in large part because, with all of the positively charged metal ions around, anions are obviously required for charge neutralization. One of the most common genetic disorders in man, cystic fibrosis (often referred to as mucoviscidosis, because of the viscous nature of bronchial secretions, resulting in frequent respiratory infections), is due to the production by epithelial cells that line the lungs, digestive tract, sweat glands and genitourinary system, of a defective form of a protein called cystic fibrosis transmembrane conductance regulator (CFTR), which is a chloride channel. Detailed structural analysis of closed and open conformations of bacterial chloride channels has shown that they can be closed by a glutamate residue, which replaces a third Cl^- ion on the extracellular site of the channel (Figure 1.4). In the closed conformation, the ion-binding sites S_{int} and S_{cen} are occupied by Cl^- ions and the ion-binding site S_{ext} is occupied by the side chain of Glu^{148}, whereas in the opened conformation Glu^{148} has moved out and the binding site is occupied by a third Cl^- anion.

Argon, an inert gas, has the useful property of being heavier than air, thus making it the ideal medium in which to work anaerobically (keep everything at the bottom of your argon-flushed glove box!).

Potassium, like sodium, is involved in ionic equilibria, and the opening and closing of sodium and potassium ion channels create the electrochemical gradients across cell membranes that transmit nerve impulses and other information and regulate cellular function.

Calcium, a crucial second messenger signalling key changes in cellular metabolism, is also important in muscle activation, in the activation of many proteases, both intra- and extracellular, and as a major component of an range of biominerals, including bone.

When Mendeleev first proposed his celebrated classification of the elements (1869), he found it necessary to leave a blank at the position now occupied by scandium. He did however predict some of its properties, and when it was discovered in Scandinavia a few years later (1879), the agreement of its properties with Mendeleev's predictions contributed greatly to the general scientific acceptance of the periodic table. Despite being very abundant in the earth's crust, titanium, the first of the otherwise biologically very important first row of the transition metals, is not essential. It has therapeutic potential in a number of anti-tumour drugs.

Vanadium is known to be essential, and is a constituent of some haloperoxidases as well as nitrogenases in some nitrogen-fixing organisms. It is particularly abundant in tunicates (a species of marine organisms) and in *Amanita* toadstools.

Chromium presents a big enigma: it appears to be essential for man, yet we do not have a clue as to what it might do—indeed most chromium compounds are toxic. The biologically relevant form, trivalent Cr^{3+}, is required for carbohydrate and lipid metabolism in mammals. Chromium has become extremely popular as a nutritional supplement, weight-loss and muscle-development agent, second only to Ca-containing products among mineral supplements[5].

Manganese is essential for man, although possibly its most significant contribution to biology is its incredible chemistry as a tetra-manganese centre in the splitting of water by photosystem II in plants, and originally at a much earlier point in geological time, in cyanobacteria. This reaction generates oxygen, which of course changed the whole pattern of the evolution of planet Earth. Perhaps this was the greatest pollution event in the history of our planet, which progressively moved us from an essentially reducing atmosphere to the oxidative world that we now know. There were, of course, advantages—respiration is almost 20 times more effective at producing ATP than fermentation. However, the oxygen paradox, as it has been termed, also resulted in the production of toxic reactive oxygen species, notably the hydroxyl radical through the well-known Fenton reaction. Another consequence of the appearance of dioxygen was that divalent copper became much more bioavailable, whereas trivalent iron underwent hydrolysis, polymerization and precipitation, making it much more difficult to extract.

The next five transition metals iron, cobalt, nickel, copper and zinc are of undisputed importance in the living world, as we know it. The multiple roles that iron can play will be presented in more detail later in Chapter 13, but we can already point out that, with very few exceptions, iron is essential for almost all living organisms, most probably because of its role in forming the amino acid radicals required for the conversion of ribonucleotides to deoxyribonucleotides in the Fe-dependent ribonucleotide reductases. In those organisms, such as *Lactobacilli*[6], which do not have access to iron, their ribonucleotide reductases use a cobalt-based cofactor, related to vitamin B_{12}. Cobalt is also used in a number of other enzymes, some of which catalyse complex isomerization reactions. Like cobalt, nickel appears to be much more extensively utilized by anaerobic bacteria, in reactions involving chemicals such as CH_4, CO and H_2, the metabolism of which was important

---

[5] Chromium picolinate generated sales of 0.5×10^9 $ in 1999.

[6] So called because they are found in milk, where the iron-binding protein lactoferrin sequesters iron so tightly that it is no longer available for microbial requirements.

before the appearance of dioxygen. In higher organisms, notably plants, the only nickel-containing enzyme is urease. Since humans, in common with most terrestrial vertebrates, excrete excess nitrogen from the metabolism of amino acids in the form of urea (i.e. they are ureotelic), they do not produce this urea-hydrolysing enzyme.

Copper, like iron, is frequently encountered in reactions involving dioxygen. The copper enzyme laccase catalyses the oxidation of uroshiol (the same poisonous substance found in poison oak and ivy) in the production of Japanese lacquer. It is the products of uroshiol oxidation, which are responsible for the lacquer's remarkable material properties.

Zinc, in addition to its use as a Lewis acid in enzyme catalysis, plays a structural role in stabilizing protein molecules. It is also involved in a characteristic motif, termed zinc finger, in a number of eukaryotic DNA-binding proteins (that regulate the transcription of DNA into RNA), first described by Aaron Klug.

Gallium is non-essential, but on account of the similarity between Ga^{3+} and Fe^{3+} it binds to iron transport and storage proteins such as transferrin and ferritin. The radioactive isotope of gallium, ^{67}Ga, concentrates to a large extent in many tumours and at sites of inflammation and infection, and since many tumours overexpress the transferrin receptor it can be used for tumour imaging.

Arsenic is highly toxic, and indeed much speculation has surrounded arsenic poisoning as the cause of death of Napoleon Bonaparte, on account of the levels of As in the Emperor's hair (perhaps derived from fungal activity on a green pigment present in the wallpaper of his apartments in St. Helena). Arsenic trioxide has been approved by the Food and Drug Administration (FDA) of the USA for the treatment of acute promyelocytic anaemia in adult patients who fail to respond to other chemotherapy, or have relapsed disease.

Selenium is essential for many species including man, on account of its presence in a number of enzymes, notably glutathione peroxidase, an important antioxidant enzyme. It is incorporated into selenoenzymes in the form of selenocysteine.

Bromine is thought to be essential for plants and animals, although no known biological role has been established. It has flame-extinguishing characteristics and is used in fireproofing agents and to make flame-resistant plastics.

Molybdenum is essential for a number of enzymes, for example xanthine oxidases in mammals and nitrogenases in nitrogen-fixing bacteria.

Tin is thought to be an essential trace element for some species, although its precise role remains unknown. Some therapeutic uses of tin compounds have been proposed, and triorganotin carboxylates are effective bacteriocides and pesticides. Tin is, of course, an important component of a number of alloys, with copper in bronze (the Bronze age began about 3500 BC), and with lead in pewter.

Iodine is an essential element with an important role in mammals in the regulation of metabolism, through the action of the two related hormones triiodothyronine (T_3) and thyroxine (T_4) produced by the thyroid gland. The biosynthesis of these two hormones (Figure 1.5) occurs through the iodination, rearrangement and subsequent hydrolysis (proteolysis) of tyrosine residues in the thyroglobulin protein. Iodine, which is relatively scarce, is actively concentrated in the thyroid gland where both T_3 and T_4 are produced.

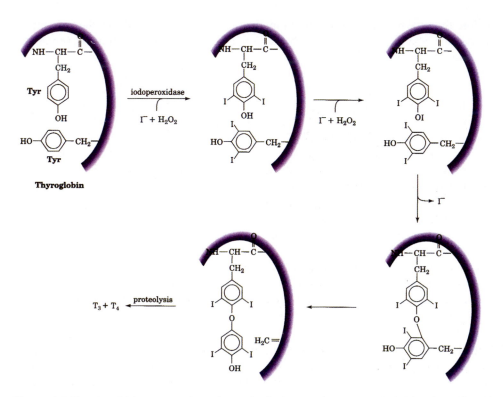

Figure 1.5 The thyroid hormones thyroxine and triiodothyronine are produced by the action of iodoperoxidase and subsequent proteolysis of thyroglobin. (From Voet and Voet, 2004. Reproduced with permission from John Wiley & Sons., Inc.)

While gadolinium, like the rest of the lanthanides, is a non-essential element, it has found wide use as a contrast agent for MRI (Figure 1.6a) because of its high paramagnetism (seven unpaired electrons) and favourable properties in terms of electronic relaxation. The presence of the contrast agent dramatically changes the water proton relaxation rates, adding an important amount of additional physiological information to the already impressive anatomical resolution of the non-contrasted images.

Tungsten, still considered to be essential for some organisms, is required as a cofactor in a number of prokaryotic enzymes. It was also the element that would replace osmium and tantalum (and before them, carbon) in the electric lamps of the early twentieth century, which gradually replaced gas lamps. Tungsten lamps, made of tightly coiled helices of finely drawn tungsten wire, in bulbs filled with argon, would provide 'Light for the Masses', the title of Chapter 5 of Oliver Sacks's delightful memories of a chemical boyhood 'Uncle Tungsten' (Sacks, 2001).

We conclude this idiosyncratic trip through the Periodic Table with four non-essential elements. Platinum, initially in the form of cisplatin, has been hugely successful in the

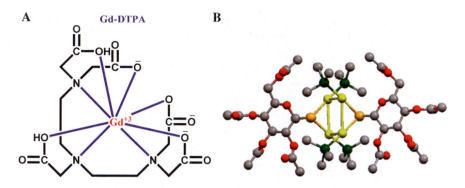

Figure 1.6 The structures of (a) the MRI contrast agent Gd-DTPA and (b) the orally active anti-rheumatoid arthritis drug Auranofin®.

treatment of testicular and ovarian cancers, and since cisplatin resistance has been encountered, new Pt anti-tumour drugs continue to be developed. The precise site of action of *cis*-platinum is discussed in detail in Chapter 20 together with a number of other striking examples of metals as drugs.

Gold might seem to be a surprising (and at first sight costly) therapeutic agent. Nonetheless gold therapy for rheumatoid arthritis, notably using the orally active derivative Auranofin (Figure 1.6b) that can be administered at doses of 3–6 mg/day without necessitating regular visits to the doctor, represents a 'second-generation' drug in the treatment of this painful condition.

The chemicals that were employed in hat making included mercurous nitrate, which is used in rendering felt more rigid (allowing the production of 'top hats'). Prolonged exposure to the mercury vapours caused mercury poisoning. Victims developed severe and uncontrollable muscular tremors and twitching limbs, called 'hatter's shakes'; other symptoms included distorted vision and confused speech. Advanced cases developed hallucinations and other psychotic symptoms. This may well explain the expression 'mad as a hatter', illustrated classically in Lewis Carrol's description in 'Alice in Wonderland' of the Mad Hatter's Tea Party (Figure 1.7).

Finally, in this last group, we include lead that causes saturnism[7], particularly among young children in socially deprived inner cities. The toxicity of environmental Pb finds its molecular explanation in the extraordinary high affinity of Pb (binding constant of 10^{15} M) for the key Zn-dependent enzyme of haem biosynthesis, porphobilinogen synthase.

---

[7] Chronic lead poisoning: Saturn was the alchemist's name for lead. The metal Pb (Latin *plumbum*) was used in domestic plumbing from Roman times, on account of it being soft and malleable. It may have been responsible for the decline of the Roman Empire, not from the plumbing but rather through the use of pewter drinking vessels. It is unlikely that modern-day plumbers ever use it (they prefer other metals or plastics). The attraction of Pb for young children is that it has a very sweet taste.

Figure 1.7 The Mad Hatter's Tea Party. (From Lewis Carroll's *Alice in Wonderland*.)

This concludes this brief introduction in which I have tried not to say too much about elements that we will encounter in greater detail later on, but to give some indications of the multiple roles, for good as well as for ill, of a number of other metal ions that play an important role in living organisms.

REFERENCES

Dutzler, R., Campbell, E.B. and MacKinnon, R. (2003) Gating the selectivity filter in ClC chloride channels, *Science*, **300**, 108–112.

Gouax, E. and MacKinnon, R. (2005) Principles of selective ion transport in channels and pumps, *Science*, **310**, 1461–1465.

Levi, P. (1985) *The Periodic Table*, Michael Joseph, London, 233 pp.

Sacks, O. (2001) Memories of a Chemical Boyhood, *Uncle Tungsten*, Picador, London, 337 pp.

– 2 –

Basic Coordination Chemistry for Biologists

INTRODUCTION

Biological inorganic chemistry is by its nature an interdisciplinary subject with linguistic and conceptual problems that render it difficult for students who have a unique background in either biology or chemistry. The major problem for the student with a background in biology is the understanding of the concepts inherent in the interactions of chemical species (charged or uncharged) with each other. Such concepts involve electronic structure and considerations of symmetry, which in turn affect the bonding between them. In this chapter we will lay out the basics of such concepts with particular reference to the interactions of metal ions with organic molecules.

The electron in its simplest description can be considered as a negatively charged cloud that occupies a definite but arbitrarily defined region of space relative to the nucleus. Such regions are called orbitals[1] and can contain a maximum of two electrons of opposing spin. The s orbitals are spherical. The p orbitals are dumb-bell shaped and there are three of them, each one lying across a Cartesian xyz-axes system. The d orbitals (apart from the d_{z^2}) are four-lobed and their orientation along the axes system is discussed, more extensively, further in this chapter. The f orbitals are seven in number but the shape and orientation are beyond the scope of this book. The shapes of the d orbitals are shown in Figure 2.1.

Atoms within the same molecule or between different molecules interact and are held together by bonds formed by electrons. The number of bonds that an atom can form is usually determined by its valency—the number of unpaired electrons in its outer shell (the valency shell). Bonding results in each atom achieving the noble gas configuration[2].

---

[1] Strictly speaking, an orbital is not a physical reality but refers to a particular solution of complicated wave equations associated with the theoretical description of atoms and they are referred to by the initial letter of the terms describing the spectral lines: sharp, principal, diffuse and fundamental.

[2] The noble gases of Group VIII of the Periodic Table all contain eight electrons in their outer shell.

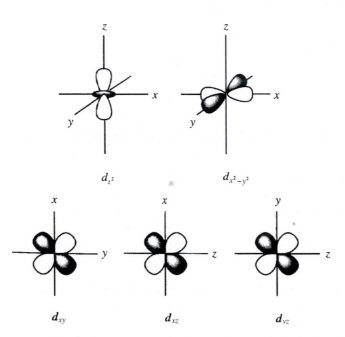

Figure 2.1 Graphic representation of the five *d* orbitals along a Cartesian *xyz*-axes system.

Ionic bonding

Electronegativity is the tendency of an atom to attract electrons in a molecule. Large differences in electronegativity between atoms in a given molecule often cause the complete transfer of an electron from the unfilled outer shell of one atom to the unfilled shell of another. The resulting charged species (ions) are held together by electrostatic forces. Such bonds are highly polarized and are referred to as ionic bonds. Ionic bonding is the simplest type of chemical bonding encountered. NaCl can be written as $[Na^+ Cl^-]$, the sodium atom giving up one electron to resemble the stable neon atom, while the chlorine atom acquiring an extra electron to resemble the stable argon atom. $MgCl_2$ $[Mg^{2+} Cl_2^-]$ and $CoBr_3$ $[Co^{3+} Br_3^-]$ are other examples of ionic compounds.

Covalent bonding

Orbital overlap, i.e. mutual sharing of one or more electrons, can occur when two atoms are in close proximity to each other. The bonding resulting from such overlap is referred to as covalent bonding. Most frequently for a significant overlap and hence a more stable bond, either both atoms have half-filled valency orbitals, as in the H_2 molecule, or one atom has a filled valency orbital not used for bonding and the other one a vacant valency orbital. Pure covalent bonding occurs in compounds containing atoms of the same element such as H_2. Most compounds however contain atoms of different elements, which have

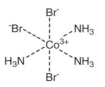

Figure 2.2 Structure of the coordination complex $CoBr_3 \cdot 3NH_3$.

different electronegativities, and hence the commonest type of bonding lies somewhere between purely ionic and purely covalent as in HCl.

Coordinate bonds are a special case of covalent bonds where the electrons for sharing are supplied by one atom. There is often a fractional positive charge on the donor atom and a fractional negative charge on the acceptor atom. $CoBr_3 \cdot 3NH_3$ of Figure 2.2 exhibits such type of bonding and hence traditionally referred to as a coordination compound.

Coordination[3] compounds consist of a central atom or ion, such as Co^{3+}, surrounded by electron-rich groups (ligands), such as NH_3. The ligands are directly bound (coordinated to) to the central atom or ion; they are usually between 2 and 9 in number and may be single atoms, ions or molecules. The ligands directly bound to the metal are said to be in the inner coordination sphere, and the counter-ions that balance out the charge are said to be outer sphere ions. Coordination compounds are usually referred to as complexes; they can be charged or uncharged and their structure is defined by the coordination number (the number of ligand atoms bonded to the central atom) and their coordination geometry (the geometrical arrangement of the ligands and the symmetry of the entire complex). The central ion can be in any oxidation state, which remains unchanged in the coordination complex. We shall endeavour in what follows to explain some of the concepts of coordination chemistry and their relevance to biological inorganic chemistry.

HARD AND SOFT LIGANDS

In 1923 the American chemist G.N. Lewis provided a broad definition of acids and bases, which covered acid–base reactions not involving the traditional proton transfer: an acid is an electron-pair acceptor (Lewis acid) and a base is an electron-pair donor (Lewis base). The concept was extended to metal–ligand interactions with the ligand acting as donor, or Lewis base, and the metal ion as acceptor, or Lewis acid.

The metal ions can be empirically sorted into two groups on the basis of their preference for various ligands: the large and polarizable ions that prefer large, polarizable ligands and the smaller, compact and less-polarizable ones that prefer compact, less-polarizable ligands. Such a correlation, coupled to the broader definition of acid–base, led to the concept of 'hard' and 'soft' acids and bases that can be useful in classifying and to some extent predicting the strength of metal–ligand bonds, and hence the stability of complexes.

---

[3] Although there is no real reason for treating coordination compounds separately from molecular ones, the historic convention will be used here for reasons of convenience.

Table 2.1

Classification of biologically important metal ions and ligands according to the 'hard–soft acid–base' concept and their general characteristics

Acid/acceptor (metal ions)		Base/donor (ligands)
Hard	High charge density	Low polarizability
	Small ionic radius	High electronegativity
	No easily excited outer shell electrons	Vacant, high-energy orbitals
	Na^+, K^+, Mg^{2+}, Ca^{2+}, Cr^{3+}, Fe^{3+}, Co^{3+}	Hard to oxidize
		H_2O, OH^-, CO_2^-, CO_3^{2-}, NO_3^-, PO_4^{3-}, $ROPO_3^{2-}$ PO_4^{3-}, $ROPO_3^{2-}$, $(RO)_2PO_2^-$, ROH, RO^-, R_2O, NH_3, RNH_2, Cl^-
Intermediate	Fe^{2+}, Co^{2+}, Ni^{2+}, Cu^{2+}, Zn^{2+}	NO_2^-, SO_3^{2-}, Br^-, N_3^-, imidazole
Soft	Low-charge density	High polarizability
	Large ionic radius	Low electronegativity
	Easily excited outer shell electrons	Low-energy vacant orbitals
	Cu^+	Easily oxidized
		RSH, RS^-, CN^-, CO

The general characteristics of each group are summarized in Table 2.1 along with a classification of metal ions and ligands of importance in biological inorganic chemistry. In general 'hard' acids prefer 'hard' ligands whereas 'intermediate' and 'soft' acids form more stable complexes with 'soft' bases. Hard–hard interactions will be primarily ionic in nature whereas soft–soft interactions will be governed by 'orbital' interactions.

A number of more specific ligand–metal ion interactions are hidden within Table 2.1. For example, Mg^{2+} is often associated with phosphate ligands (see Chapter 10), Ca^{2+} is most commonly coordinated by carboxylate ligands as in proteolytic enzymes of the blood coagulation cascade where Ca^{2+} is often bound to γ-carboxyglutamate residues, Cu^{2+} is often bound to histidine residues. Non-biological metal ions, which are of importance in medicine or as environmental pollutants, can also use the same ligands. Thus, Al^{3+} and Ga^{3+} fall into the 'hard' category, while Cd^{2+}, Pt^{2+}, Pt^{4+}, Hg^{2+} and Pb^{2+} are classified as 'soft'.

Ligands are also classified electronically (according to the number of electrons donated to the central atom) and structurally (by the number of connections they make to the central atom). The structural classification of the ligands refers to their denticity i.e. the number of donor atoms from each molecule. A ligand attached by one atom is described as monodentate, by two bidentate, by three tridentate and so on. Multi-dentate ligands bound directly to one atom are known as chelating agents and a central metal atom bound to one or more ligands is called a chelate.

The chelate effect

Metal ions dissolved in water are effectively complexed to water molecules. Displacing the set of water ligands, partially or entirely by another set, in such aqua metal ions results in

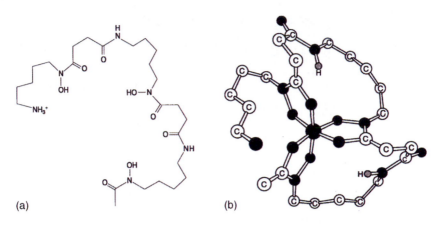

Figure 2.3 (a) The metal chelator desferrioxamine (DFO) and (b) its complex with iron.

forming what is more conventionally known as complexes. Displacement of water molecules by multi-dentate ligands results in more stable complexes than similar systems with none or fewer chelates.

Such enhanced stability, referred to as the chelate effect, is due to a favourable entropic contribution irrespective of the associated enthalpy changes. The large increase in entropy is the result of the net increase in the number of unbound molecules, i.e. released non-chelating ligands, usually water, from the coordination sphere of the metal ion. The chelate effect decreases in magnitude with increasing size of the chelate ring. The complexing capacity of chelators is best expressed by using the pM[4] where M is the central metal ion. The term allows for comparisons between ligands of different denticity. The larger the value of pM for a particular ligand, the more stable is the metal complex.

Chelation is important in medicine. Treatment of the hereditary disease thalassaemia[5] requires regular blood transfusion and the excess iron can be removed by the hexadentate chelator desferrioxamine (Desferal®, DFO) with pFe of the order of 27, depicted in Figure 2.3. DFO loses three protons when it binds to Fe^{3+}. This illustrates an important aspect of coordination chemistry, namely that the positive charge on the metal ion stabilizes the acid anion (i.e. the conjugate base) of protonated ligands. The same thing is true for other biological ligands, such as water, alcohols, carboxylic acids, imidazole, phenols, phosphoric acid and thiols. In the particular case of water, deprotonation to form a hydroxy ligand is presumed to be involved in a number of metalloenzyme-catalysed hydrolytic reactions (e.g. the role of Zn^{2+} in carbonic anhydrase).

The corrins and porphyrins are another important class of natural chelator molecules (Figure 2.4). They are thermodynamically very stable and have four nearly coplanar pyrrole rings, the nitrogen atoms of which can accommodate a number of different metal ions in different oxidation states like Fe^{2+} in haem, Mg^{2+} in chlorophyll and Co^{3+} in vitamin B_{12}.

---

[4] This has been defined by Ken Raymond as the negative logarithm of the concentration of the free or uncomplexed metal ion M^{n+}_{aq} ($pM^{n+} = -\log[M^{n+}_{aq}]$) and it is calculated from the formation constant for a total ligand concentration of 10^{-5} M and $[M^{n+}_{aq}]_{tot}$ of 10^{-6} M under standard conditions (i.e. pH 7.4 and 25 °C).

[5] The blood of the patients with thalassaemia contains an abnormal form of haemoglobin.

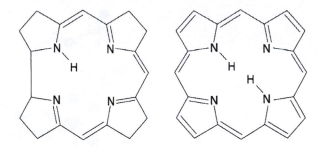

Figure 2.4 The structures of corrin (left) and porphyrin (right).

The chelate effect in proteins is also important, since the three-dimensional (3-D) structure of the protein can impose particular coordination geometry on the metal ion. This determines the ligands available for coordination, their stereochemistry and the local environment, through local hydrophobicity/hydrophilicity, hydrogen bonding by nearby residues with bound and non-bound residues in the metal ion's coordination sphere, etc. A good example is illustrated by the Zn^{2+}-binding site of Cu/Zn superoxide dismutase, which has an affinity for Zn^{2+}, such that the non-metallated protein can extract Zn^{2+} from solution into the site and can displace Cu^{2+} from the Zn^{2+} site when the di-Cu^{2+} protein is treated with excess Zn^{2+}.

COORDINATION GEOMETRY

The shape of a molecule, i.e. its geometry, is generally defined by the bonds within the molecule, which are disposed in a 3-D array. The different pairs of electrons involved in bonding are attracted by two nuclei and they will tend to stay as far from each other as possible to minimize electrostatic repulsions. The shape of a molecule can be predicted on the basis of the number of electron pairs in the valence shell of the central atom. Two pairs result in a linear arrangement, three a triangular one and four occupy the vertices of a tetrahedron. Five pairs give two possible arrangements: the stable trigonal bipyramid and the less-stable square pyramid. The predicted stable geometries are shown in Table 2.2.

Hence diatomic molecules are linear and so are triatomic ones in the absence of unshared pairs (lone pairs). However, deviations from linearity are observed when lone

Table 2.2

Predicted arrangements of electron pairs in the valence shell of the central atom

Number of pairs	Predicted stable geometry
2	Linear
3	Equilateral triangle
4	Tetrahedron
5	Trigonal bipyramid square pyramid (less stable)
6	Octahedron

Table 2.3

The most commonly encountered types of hybridization

Overlapping orbitals	Name of hybrid	Geometry	Example
One *s* and three *p*	sp^3	Tetrahedral	Carbon
One *s* and two *p*	sp^2	Trigonal	Boron
One *s* and one *p*	sp	Linear	Beryllium
One *s*, three *p* and one *d*	sp^3d	Trigonal bipyramid	Platinum
One *s*, three *p* and two *d*	sp^3d^2	Octahedral	Titanium

pairs are present. Lone pairs are attracted by one nucleus instead of two (as in the case of the shared pairs) and hence occupy more space than shared pairs. For example the water molecule is a 'bent' molecule with a bond angle H–O–H less than 180° due to the repulsions between the two lone pairs of the oxygen atom.

The charge cloud model gives reasonable predictions concerning the shapes of molecules. However, it does not account for the positions of the non-bonding electron pairs (lone pairs) in the molecules. A limitation of such treatment is apparent in the implication that the valency of an element is equal to the number of unpaired electrons in the valency shell. For example oxygen has two unpaired electrons and hence a valency of 2. However, carbon has two unpaired electrons but a valency of 4 instead of the expected 2. This higher than expected valency can be explained by the reorganization of the valence orbitals into new ones possessing a spatial orientation other than the ones discussed above i.e. the atomic orbitals combine to give new orbitals of different shape and orientation: the hybrid orbitals. A summary of the types of hybridization and the geometry of the resulting hybrid orbitals is shown in Table 2.3.

The geometry of the coordination compounds can be similarly predicted based on the coordination number of the central atom. Coordination numbers 2 and 3 are both relatively rare and give linear and planar or pyramidal geometries, respectively. The most important coordination numbers are 4, 5 and 6 with the latter being the most important one as nearly all cations form 6-coordinate complexes. Table 2.4 shows the geometries corresponding to the commonest coordination numbers in biological systems.

The nature of the ligand donor atom and the stereochemistry at the metal ion can have a profound effect on the redox potential of redox-active metal ions. The standard redox potentials of Cu^{2+}/Cu^+, Fe^{3+}/Fe^{2+}, Mn^{3+}/Mn^{2+}, Co^{3+}/Co^{2+}, can be altered by more than 1.0 V by varying such parameters. A simple example of this effect is provided by the couple Cu^{2+}/Cu^+. These two forms of copper have quite different coordination geometries, and ligand environments, which are distorted towards the Cu(I) geometry, will raise the redox potential, as we will see later in the case of the electron transfer protein plastocyanin.

CRYSTAL FIELD THEORY AND LIGAND FIELD THEORY

The crystal field theory (CFT) was developed for crystalline solids by the physicist Hans Bethe in 1929. The model takes into account the distance separating the positively and

Table 2.4

Common geometries for 4- and 6-coordinate metal ions with examples for each case

Coordination number	Geometry of coordination compound	Example

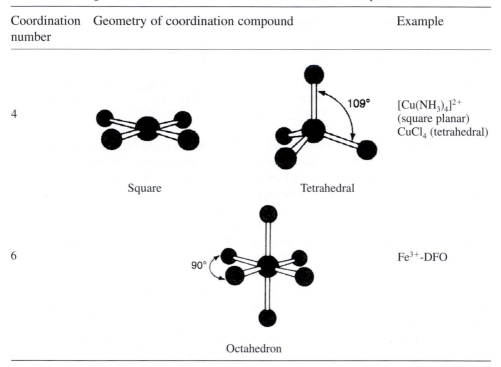

4	Square Tetrahedral	$[Cu(NH_3)_4]^{2+}$ (square planar) $CuCl_4$ (tetrahedral)
6	Octahedron	Fe^{3+}-DFO

negatively charged ions and treats the ions simply as point charges with the attractive and repulsive interactions between them as purely electrostatic/ionic ones. In the case of neutral ligands, such as water and ammonia, dipolar charge separations are considered. The central point in this theory is the effect of the symmetry of the arrangement of ligands on the energy of the d orbitals of a central metal atom. Imagine a cube with a metal ion occupying its centre and a Cartesian system of xyz-axes going through it. There are five d orbitals for the metal ion: two aligned along the principle axes hence referred to as d_{z^2} and $d_{x^2-y^2}$ and three distributed between the axes and hence referred to as d_{xy}, d_{xz} and d_{yz}. In the absence of any ligand, the d orbitals are all of equal energy. We describe such orbitals as degenerate. Imagine negatively charged ligands approaching the cube along the xyz-axes. For an octahedral compound that means six ligands moving towards the centres of the faces of the cube. The ligands have a negative field around them, which will be at a maximum along the direction of the approach, i.e. the xyz-axes. For s and p electrons this is of little consequence but for any d electrons this is of great importance. Not only will such electrons be repelled, but also those in the orbits along one of the Cartesian axes will experience a greater repulsion than those in an orbit between the axes, since such electrons will be pointing towards where the ligand negative field is at its maximum value. Such unevenness in the repulsion will lift the degeneracy of the orbitals and will create preferences for

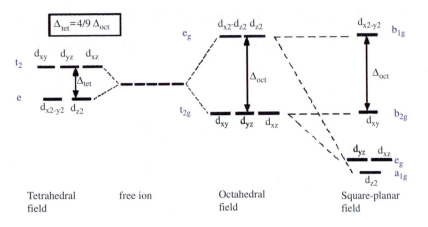

Figure 2.5 Crystal field d-orbital splitting diagrams for common geometries.

occupation along the orbitals of the lowest energy: the electrons will occupy the orbitals in between the xyz-axes, i.e. the d_{xy}, d_{yz} and d_{xz} rather than the orbitals along the axes, i.e. d_{z^2} and $d_{x^2-y^2}$, which lie along the direction of approach of the ligands. In other words the field associated with the ligands splits the previously homogenous spherical field of the central ion into two groups of different energy levels: the e_g group of the d_{z^2} and $d_{x^2-y^2}$ orbitals of relatively high energy and the t_{2g} group of the d_{xy}, d_{xz} and d_{yz} orbitals of relatively low energy. The notation/symbol used for each subset of orbitals indicates its symmetry: e is used for doubly degenerate orbitals, t for triply degenerate ones. The energy splitting is shown schematically in Figure 2.5.

The above treatment considers the ligands in an octahedral geometry (i.e. with the ligands placed at the centre of the faces of the cube). The square planar case is simply an extension of the octahedral where two ligands are removed from the z-axis. The repulsion of electrons in d_{z^2} and $d_{x^2-y^2}$ orbitals will not be the same and the result is a square planar shape.

Consider now the cube and the ligands fitting into a tetrahedral geometry (i.e. the ligands are placed at four corners of the cube). The energy of the d orbitals that point towards the edges should now be raised higher than those that point towards the faces. The tetrahedral ligand-field splitting is exactly the opposite to that of the octahedral field.

A splitting of magnitude Δ^6 is produced and it depends on the nature of both the metal ion and the ligand. In the case of the octahedral field each electron placed in one of the t_{2g} orbitals is stabilized by a total of $-2/5\Delta$, while electrons placed in the higher energy e_g orbitals are destabilized by a total of $3/5\Delta$. The splitting for a tetrahedral complex, Δ_{tet} is less than that for an octahedral one and algebraic analysis shows that Δ_{tet} is about $4/9\Delta_{oct}$.

---

[6] This is referred to as the crystal field splitting in CFT measuring the magnitude of the electrostatic interaction or the ligand field stabilization energy in LFT measuring the strength of the ligand field. In this book we shall simply refer to it as the energy difference or 'splitting'.

Two factors shall determine how electrons will redistribute themselves among the d orbitals: (a) the tendency for electrons to repel each other results in the half-filling of the d orbitals by single electrons before electron pairing occurs and (b) the orbitals of low energy will be filled before higher energy orbitals are occupied. The difference in energy Δ will determine which of those two factors will become important. A strong ligand field will result in a large Δ which in turn will result in the low energy t_{2g} orbitals to be occupied in preference to the e_g orbitals leaving the e_g vacant and hence available for bonding. The resultant electronic configuration of the central ion is then known as spin-paired. A weak ligand field will result in half-filled orbitals and the spin-free configuration.

Different ligands will cause different separation of the d-orbitals. This is evident in the multitudes of colours available for a given metal ion when the ligand or stereochemistry varies. The ability of the ligands to cause a large splitting of the energy between the orbitals is independent of the metal ion, its oxidation state and the geometry of the molecule. The ranking of the ligands in order of their ability to cause large orbital separations gives rise to spectrochemical series, a shortened version of which is the following:

$$I^- < Br^- < SCN^- < S^{2-} < Cl^- < NO_3^- < F^- < OH^- \sim RCOO^- < H_2O \sim RS^- < NH_3$$
$$\sim Im \text{ (imidazole)} < bpy \text{ (2,2'-bipyridine)} < CN^- < CO$$

Thus, iodide is a weak field ligand and gives small ligand-field splitting, while carbon monoxide gives a strong field and a large Δ. The energy difference Δ over this range of ligands increases by a factor of about two. It must be noted that this series must be used as a simple and useful rule and not taken as universally accepted, as it has been built on experimental data for metal ions in common oxidation states.

Metals can also be arranged according to a spectrochemical series:

$$Mn^{2+} < Ni^{2+} < Co^{2+} < Fe^{2+} < V^{2+} < Fe^{3+} < Co^{3+} < Mn^{3+} < Mo^{3+} < Rh^{3+} < Ru^{3+}$$
$$< Pd^{4+} < Ir^{3+} < Pt^{4+}$$

When CFT is applied to metal ions of a symmetrical spherical charge, such as the alkali metal ions K^+ and Na^+, the energy calculations show that large cations of low charge should form few coordination compounds. Transition metal cations however contain electrons in orbitals, which are not spherically symmetric and affect bond energies and properties of the metal concerned. The weakness of the CFT is further highlighted by the spectrochemical series. One would expect negatively charged ligands to give stronger crystal fields than neutral ones if only pure electrostatic repulsions were in operation. The position of the negatively charged halide ions as weak field ligands therefore seems odd, as does the fact that hydroxide ion is a weaker field ligand than its parent acid water, despite having the same donor ion. CFT is incapable of explaining the differences in magnetic and spectral properties of coordinated metal ions compared to the free metal ion, and indeed in explaining why these properties depend on the nature of the ligand. For example, $[FeF_6^{3-}]$ has magnetic properties corresponding to five unpaired electrons, whereas those of $[Fe(CN)_6]^{3-}$ correspond to only one unpaired electron (for a more detailed discussion on the magnetic properties and their applications see Chapter 6).

Such discrepancies between empirical observations and theory eventually prescribed a need to describe the bonding in complexes of various symmetries taking into account not only the electrostatic interactions but also the overlap interactions of the molecular orbitals (for further reading see Constable 1996; Cotton and Wilkinson, 1980; Huheey et al., 1993; Mackay and Mackay, 1989). This theory is referred to as the ligand field theory (LFT). Consider the shapes of the *s*, *p* and *d* orbitals and the same symmetry arrangements as in CFT but with the additional use of the molecular orbital theory (MOT) of chemical bonding. MOT combines the approximate energies and wave functions of all the component atomic orbitals to obtain the best approximations for the energies and wave functions of the molecule. In other words, it makes use of covalency in the metal–ligand interactions.

During the formation of a molecule the atomic orbitals of the individual nuclei interact. Such interactions may be constructive or destructive depending on whether their wave functions add or subtract in the region of overlap. Which orbitals can overlap effectively is dictated by symmetry considerations, and only orbitals of matching symmetry may interact. A constructive interaction will result in the formation of two types of bonding molecular orbital: the σ and the π molecular orbitals[7] with a build-up of electron density between the two nuclei. Destructive interactions will give rise to anti-bonding orbitals called σ^* and π^* with an associated decrease in electron density. The bonds associated with σ and π orbitals are called σ and π bonds, respectively. In simplistic terms, direct 'head-on' overlap of two suitably oriented orbitals results in a σ-bond with uniform distribution of charge density that abounds the axis of the bond whereas 'side-ways' overlap will give rise to a π-bond with distribution of the charge density above and below a plane crossing the axis of the bond. The electrons involved in the latter type of bonding are spread out over a greater volume than those involved in the former type. A π-bond will hence be more readily polarized than σ-bond and such bonds are said to be delocalized as sideways overlap occurs between all orbitals in the vertical plane and all those in the horizontal plane. This is the case of alkynes and nitriles, both possessing two sets of π-bonds perpendicular to each other. Delocalization gives additional stability to a molecule as the increase in the volume of the space occupied by the electrons involved lowers the potential energy of the system.

The bonds involved in coordination complexes can then be described as σ-bonds (any lone pair donation from a ligand to the metal) and π-bonds (any donation of electron density from filled metal orbitals to vacant π orbitals of the ligand or from the *p* orbitals of the ligand to the metal *d* orbitals). In the octahedral environment of a central metal atom with six surrounding ligands, the s, p_x, p_y, p_z, d_{z^2} and $d_{x^2-y^2}$ valence shell orbitals of the central metal atom have lobes lying along the metal–ligand bond directions and hence are suitable for σ-bonding. The orientation of the d_{xy}, d_{xz} and d_{yz} makes such orbitals appropriate only for π-bonding. It is assumed that each ligand possesses one σ orbital[8]. Each of the metal orbitals will be combined with its matching symmetry of the ligand system to give a bonding (maximum positive overlap) and an anti-bonding (maximum negative overlap)

---

[7] Pronounced as *sigma* and *pi* from the Greek letters.
[8] If the ligands possess also π orbitals, these have to be taken into account.

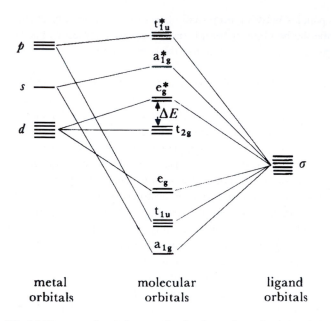

metal
orbitals

molecular
orbitals

ligand
orbitals

Figure 2.6 Simplified MO energy-level diagram for the formation of a σ-bonded octahedral ML_6 complex in which there are no π-bonding interactions between metal and ligand.

molecular orbital. The simplified MO diagram[9] for the formation of a σ-bonded octahedral ML_6 complex is shown in Figure 2.6.

If a molecular orbital is closer in energy to one of the atomic orbitals used to construct it than to the other one, it shall have more the character of the first one than the second one. Hence the electrons occupying the six bonding σ molecular orbitals will be largely 'ligand' electrons with some metal ion character. Electrons occupying the anti-bonding orbitals will be mainly 'metal' electrons. During the complex formation, the metal d electrons will go either only to t_{2g} or to both t_{2g} and e_g^*. In the absence of any π-bonding any electrons in the t_{2g} (which could contribute to π-bonding) will be purely metal electrons and the level is essentially non-bonding, whereas the e_g^* levels participate in σ-bonding with the ligand. In other words the central portion of the diagram closely resembles the t_{2g} and e_g orbitals derived from CFT (Figure 2.5), with one difference: the e_g orbital is now e_g^*.

In terms of CFT, the larger gap between t_{2g} and e_g^* energy levels in strong field ligands is essentially a consequence of the raising of the e_g^* energy levels by electrostatic interactions between the ligand and the d electrons of the metal. However, the molecular orbital model shows how the difference in energy Δ could also be increased by lowering the

---

[9] a_{1g}, e_g, t_{1u} encountered in the diagram are symmetry symbols for the associated orbitals: a_{1g} represents a single orbital which has the full symmetry of the molecular system, e_g and t_{1u} represent a set of two and three orbitals respectively, which are equivalent within the individual set apart from their orientation in space. Subscripts g and u indicate whether the orbital is centrosymmetric or anti-centrosymmetric.

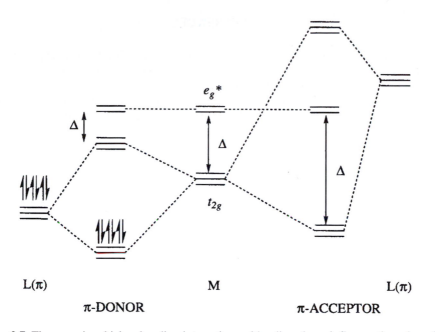

π-DONOR π-ACCEPTOR

Figure 2.7 The ways in which π-bonding interactions with a ligand can influence the value of the energy difference, Δ for an octahedral complex. High energy, poorly populated π-orbitals in the ligand increase the splitting (i.e. are π-acceptors), whereas filled, low-energy π-orbitals decrease the splitting (they are π-donors).

energy of the t_{2g} orbitals. Figure 2.7 shows the situation when π-bonding interactions between the metal t_{2g} orbitals and the p orbitals of the ligand are considered. π-bonds are generally weaker than σ-bonds so the effect is to modify rather than dramatically alter the description. Two orbitals from each ligand are combined to give a total of 12, which are subdivided into four sets with three ligand group orbitals in each set. The metal t_{2g} orbitals are the most suitable for interaction. There are two cases to be considered. (a) The metal t_{2g} orbitals are vacant and the ligand π orbitals are full and at a lower energy than the metal t_{2g}. In this case there is a decrease in the magnitude of Δ. The electron density will be transferred from the ligand to the metal (the ligand is now a π donor). (b) The metal t_{2g} orbitals are filled, and the ligand π orbitals unfilled and at higher energy than the metal t_{2g}. This causes an increase in Δ and the transfer of electron density will now be from the metal to the ligand (the ligand is now a π acceptor). This explains in a satisfactory way the position of CO and CN^- in the spectrochemical series: because they have vacant orbitals of π-symmetry. This additional π-bonding is responsible for the stabilization of the low oxidation states of metals by strong field ligands as the high electron density on such metal ions can be delocalized onto the ligands. In contrast, weak field ligands, such as F^- and OH^-, are π donors, and would be expected to stabilize high oxidation states. The essential take-home message is that the metal ion has its properties influenced by the ligands, and vice versa.

REFERENCES

Constable, E.C. (1996) *Metals and Ligand Reactivity*, Chapter 1, VCH, Weinheim, pp. 1–21.

Cotton, F.A. and Wilkinson, G. (1980) *Advanced Inorganic Chemistry: A Comprehensive Text*, 4th edition, Wiley, New York, Chichester, 1396 pp.

Huheey, J.E., Keiter, E.A. and Keiter, R.L. (1993) *Inorganic Chemistry: Principles of Structure and Reactivity*, 4th edition, HarperCollins College Publishers, New York; 964 pp.

Mackay, K.M. and Mackay, R.A. (1989) *Introduction to Modern Inorganic Chemistry*, 4th edition, Blackie, Glasgow and London, 402 pp.

– 3 –

Biological Ligands for Metal Ions

INTRODUCTION

In the previous chapter we have explained the basic notions involved in the coordination chemistry of metal ions. We now consider the potential ligands that could be involved in binding metals in metalloproteins. We can divide them into those which are naturally occurring amino acids in the protein itself; amino acids that have been chemically modified in order to bind specific metal ions, such as Ca^{2+}; low-molecular weight inorganic ligands, such as carbonate, cyanide and carbon monoxide; metal-binding organic cofactors that have been introduced into the protein (such as porphyrins, corrins and iron–sulfur (Fe–S) clusters, the molybdenum cofactor, MoCo, the CuZ centre of nitrous oxide reductase and the FeMoCo and P-clusters of nitrogenase); and, finally, metal-binding molecules excreted from the cell and then taken up as the metal chelate (such as siderophores).

As was pointed out in the previous chapter, biologically important metal ions and their ligands can be classified according to the hard–soft theory of acids and bases (Table 2.1). While there are exceptions, most metal ions bind to donor ligands as a function of preferences based on this concept, with hard acids (metal ions) binding preferentially to hard bases (ligands) and soft acids to soft bases.

PROTEIN AMINO ACID RESIDUES (AND DERIVATIVES) AS LIGANDS

Of the 20 amino acids present in proteins, only a relatively small number are potential metal ligands. The ligand groups, which are encountered most often, are the thiolate of Cys, the imidazole of His, the carboxylates of Glu and Asp, and the phenolate of Tyr (Figure 3.1). Less frequently we encounter the thioether group of Met, the amino group of Lys and the guanidino group of Arg, and the amide groups of Asn and Gln. Metal ions can also bind to peptide bonds, through the carbonyl or the deprotonated amide nitrogen, and to the terminal amino and carboxyl groups of the protein.

Cysteine can bind to either one or two metal ions, and is frequently found as a ligand to iron (in Fe–S clusters—see later) and to Cu^+ (for example in the copper chaperones, which transfer copper to specific copper-binding proteins). Histidine can bind metal ions in two

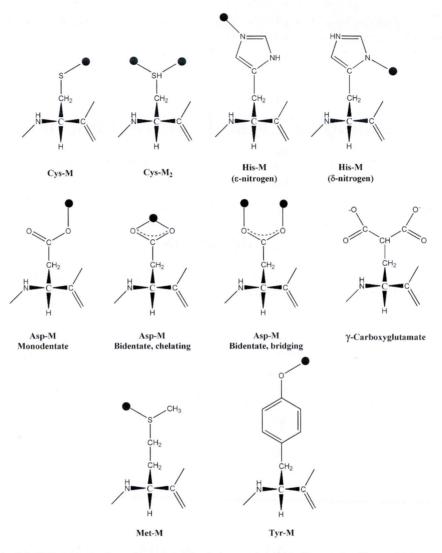

Figure 3.1 Principal protein amino acid side-chain metal–ion binding modes (the metal ion represented as a dark filled circle) and (right) the structure of the Ca^{2+}-binding γ-carboxyglutamate found in proteins of the blood-clotting cascade.

positions, and has a strong preference for Cu^{2+}. The carboxylate oxygens of aspartate (and glutamate, which is not included in Figure 3.1) are preferential ligands for the alkali and alkaline earth metals such as Ca^{2+}. They can bind a single metal ion in either a mono- or bi-dentate (chelating) mode, or bind two metal ions in a bi-dentate, bridging mode. Many proteins of the blood-clotting cascade contain γ-carboxyglutamic acid (Figure 3.1), which binds Ca^{2+} more strongly than glutamate itself. Vitamin K is a cofactor for the enzymatic conversion of glutamic acid to γ-carboxyglutamic acid in these vitamin K-dependent proteins after their biosynthesis (an example of post-translational modification). Vitamin K antagonists (such as Warfarin, originally developed as a rat poison) are used therapeutically as anti-coagulants.

Fe^{3+} also shows a strong affinity for the oxygen donor atoms of carboxylates as well as the phenoxide of tyrosine. Like cysteine, the sulfur ligand of methionine is often found bound to iron, for example in electron-transfer haemoproteins such as cytochrome c.

AN EXAMPLE OF A NON-PROTEIN LIGAND: CARBONATE AND PHOSPHATE

In addition to the amino acid side chains mentioned above, a number of other low molecular weight ligands are found in metalloproteins. These include cyanide and carbon monoxide, which we will describe later in this chapter. Here we consider carbonate and phosphate anions in the context of the super family of iron-binding proteins, the transferrins.

Transferrins, which transport iron in serum and other extracellular fluids in mammals, are part of a super family of proteins, all of which function by a 'Venus fly trap' mechanism. They are made up of two homologous lobes, termed N- and C-lobes, each of which binds a single atom of ferric iron together with a 'synergistic' carbonate anion. Each lobe is composed of two domains indicated in Figure 3.2, which close together upon iron and carbonate binding[1]. The ferric iron is bound in almost ideal octahedral geometry with four

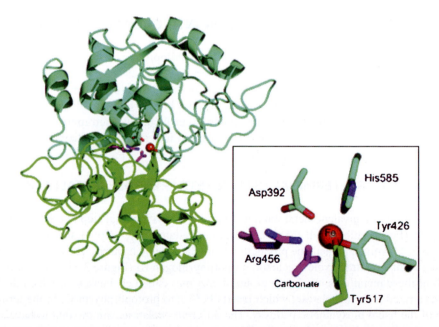

Figure 3.2 Ribbon diagram of the C-lobe of human transferrin with the two domains shown in different colours (cyan for C1 and green for C2). The inset shows the four protein ligand residues together with the arginine residue which stabilizes binding of the synergistic carbonate ion (both in magenta). (Reprinted with permission from Mason et al., 2005. Copyright (2005) American Chemical Society.)

---

[1] The Venus fly trap is a carnivorous swamp flower of the sundew family, native to the Carolinas; it has leaves with two-hinged blades that snap shut to trap insects.

ligands supplied by the protein and the remaining two contributed by the carbonate anion. The charge on the ferric ion is matched by the three anionic ligands contributed by the protein itself (two tyrosines and one aspartate), while the charge on the carbonate is stabilized by the positive charge on an arginine residue, which is highly conserved in each lobe of all mammalian transferrins. In the open configuration, the carbonate can bind in the bottom of the metal-binding pocket. Upon binding of the Fe^{3+}, two of the four protein ligands are already in place, and closure of the domains brings the other two into place. Iron release from transferrin is characterized by a large conformational change, in which the two domains move away from each other and is thought to involve protonation of the carbonate and movement of the arginine residue, destabilizing the coordination of the iron atom.

In pathogenic bacteria, such as *Haemophilus influenzae* and various species of *Neisseria*, Fe^{3+} is transported by ferric-binding proteins, which have an overall structure almost identical to that of one of the iron-binding lobes of transferrin. Again there are four protein ligands to the metal, almost identical to transferrin (except that aspartate is replaced by glutamate): However the carbonate is replaced by a phosphate, and only one of the phosphate oxygens (alternatively oxygen atoms of the phosphate) binds the metal with the sixth coordination position taken by a water molecule.

ENGINEERING METAL INSERTION INTO ORGANIC COFACTORS

As we will see in subsequent chapters, many metalloproteins have their metal centres located in organic cofactors (Lippard and Berg, 1994), such as the tetrapyrrole porphyrins and corrins, or in metal clusters, such as the Fe–S clusters in Fe–S proteins or the FeMo-cofactor of nitrogenase. Here we discuss briefly how metals are incorporated into porphyrins and corrins to form haem and other metallated tetrapyrroles, how Fe–S clusters are synthesized and how copper is inserted into superoxide dismutase.

CHELATASE: TERMINAL STEP IN TETRAPYRROLE METALLATION

The insertion of a divalent metal ion into a tetrapyrrole is the final step in the biosynthesis of haem (Fe^{2+}), chlorophyll (Mg^{2+}), cobalamin (vitamin B_{12}—Co^{2+}) and the coenzyme F_{430} (Ni^{2+}) involved in methane production in methanogenic bacteria. These are all derived from a common tetrapyrrole precursor, uroporphyrinogen III (Figure 3.3). The insertion of each of these metal ions involves a group of enzymes called chelatases, of which the best characterized is ferrochelatase, which inserts Fe^{2+} into protoporphyrin IX in the terminal step of the haem biosynthetic pathway. The different chelatases are thought to have similar mechanisms, which involve, as the first step, the distortion of the tetrapyrrole porphyrin upon binding to the enzyme to give a saddled structure (Figure 3.4a) in which two opposite pyrrole rings are slightly tilted upwards while the other two pyrrole rings are tilted slightly downwards. In the figure, the two unprotonated nitrogen atoms of the pyrrole rings point upwards, while the two protonated nitrogens point downwards with respect to the porphyrin ring. Subsequent to the distortion of the porphyrin ring, the first

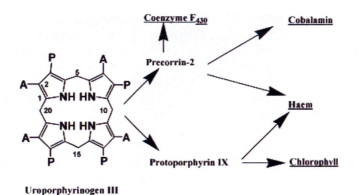

Uroporphyrinogen III

Figure 3.3 The tetrapyrrole biosynthetic pathways. Chelatases selectively insert Fe^{2+} to form haem, Mg^{2+} to form chlorophyll, Co^{2+} to form cobalamin and in methanogenic bacteria Ni^{2+} to form coenzyme F_{430}.

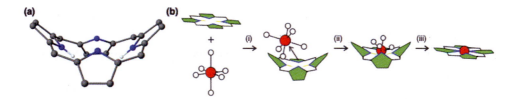

Figure 3.4 Mechanism of porphyrin metallation. (a) Out-of-plane saddle structure in which two pyrrole rings with unprotonated nitrogens (blue spheres) point upwards, while the other two, protonated (blue and white spheres) point downwards. (b) Steps in the mechanism for incorporation of the metal ion (red) into the porphyrin (pyrrole rings in green), described in the text. (From Al-Karadaghi et al., 2006. Copyright 2006, with permission from Elsevier.)

metal–porphyrin bond is formed (Figure 3.4b), followed by other ligand exchange steps leading to the formation of a complex in which the iron atom is sitting on top of the porphyrin, with two of its nitrogen atoms coordinated to the metal while the other two are still protonated. This is followed by the sequential deprotonation of the two pyrrole nitrogen atoms coupled with formation of the metallated porphyrin. The saddling of the porphyrin is an out-of-plane deformation, which exposes both the protons and the lone pairs of the nitrogen atoms of the porphyrin molecule in an appropriate arrangement for metal insertion.

The structure of several ferrochelatases has been determined, and it is clear that the porphyrin rings B, C and D are held in a very tight grip by conserved amino acids, whereas the A ring is distorted. Two metal–ion binding sites have been identified, one located at the surface of the molecule, occupied by a fully hydrated Mg^{2+} ion, and the other located in the porphyrin-binding cleft, close to the distorted porphyrin ring A, with its nitrogen pointing towards His183 and Glu264 (Figure 3.5). It has been proposed that the metal ion on the outermost site, by ligand exchange with a series of acidic residues arranged along

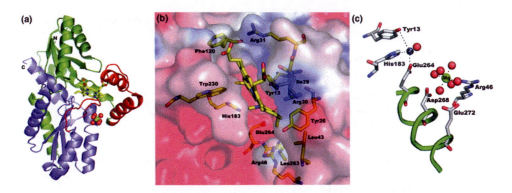

Figure 3.5 Structure (a), porphyrin (b) and metal ion-binding sites (c) in *Bacillus subtilis* ferrochelatase. In (c) the two metal ions are a Zn^{2+} ion (grey) and a fully hydrated Mg^{2+} ion. The side chains of Glu272, Asp 268 and Glu264 are aligned along a π-helix (green). (From Al-Karadaghi et al., 2006. Copyright 2006, with permission from Elsevier.)

the helical edge of a π-helix[2], would be shuttled to the inner site, to be exchanged with the pyrrole nitrogens, resulting in insertion of the metal ion into the porphyrin. The two sites, occupied respectively by a Zn^{2+} ion and a fully hydrated Mg^{2+} ion, are ~7 Å apart. Two of the ligands to the Zn^{2+} ion in the outer site, His183 and Glu264 are invariant in all ferrochelatases. The side chains of Glu272, Asp268 and Glu272 are aligned along the π-helix, in a line connecting the two metal sites. Only a π-helix can provide such an alignment of side chains. This is reminiscent of several other metalloproteins, such as nitrogenase, in which residues in π-helices function to coordinate metal ions involved in enzymatic activity.

IRON–SULFUR CLUSTER CONTAINING PROTEINS

For the first billion years of evolution the environment was anaerobic; this meant that, since iron and sulfur were abundant, proteins containing Fe–S clusters were probably abundant, and therefore were among the first catalysts that Nature had available to it. They are distributed in virtually all the living organisms, but their recognition as a distinct class of metalloproteins only occurred after the discovery of their characteristic EPR spectra in the oxidized state in the 1960s. This second class of iron-containing proteins have iron atoms bound to sulfur, either bound to the polypeptide chain by the thiol groups of cysteine residues or else with both inorganic sulfide and cysteine thiols as ligands. The biochemical utility of these Fe–S clusters resides not only in their possibility to easily transfer electrons, but also in their tendency to bind the electron-rich oxygen and nitrogen atoms of organic substrates.

Fe–S proteins contain four basic core structures, which have been characterized crystallographically both in model compounds (Rao and Holm, 2004) and in iron–sulfur proteins. These are (Figure 3.6), respectively, (A) rubredoxins found only in bacteria, in which the [Fe–S] cluster consists of a single Fe atom liganded to four Cys residues—the iron atom

---

[2] For more information concerning this unusual type of helix see Chapter 4.

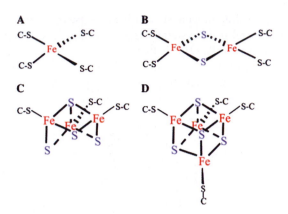

Figure 3.6 Structures of common iron–sulfur centres (C - Cys).

can be in the +2 or +3 valence; (B) rhombic two iron–two sulfide [Fe$_2$–S$_2$] clusters—typical stable cluster oxidation states are +1 and +2 (the charges of the coordinating cysteinate residues are not considered); (C) cuboidal three-iron–four sulfide [Fe$_3$–S$_4$] clusters—stable oxidation states are 0 and +1; and (D) cubane four iron–four sulfide [Fe$_4$–S$_4$] clusters—stable oxidation states are +1 and +2 for ferredoxin-type clusters and +2 and +3 for 'Hipip'[3] clusters. Electrons can be delocalized, such that the valences of individual iron atoms lie between ferrous and ferric forms. Low molecular weight proteins containing the first and the last three types are referred to as rubredoxins (Rd) and ferredoxins (Fd), respectively. The protein ligands are frequently Cys residues, but a number of others are found, notably His, which replace two of the thiol ligands in the high-potential [Fe$_2$–S$_2$] Rieske proteins.

IRON–SULFUR CLUSTER FORMATION

Numerous Fe–S proteins are known in each of the three kingdoms of living organisms, i.e. in Eubacteria, Archaebacteria and Eukaryotes, and their multiple functions in electron transport and catalysis are reviewed in Chapter 13. In contrast to most other cofactors, they are essentially of an inorganic nature consisting simply of iron cations (Fe^{2+} or Fe^{3+}) and inorganic sulfide anions (S^{2-}). Our understanding of the way in which these clusters are assembled has evolved rapidly in the last few years (Lill and Muhlenhoff, 2005, 2006 ; Lill et al., 2006) and we summarize our current understanding of the eukaryotic mitochondrial iron-cluster assembly (ICA) machinery here. The mitochondrial ISC assembly system is strikingly similar to that in bacteria, and it is now clear that mitochondria play a prime role in Fe–S protein biogenesis, since they are not only responsible for the maturation of Fe–S proteins inside but also outside of the organelle. The current view of Fe–S protein biogenesis in eukaryotes (Figure 3.7) involves the interplay of three complex multi-protein systems, referred to as ISC assembly, ISC export and CIA (cytosolic iron–sulfur protein assembly machinery).

---

[3] Hipip: high potential iron–sulfur protein.

The general concept of ISC biogenesis involves the transient *de novo* synthesis of an ISC, with the sulfur derived from cysteine being delivered as sulfide to a so-called scaffold protein complex (Isu1/2). The ISC is then finally transferred to apoproteins with the help of additional ISC proteins (Figure 3.8). Iron, as Fe^{2+}, is imported in a membrane potential-dependent process facilitated by the carrier proteins Mrs3 and Mrs4, together

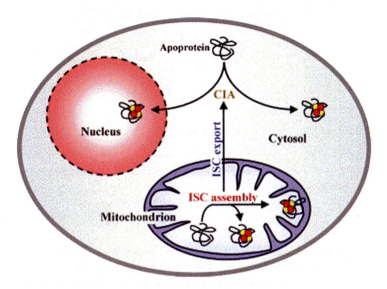

Figure 3.7 The three systems involved in the generation of Fe–S proteins in eukaryotes. (From Lill et al., 2006. Copyright 2006, with permission from Elsevier.)

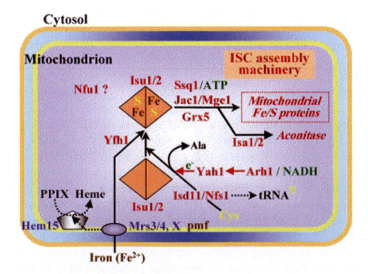

Figure 3.8 Current working model for the mechanism of ISC assembly in mitochondria. (From Lill et al., 2006. Copyright 2006, with permission from Elsevier.)

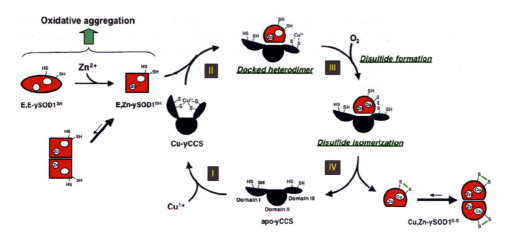

Figure 3.9 Proposed mechanism of copper insertion into SOD1 by its metallochaperone CCS. (From Culotta et al., 2006. Copyright 2006, with permission from Elsevier.)

with further unknown proteins (X). The sulfide required for ISC synthesis is supplied by the pyridoxal phosphate-dependent cysteine desulfurase, Nfs1. In this reaction, the sulfur atom of free cysteine is transferred to a conserved cysteine of Nfs1 to form a persulfide as a reaction intermediate. The sulfur is then transferred directly to the scaffold protein complex, a necessary process to avoid the potential unregulated release of toxic sulfide. The reduction of sulfur to sulfide requires electrons derived from NADH via an electron-transfer chain involving the ferredoxin reductase Arh1 and the ferredoxin Yah1. The binding of iron to the scaffold protein complex Isu1/Isu2 seems to require the protein Yfh1 (known in man as frataxin[4]) as an iron chaperone, which donates its iron to Isu1.

Additional ISC proteins are required later in the process for the insertion of ISC into mitochondrial Fe–S proteins. Isa1 and Isa2 are required specifically and in addition for the maturation of aconitase-type Fe–S proteins.

COPPER INSERTION INTO SUPEROXIDE DISMUTASE

As we will discuss later, in Chapter 8, free copper levels are extremely low within cells because the copper is bound to a family of metallochaperones, which are subsequently involved in the incorporation of copper into copper-containing proteins. The mechanism proposed for copper insertion into the Cu/Zn superoxide dismutase, SOD1, is presented in Figure 3.9. The copper chaperone, CCS, acquires copper as Cu[+] from a copper transporter and then docks with the reduced dithiol form of SOD1 (Steps I and II) to give a docked

---

[4] Frataxin is the protein involved in Friedreich's ataxia, the most common ataxia found in man, associated with massive iron accumulation in mitochondria.

heterodimer. When exposed to oxygen or superoxide (Step III) a disulfide is formed within the heterodimer, which subsequently undergoes disulfide isomerization to an intramolecular disulfide in SOD1 (Step IV). At some point after the introduction of oxygen, copper is transferred from the chaperone to the SOD1 and the mature monomeric SOD1 is released from the CCS chaperone.

MORE COMPLEX COFACTORS: MoCo, FeMoCo, P-CLUSTERS, H-CLUSTERS AND CuZ

Our understanding of metal incorporation into metalloporphyrins and Fe–S clusters has advanced greatly in recent years. These are cofactors, which are widely distributed in great many metalloproteins. However, it has also become apparent that there are a growing number of more complex cofactors with a more specific distribution. The transition metal molybdenum (Mo) is found as an essential part of the active site in a wide range of metalloenzymes in bacteria, fungi, algae, plants and animals. However, the metal itself is biologically inactive unless it is incorporated into a special MoCo. In all organisms studied to date, MoCo is synthesized by a highly conserved biosynthetic pathway that can be divided into four steps (Figure 3.10). The six enzyme activities involved in MoCo biosynthesis (and their corresponding genes) have been identified in plants, fungi and humans, and are homologous to their counterparts in bacteria. The human genes are indicated in the figure. In the first step of molybdenum cofactor synthesis MOCS1A and MOCS1B catalyse the circularization of guanosine triphosphate (GTP) to precursor Z. Three enzymes MOCS2A, MOCS2B and MOCS3 are then responsible for the formation of the dithiolene group. The final steps of the pathway, transfer and insertion of Mo into MTP are catalysed by the individual domains, Geph-G and Geph-E of a two-domain protein called gephyrin[5] in mammals. Since MoCo is labile and oxygen sensitive, it comes as no surprise that in order to buffer the cellular supply and demand of MoCo, all cells contain an MoCo carrier protein that binds the cofactor, and protects it from oxidation. It is not known how MoCo is inserted into Mo enzymes, but for some bacterial Mo enzymes, specific chaperones are required for MoCo insertion and protein folding.

Nitrogen fixation is carried out by a small number of microorganisms, called diazotrophs, some of which (of the genus *Rhizobium*) function symbiotically in the root nodules of nitrogen-fixing legumes (such as peas, clover). The reduction of the triple bond of dinitrogen to ammonia is carried by nitrogenases, which are typically composed of two proteins, the *Fe-protein*, which contains one [4Fe–4S] cluster and two ATP binding sites, and the *MoFe-protein*, which contains both Fe and Mo. The MoFe protein contains two complex metallo-clusters. Both clusters contain eight metal ions. The *P-cluster* (Figure 3.11a,b) can be considered as two [4Fe–3S] clusters linked by a central sulfide ion which, in the reduced form (a) of the cluster, forms the eighth corner of each of the two cubane-like structure, coordinated to two iron atoms of each [4Fe–3S] unit. Two cysteine thiols serve as bridging ligands, each coordinating one Fe atom from each cluster—these are the

---

[5] Gephyrin was initially found in the central nervous system, where it is essential for the clustering of inhibitory neuroreceptors in the postsynaptic membrane.

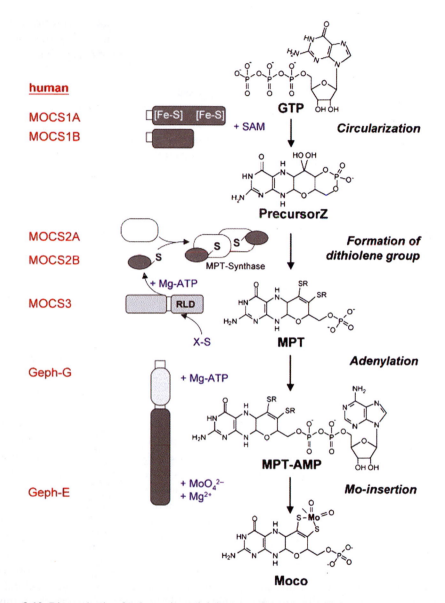

Figure 3.10 Biosynthesis of eukaryotic molybdenum cofactor occurs in four steps. (Adapted from Mendel and Bittner, 2006. Copyright 2006, with permission from Elsevier.)

same four Fe atoms that are also coordinated to the central sulfide ion. In the two-electron oxidized cluster (b), two of the Fe atoms that were bridged to the central sulfide have moved away from it, leaving it tetracoordinate. The two Fe atoms in question remain four-coordinate, through coordination to the amide nitrogen of Cys 87α and the side-chain hydroxyl of Ser 186β. The *FeMo-cofactor* (Figure 3.11c) consists of a [4Fe–3S] cluster and a [1Mo–3Fe–3S] cluster, bridged by three sulfide ions, such that its overall inorganic

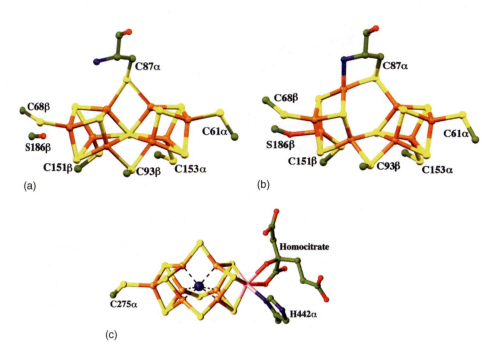

(a)

(b)

(c)

Figure 3.11 (a) and (b) the P-cluster of nitrogenase in its reduced and oxidized state and (c) the FeMo-cofactor. The molecules are represented with C green, N blue, O red, S yellow, Fe orange and Mo pink. (From Voet and Voet, 2004. Reproduced with permission from John Wiley & Sons., Inc.)

composition is [1Mo–7Fe–9S]. The cofactor is bound to the protein by one Cys and one His residue at either end of the structure, and the Mo ion is coordinated approximately octahedrally by three sulfide ions from the cofactor itself, the terminal imidazole nitrogen of the histidine residue of the protein and two oxygens from a molecule of the unusual tricarboxylic acid homocitrate, which is an essential component of the cofactor. The complexity of the enzyme systems required to synthesize both the P-cluster and the FeMoCo cluster, together with the proteins required for their insertion into functionally active nitrogenase, have combined to render the biotechnological dreams of cloning nitrogen fixation into other crop plants an illusion.

Yet another organic cofactor of extraordinary complexity is represented by the H-cluster of microbial hydrogenases. In this case, an unusual coordination of cyanide and carbon monoxide ligands to a metal centre was found, notably by spectroscopic methods, since the electron density of carbon, nitrogen and oxygen cannot easily be differentiated by X-ray crystallography. This is illustrated in Figure 3.12 for the Fe-only hydrogenase from *Desulfovibrio desulfuricans*. The two active site Fe atoms are each coordinated by one CO and one CN ligand and are bridged by an unusual organic 1,3-dithiolate ligand. The Fe atom designated Fe^P is bound to the protein by a cysteine residue, which is itself bridged to a 4Fe–4S cluster. The other, Fe^D has a second CO ligand bound in the reduced form, which changes to become a bridging ligand upon oxidation. The nature of the bridgehead atom (C, N or O) in the 1,3-dithiolate ligand could not be determined unequivocally from the X-ray data, and is therefore shown as X in the figure.

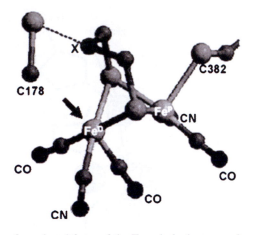

Figure 3.12 Active site of a reduced form of the Fe-only hydrogenase from *Desulphovibrio desulphuricans*. The Fe atom on the right is defined as the proximal Fe (relative to the neighbouring [Fe–S] cluster), Fe^P; the Fe atom on the left is defined as the distal Fe, Fe^D. The arrow indicates the potential hydron-binding site on Fe^D that is occupied by either H_2O or an extrinsic CO in the structure of *Cp* I. Also shown is a close contact between the bridgehead atom X of the exogenous dithiolate ligand and the S atom of cysteine-178. (Reprinted with permission from Parkin et al., 2006. Copyright (2005) American Chemical Society.)

Finally in this gallery of extraordinary ligands, we have the Cu_Z cluster of nitrous oxide reductase. This enzyme is found in denitrifying bacteria where it catalyses the final step in the nitrogen cycle, the reduction of nitrous oxide (N_2O) to dinitrogen, thereby returning fixed nitrogen to the atmosphere. Nitrous oxide reductase contains two types of copper centres, Cu_A and Cu_Z. The Cu_A centre (which is described in Chapter 14) serves as an electron-transfer centre, while the Cu_Z centre is associated with the active site of nitrous oxide reduction. The Cu_Z site comprises four copper ions arranged in a tetranuclear cluster bound to a central sulfide (Figure 3.13) with a bridging oxygen ligand (not shown) assigned between Cu_1 and Cu_4. Seven histidine residues complete the coordination of the cluster, three of the Cu atoms bound to two His residues while Cu_4 is liganded by a single His ligand. It has been suggested that Cu_4 is the binding site for N_2O, since it has only one His ligand and coordinates the bridging oxygen species. In addition to the structural gene for the N_2O reductase protein, a number of other gene products are required for cofactor assembly and insertion, which have been well characterized.

SIDEROPHORES

In the previous chapter, we mentioned the Fe(III) chelator, desferrioxamine B, which is a member of a large class of iron-binding molecules called siderophores. Siderophores are iron-complexing molecules of low molecular weight (typically less than 1000), specifically designed to complex Fe^{3+}, which are synthesized by bacteria and fungi and serve to deliver iron to the microbes (described in Chapter 7). They can be classified into several groups according to their chemical structures: hydroxamates, catecholates, carboxylates and heterocyclic compounds. All of the natural siderophores are designed to

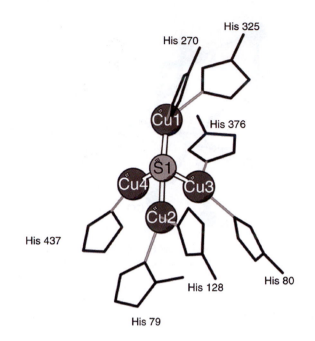

Figure 3.13 Structure of the Cu_Z centre in nitrous oxide reductase. The central sulfide interacts with all four copper atoms. (From Rees, 2002. Copyright 2002 Annual Reviews.)

selectively chelate Fe(III), which under aerobic conditions is the predominant and potentially bioinaccessible form of iron in the environment. This means that they usually contain hard O-donor atoms as ligands, and form thermodynamically extremely stable complexes with Fe(III). Examples are given in Figure 3.14. Ferrichrome (*p*Fe = 25.2)[6], first isolated from the smut mould *Ustilago* in 1952, the best characterized of the hydroxamate siderophores, has a cyclic hexapeptide backbone to which are attached three molecules of *N*-acyl-*N*-hydroxy-L-ornithine. Enterobactin (*p*Fe = 35.5), the prototype of the catecholate siderophores, is the principal siderophore produced by *Eschericia coli*. It is a cyclic triester of dihydroxybenzoyl-serine. When enterobactin binds iron, the six deprotonated hydroxyl groups of the dihydroxybenzoyl (or catecholate) functions wrap around the metal ion at the centre of the molecule (Figure 3.15). It is clear from this figure that recognition of the ferric enterobactin does not involve recognition of the metal, and that the receptor will have no difficulty distinguishing between the apo- and the ferri-enterobactin. Staphyloferrin A, the iron-transporting siderophore of *Staphylococci*, contains a D-ornithine backbone to which two citric acid residues are linked, which are clearly involved in Fe(III) binding. Yersiniabactin, an example of a heterocyclic siderophore, is from the highly pathological *Yersinia* family. In ferric–yersiniabactin the iron atom is coordinated by the three nitrogens and three negatively charged oxygen atoms, arranged in a distorted octahedral arrangement.

---

[6] pFe as defined in Chapter 2.

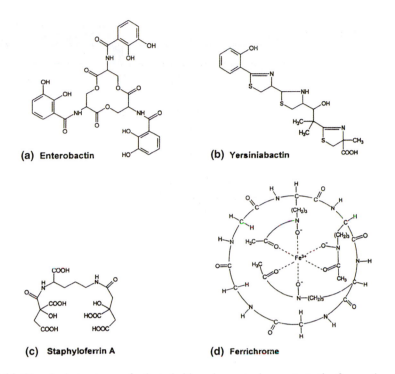

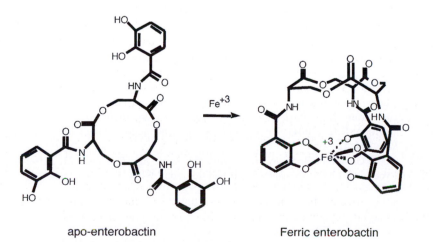

Figure 3.14 Chemical structures of selected siderophores to demonstrate the four major structural classes.

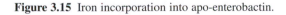

apo-enterobactin   Ferric enterobactin

Figure 3.15 Iron incorporation into apo-enterobactin.

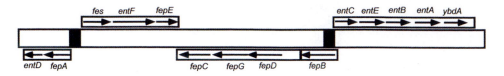

Figure 3.16 The genes involved in enterobactin-mediated iron uptake in *E. coli*.

The importance of iron for a bacteria-like *E. coli* can be illustrated by fact that 14 genes alone are required for enterobactin-mediated iron uptake, including those for its synthesis, export, transport of the ferric–enterobactin back into the cell and iron release (Figure 3.16). In total, *E. coli* has at least 8 uptake systems for iron, encoded by some 50 genes.

REFERENCES

Al-Karadaghi, S., Franco, R., Hansson, M., Shelnutt, J.A., Isaya, G. and Ferreira, G.C. (2006) Chelatases: distort to select? *TIBS*, **31**, 135–142.

Culotta, V.C., Yang, M. and O'Halloran, T.V.O (2006) Activation of superoxide dismutases: putting the metal to the pedal, *Biochim. Biophys. Acta*, **1763**, 747–758.

Lill, R., Duftkiewitz, R., Elsässer, H.-P., Hausmann, A., Netz, D.J.A., Pierik, A.J., Stehling, O., Urzika, E. and Mühlenhoff, U. (2006) Mechanisms of iron–sulfur protein maturation in mitochondria, cytosol and nucleus of eukaryotes, *Biochim. Biophys. Acta*, **1763**, 652–667.

Lill, R. and Mühlenhoff, U. (2005) Iron–sulfur protein biogenesis in eukaryotes, *TIBS*, **30**, 133–141.

Lill, R. and Mühlenhoff, U. (2006) Iron–sulfur protein biogenesis in eukaryotes: components and mechanisms, *Annu. Rev. Cell Dev. Biol.*, **22**, 457–486.

Lippard, S.J. and Berg, J.M. (1994) *Principles of Bioinorganic Chemistry*, University Science Books, Mill Valley, CA, 411 pp.

Mason, A.B., Halbrooks, P.J., James, N.G., Connolly, S.A., Larouche, J.R., Smith, V.C., MacGillivray, R.T.A. and Chasteen, N.D. (2005) Mutational analysis of C-lobe ligands of human serum Transferrin: insights into the mechanism of iron release, *Biochemistry*, **44**, 8013–8021.

Mendel, R.R. and Bittner, F. (2006) Cell biology of molybdenum, *Biochim. Biophys. Acta*, **1763**, 621–635.

Parkin, A., Cavazza, C., Fontecilla-Camps, J.C. and Armstrong, F.C. (2006) Electrochemical investigations of the interconversions between catalytic and inhibited states of the [FeFe]-hydrogenase from *Desulphovibrio desulfuricans*, *JACS*, **128**, 16808–16815.

Rao, P.V. and Holm, R.H. (2004) Synthetic analogues of the active sites of iron–sulfur proteins, *Chem. Rev.*, **104**, 527–559.

Rees, D.C. (2002) Great metalloclusters in enzymology, *Annu. Rev. Biochem.*, **71**, 221–246.

Voet, D. and Voet, J.G. (2004) *Biochemistry* 3rd edition, John Wiley & Sons, Chichester.

– 4 –

Structural and Molecular Biology
for Chemists

INTRODUCTION

We have included introductory chapters to enable readers who come from a more biological background to understand the notions of inorganic chemistry, and vice versa to explain chemists the important notions of structural and molecular biology, which will be necessary to follow our path through the diverse roles of metals in biological systems.

We begin our introduction to structural and molecular biology with a brief survey of the building blocks used to build up the macromolecular structures, which constitute the principal elements of biological systems, and then discuss in greater detail the hierarchy of structural organization in proteins, nucleic acids, carbohydrates and the lipid constituents of biological membranes. In the second part of the chapter we describe how genetic information encoded in DNA is transcribed into RNA and then into proteins, followed by a short analysis of how we can use the tremendous tools of modern molecular biology to approach the much more difficult problems of how these gene products function in a biochemical and physiological context.

THE STRUCTURAL BUILDING BLOCKS OF PROTEINS

Proteins are formed from α-amino acids that, with the exception of proline, have a primary amino group and a carboxylate group on the same carbon (proline is an α-imino acid, with a secondary amino group). They all, with the exception of glycine, have the same configuration and are joined by peptide (amide) bonds (Figure 4.1) to form polypeptide chains. Since, at pH 7, they are present as dipolar ions, it follows that proteins will have a positively charged terminal amino group and a negatively charged terminal carboxyl group. The charge properties of the protein will be determined by the five amino acids with potentially charged R-groups: of these Glu and Asp will have a negative charge at pH 7, Lys and Arg a positive charge, while His residues will be ~10% positively charged[1].

---

[1] Assuming that the pKs are not influenced by their environment in the protein (which they often are).

43

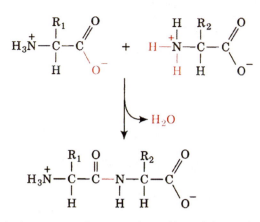

Figure 4.1 The zwitterionic structure of two α-amino acids, and the condensation reaction to form a dipeptide, linked by a peptide bond.

The basic assembly kit of 20 amino acids (Table 4.1) can be divided notionally into those which have non-polar, uncharged and generally hydrophobic side chains: those which have polar, but uncharged side chains and those which have polar, charged side chains.

When we come to consider the individual amino acids, represented in Table 4.1 by their one-letter symbols and the structure of their R-groups, it is clear that those with non-polar, aliphatic side chains, Ala, Ile, Leu, Met and Val, are sufficiently hydrophobic that they are most often buried in the generally hydrophobic core of non-membrane-embedded proteins. Note that Ile and Val have particularly sterically hindered β-carbons. Of the aromatic amino acids His, with a *p*K of around 6, will mostly be in the uncharged form at physiological pH values (therefore more often hydrophobic than polar) and will be a likely choice for reactions that involve proton transfer. Phe and Trp are clearly hydrophobic, but when we examine the free energy required to transfer Tyr from water to an organic solvent, it is not very different from Phe, so we should not be surprised to find it in hydrophobic environments. Gly is an unusual amino acid, with only hydrogen as side chain, and this results in peptide bonds involving Gly residues taking up all sorts of conformations. Pro, which does not have a proton when in a peptide bond, clearly is hydrophobic, cannot participate in hydrogen bonding and has the unique capacity among the protein amino acids to be able to form *cis*-peptide bonds (there is a *cis/trans* prolyl isomerase to restore the *trans* form).

The polar, charged residues Asp, Glu, Lys, Arg and, in its protonated form, His, will often be found at the surface of proteins, where they may not only interact with the polar layers of ordered water molecules surrounding the protein, but may also participate in hydrogen bonds and salt bridges with other polar charged residues.

Of the polar, uncharged residues, as pointed above, Tyr is not particularly polar, although it can participate in hydrogen bonding through its phenol group. The other hydroxylated amino acids, Ser and Thr, and the amides, Asn and Gln, can be at surface positions interacting with water or in the interior where they may participate in hydrogen bonding with other polar residues. A word of caution concerning Ser: there are an important number of enzymes—collectively known as serine proteases and esterases—in which a particular local environment renders the oxygen of the serine residue particularly nucleophilic.

Table 4.1

Protein amino acids

Name of amino acid		Structure of R-group	Properties	pK of R-group
Alanine	A	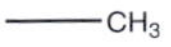 —CH$_3$	Hydrophobic	
Cysteine	C	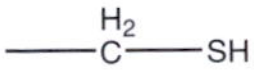 —C(H$_2$)—SH	Polar, forms disulfide bridges	8.37
Aspartate	D	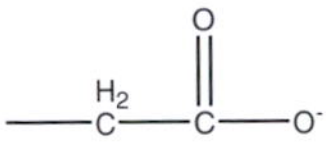 —C(H$_2$)—C(=O)—O⁻	Polar, charged	3.90
Glutamate	E	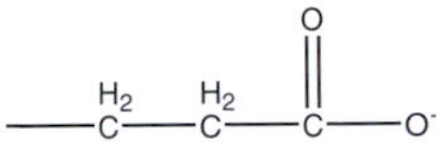 —C(H$_2$)—C(H$_2$)—C(=O)—O⁻	Polar, charged	4.07
Phenylalanine	F	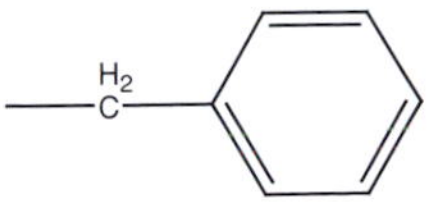 —C(H$_2$)— (phenyl ring)	Hydrophobic	
Glycine	G	 —H	Highly flexible	
Histidine	H	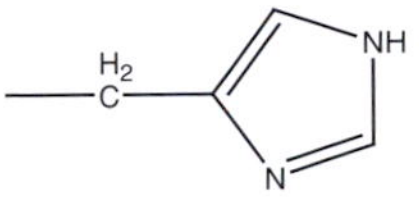 —C(H$_2$)— (imidazole ring, NH, N)	Hydrophobic/polar donor/acceptor of H⁺	6.04
Isoleucine	I	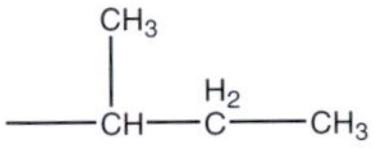 —CH(CH$_3$)—C(H$_2$)—CH$_3$	Hydrophobic, sterically hindered β-carbon	
Lysine	K	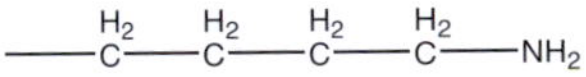 —C(H$_2$)—C(H$_2$)—C(H$_2$)—C(H$_2$)—NH$_2$	Polar, flexible side chain	10.54
Leucine	L	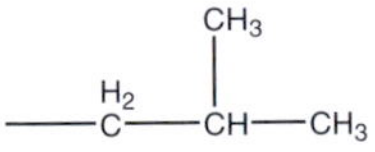 —C(H$_2$)—CH(CH$_3$)—CH$_3$	Hydrophobic	
Methionine	M	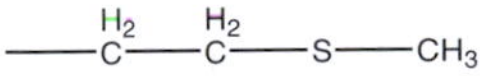 —C(H$_2$)—C(H$_2$)—S—CH$_3$	Hydrophobic	
Asparagine	N	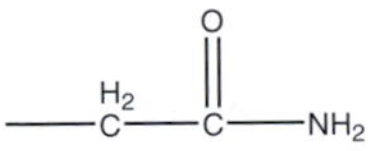 —C(H$_2$)—C(=O)—NH$_2$	Polar, uncharged	
Proline	P	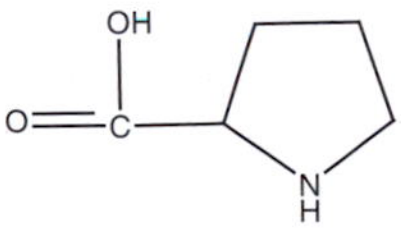 O=C— (pyrrolidine ring, OH, N–H)	Imino acid, can form cis-peptide bonds	
Glutamine	Q	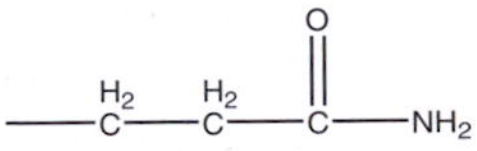 —C(H$_2$)—C(H$_2$)—C(=O)—NH$_2$	Polar, uncharged	
Arginine	R	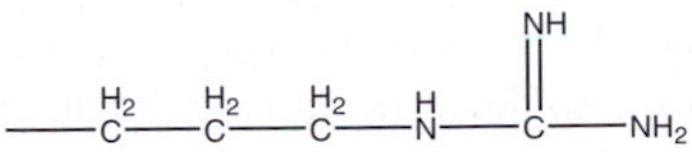 —C(H$_2$)—C(H$_2$)—C(H$_2$)—N(H)—C(=NH)—NH$_2$	Polar, charged	12.48
Serine	S	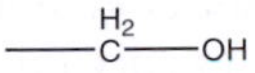 —C(H$_2$)—OH	Polar, uncharged	
Threonine	T	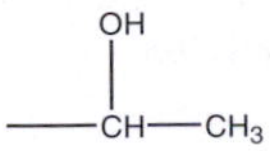 —CH(OH)—CH$_3$	Polar, uncharged, rather hydrophobic	
Valine	V	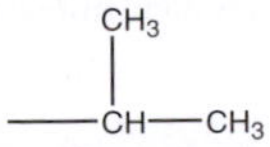 —CH(CH$_3$)—CH$_3$	Hydrophobic, sterically hindered β-carbon	

(Continued)

Table 4.1

Protein amino acids—Continued

Name of amino acid	Structure of R-group	Properties	pK of R-group
Tryptophan W		Hydrophobic, very bulky	
Tyrosine Y		Polar, but almost as hydro-phobic, as phenylalanine	10.46

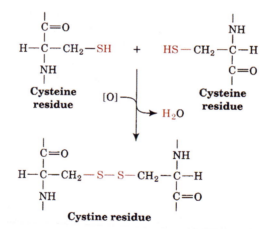

Cystine residue

Figure 4.2 Disulfide formation between two cysteine residues. The product of the oxidation reaction, stable to acid hydrolysis is called cystine.

This is, however, the exception rather than the rule, and in most contexts the Ser hydroxyl group is no more reactive than that of ethanol.

Finally, special mention must be made of Cys, which, when present alone, can be considered to belong to the polar uncharged group described above. It can, however, when correctly positioned within the three-dimensional (3-D) structure of a protein, form disulfide bridges with another Cys residue (Figure 4.2). These are the only covalent bonds, apart from the peptide bond of course, that we usually find in proteins[2].

---

[2] Some proteins, like collagen and elastin, have covalent cross-links between their fibres, which are formed after their synthesis (post-translational modification).

PRIMARY, SECONDARY, TERTIARY AND QUATERNARY STRUCTURES OF PROTEINS

We can distinguish several levels of structural organization in proteins. These are typically described as primary, secondary, tertiary and quaternary, as shown in the well-known illustration of Irving Geis (Figure 4.3). The *primary* structure is quite simply the linear amino acid sequence of the polypeptide chain. We know, from the classic experiments of Christian B. Anfinsen (1973), that the amino acid sequence inherently contains all the information required for the overall 3-D structure of the protein (but we do not yet know how to predict the latter accurately from the former). The *secondary* structure of the protein consists of local, regular structures stabilized by hydrogen bonds involving the amide backbone of the polypeptide chain, such as α-helices, β-pleated sheets and reverse turns. The *tertiary* structure is formed by the packing of such structural elements into one or more compact globular units, often referred to as either supersecondary structures or domains. This compact structure is described by the 3-D localization of all atoms of the protein's amino acid sequence, both main and side chains. This is in contrast to the secondary structure, which only involves the polypeptide backbone and brings together amino acid residues that are far apart in the structure to form a functional region, known as the active site. Some proteins, such as haemoglobin illustrated in Figure 4.3, contain several polypeptide chains, each with their tertiary structure, arranged together in a *quaternary* structure.

When Linus Pauling and Robert Corey carried out their pioneering X-ray crystallographic studies on a number of amino acids and dipeptides in the 1930s and 1940s, they

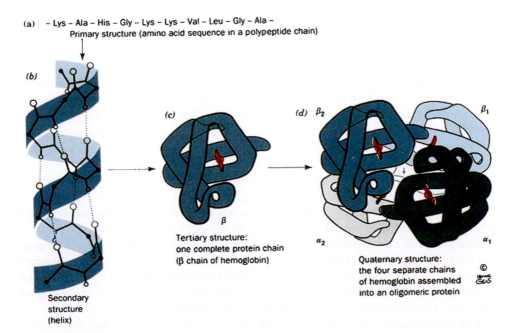

(a)   – Lys – Ala – His – Gly – Lys – Lys – Val – Leu – Gly – Ala –
Primary structure (amino acid sequence in a polypeptide chain)

(b)

Secondary structure (helix)

(c) β

Tertiary structure: one complete protein chain (β chain of hemoglobin)

(d) β₂   β₁   α₂   α₁

Quaternary structure: the four separate chains of hemoglobin assembled into an oligomeric protein

Figure 4.3 Structural organization of proteins. (From Voet and Voet, 2004. Reproduced with permission from John Wiley & Sons., Inc.)

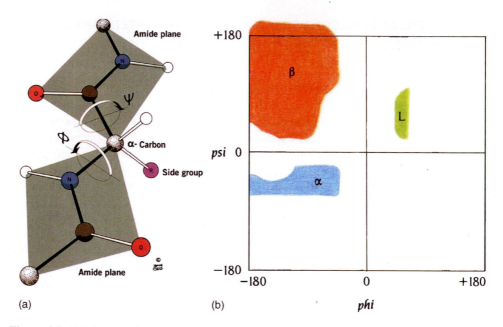

Figure 4.4 (a) Diagram of a polypeptide chain with the peptide units represented as rigid structures showing the conformational angles ϕ and ψ (from Voet and Voet, 2004) and (b) a Ramachandran plot, showing the sterically allowed angles for ϕ and ψ. (From Branden and Tooze, 1991. Reproduced by permission of Garland Publishing, Inc.)

arrived at three very important conclusions and constraints: (i) the most important constraint was that all six atoms of the amide (or peptide) group lie in the same plane. Pauling had predicted planar peptide groups because of the resonance of electrons between the double bond of the carbonyl group and the amide C–N bond of the peptide bond, which results in partial double-bond character in the C–N bond and partial single-bond character of the C=O bond, (ii) the peptide bond is usually trans and (iii) the maximum amount of hydrogen bonding potential is realized between amide functions. Since the peptide units are essentially rigid structures, linked by covalent bonds at the α-carbons, the only degree of freedom that they have are the rotations around these bonds—defined by the angles ϕ (*phi*) around the N–C$_\alpha$ bond and ψ (*psi*) around the C$_\alpha$–C′ bond (Figure 4.4a). If we can define the angles ϕ and ψ for each amino acid, we can describe the conformation of the main chain of the protein.

The Indian biophysicist G.N. Ramachandran made calculations of the sterically allowed values of ϕ and ψ, which would avoid steric collisions, either between atoms in different peptide groups or between a peptide unit and the side chains attached to C$_\alpha$. The representation of the allowed values of ϕ and ψ in Figure 4.4b is called a Ramachandran plot, and it is clear that only a few regions of the diagram, represented by the coloured regions, are sterically allowed. The areas labelled α, β and L correspond to right-handed α-helices, β-strands and left-handed α-helices, respectively. Observed values for all residue types, except glycine, for amino acid residues in well-refined X-ray structures of proteins which have been determined to high resolution, are shown in Figure 4.5a and the observed values for Gly residues in these same proteins in Figure 4.5b. This underlines the remark made earlier that Gly plays a structurally important role by allowing unusual

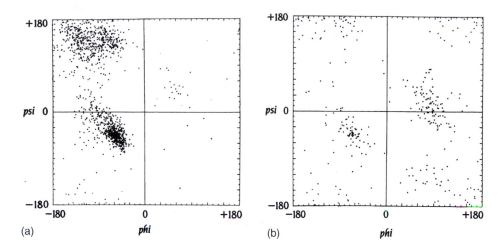

Figure 4.5 (a) Observed values for ϕ and ψ angles in protein structures for all residues except Gly and (b) observed values for Gly. (From Branden and Tooze, 1991. Reproduced by permission of Garland Publishing, Inc.)

main chain conformations. This may explain why a high proportion of Gly residues in homologous protein sequences are conserved.

On the basis of the observations described above, together with model-building studies, Pauling and his colleagues proposed two important structures, namely the α-helix and β-pleated sheets, which they predicted would be found in proteins (Eisenberg, 2003). The α-helix has 3.6 residues per turn, with hydrogen bonds between the C=O of residue n and the NH of residue $n+4$ (Figure 4.6), corresponding to the allowed ϕ and ψ angles of 60° and $-50°$, respectively. Some other helical structures are also found in proteins, notably the 3_{10} helix with three residues per helical turn, which is often found as a single turn at the C-terminus of α-helices, enabling the polypeptide chain to change direction. Although Pro cannot participate in hydrogen bonding, Pro residues do turn up in the first turn of α-helices (the first hydrogen bond is between the C=O of the first and the NH of the fifth residue). When Pro does occur elsewhere in a helix, it usually produces a bend in the helix. Since the peptide unit has a dipole moment (due to the different polarity of NH and C=O groups), it follows that α-helices have a significant dipole moment, with a partial positive charge at the amino end and a partial negative charge at the carboxyl terminus. α-Helices can contain from 5 to over 40 residues, with an average length of around 10 residues.

The second structural element to be proposed by Pauling and Corey was the β-pleated sheet (Figure 4.7). These sheets are made up of β-strands, typically from 5 to 10 residues long, in an almost fully extended conformation, aligned alongside one another with hydrogen bonds formed between the C=O bonds of one strand and the NH of the other, and vice versa. The β-sheets are pleated (i.e. they undulate) with the C_{α} atoms alternatively a little above, or a little below the plane of the β-sheet, which means that the side chains project alternatively above and below the plane. β-Strands can interact to form two types of pleated sheets.

(i)   Parallel β-pleated sheets, in which the polypeptide chains run in the same direction, have less-stable hydrogen bonds than anti-parallel β-sheets. This is reflected

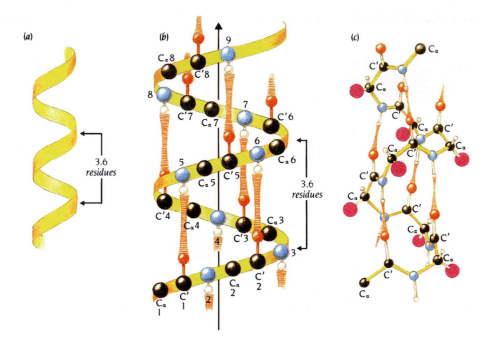

Figure 4.6 (a) Idealized diagram of the path of the main chain in an α-helix; (b) approximate positions of main chain atoms and hydrogen bonds; the arrow indicates the direction from *N*- to *C*-terminal; (c) schematic diagram of an α-helix; H-bonds are red and striated, as in (b), side chains represented by purple circles. (From Branden and Tooze, 1991. Reproduced by permission of Garland Publishing, Inc.)

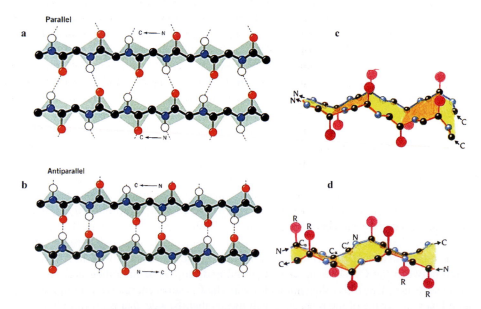

Figure 4.7 (a) Parallel and (b) anti-parallel β-sheets showing hydrogen bonds, but omitting side chains (from Voet and Voet, 2004); (c) parallel and (d) anti-parallel β̃-sheets illustrating the pleated nature of the sheet. (From Branden and Tooze, 1991. Reproduced by permission of Garland Publishing, Inc.)

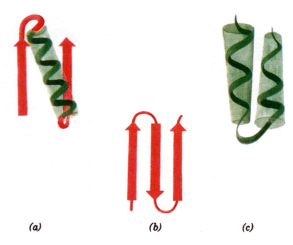

(a)   *(b)*   *(c)*

Figure 4.8 Super-secondary structures found in proteins (a) β-α-β motifs; (b) anti-parallel β-sheets connected by hairpin loops; (c) α-α motifs. (From Voet and Voet, 2004. Reproduced with permission from John Wiley & Sons., Inc.)

in the finding that parallel β-sheets of less than five strands are rare. However, the C_α-carbon atoms are all at the same distance, decreasing the restrictions on the amino acid sequences that can take up this secondary structure. In contrast to anti-parallel β-pleated sheets, which often involve contiguous amino acid sequences, parallel β-pleated sheets require a substantial number of amino acids between the individual β-strands, and this is frequently furnished by an α-helix (Figure 4.8a). The resulting β-α-β motifs constitute one of the frequently encountered super-secondary structures found in many proteins.

(ii)   Anti-parallel β-pleated sheets, in which the polypeptide chains run in opposite directions. The hydrogen bonds are well oriented, but the C_α carbons occupy two distinct positions—one relatively close, the other further apart. This places important restrictions on the amino acid residues that can occupy the former positions (the principal protein of silk, fibroin, has a repetitive sequence GAGA(S), with the Gly residues in the close positions and the Ala (or Ser) in the distance). Adjacent anti-parallel β-sheets can be joined by hairpin loops (often referred to as β-bends) with a hydrogen bond between the C=O of the first amino acid and the NH of the fourth (Figure 4.8b). These anti-parallel β-sheets, made up of hairpin loops from 4 to 5 up to more than 10, represent the second class of super-secondary structures.

The association of secondary structures to give super-secondary structures, which frequently constitute compactly folded domains in globular proteins, is completed by the α-α motifs in which two α-helices are packed in an anti-parallel fashion, with a short connecting loop (Figure 4.8c). Examples of these three structural domains, often referred to as folds, are illustrated in Figures 4.9—4.11. The schematic representation of the main chains of proteins, introduced by Jane Richardson, is used with the polypeptide backbone

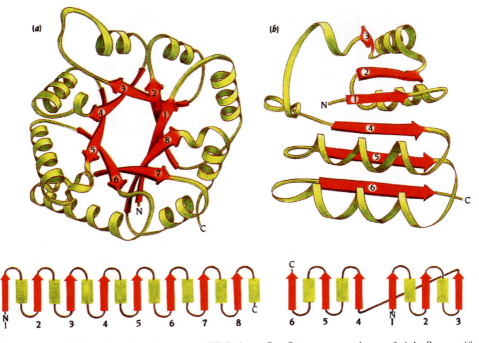

Figure 4.9 (a) Triose phosphate isomerase (TIM), has a β-α-β structure made up of eight β-α motifs terminating in a final α-helix, which form a barrel-like structure. (b) An open twisted β-sheet with helices on both sides, such as the coenzyme-binding domain of many dehydrogenases. (From Branden and Tooze, 1991. Reproduced by permission of Garland Publishing, Inc.)

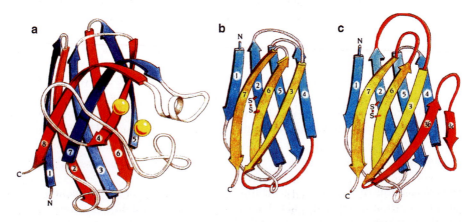

Figure 4.10 (a) The Cu–Zn superoxide dismutase is made up of eight anti-parallel β-strands; both the constant (b) and variable (c) domains of immunoglobulins are made up of seven anti-parallel β-strands with the same topology: the variable domain contains two additional β-strands. (From Branden and Tooze, 1991. Reproduced by permission of Garland Publishing, Inc.)

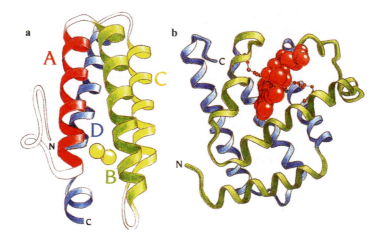

Figure 4.11 (a) Four helix bundle domain proteins, illustrated by myohaemerythrin. The oxygen-binding site is located at the di-iron centre within the hydrophobic core of the helical bundle. (b) The globin fold, represented here by myoglobin. (From Branden and Tooze, 1991. Reproduced by permission of Garland Publishing, Inc.)

represented by ribbons, α-helices as coils and β-sheets as arrows pointing towards the C-terminus, joined by loops, usually on the outer surface of the protein. This enables the characteristic folds of proteins to be more easily identified than when all of the side chains are included (Branden and Tooze, 1991; Creighton, 1993; Fersht, 1999).

Triose phosphate isomerase (TIM), a 247-residue enzyme of the glycolysis pathway, is an example of the first of the two main classes of α/β proteins, made up of β-α-β motifs. It consists of a core of eight twisted parallel β-strands, arranged close together to form a cylindrical structure, known as a β-barrel viewed in Figure 4.9a from the top. The β-strands are connected by α-helices, located on the outside of the β-barrel. This domain structure is often referred to as the TIM barrel, since it was first found in TIM. In the topological representation at the bottom of the figure, the α-helices are represented by rectangles and the β-sheets as arrows. The second class of α/β proteins contains an open-twisted β-sheet surrounded by α-helices on both sides of the β-sheet. A typical example (Figure 4.9.b) is the nucleotide-binding domain found in a number of dehydrogenases and kinases.

The second family of protein super-secondary structures, the anti-parallel β-structures, represent a very diverse range of function, including enzymes, transport proteins, antibodies and virus coat proteins. Their cores are built up of a number of β-strands, which can vary from four or five to over ten, arranged in such a way that they form two β-sheets that are joined together and packed against each other. The enzyme superoxide dismutase (Figure 4.10) is made up of eight anti-parallel β-strands, arranged in such a way that they form two β-sheets that are joined together and packed against each other to form a β-sandwich. When two such twisted β-sheets are packed together, they form a barrel-like structure. The immunoglobulins are a family of molecules, which function in the molecular recognition of foreign antigens. Though they are made up of a large number of domains, most of them (12 out of 13 in immunoglobulin G, IgG) have the ancestral immunoglobin fold (Figure 4.10b). In the constant domains of immunoglobulins, this consists of seven β-strands with four strands forming one β-sheet and three strands forming a second sheet. The sheets are closely packed against each other and are joined by a disulfide bond between them. The variable domains (Figure 4.10c) have a very similar structure to the constant domains, but have nine β-strands instead of seven. The two additional strands are inserted in the loop region between strands 3 and 4 and are functionally important in that they contain one of the hyper-variable regions (in red), which is placed close to the other two hyper-variable domains. These three hyper-variable regions together constitute the part of the immunoglobulin molecule that determines its specificity in the recognition of antigens.

The simplest way to pack a pair of adjacent helices is to place them anti-parallel to one another connected by a short loop. A frequently encountered domain structure in proteins is a bundle of four helices with their long axes aligned around a central-hydrophobic core. This is illustrated by myohaemerythrin (Figure 4.11), a non-haem oxygen transport protein in marine worms.

Another important helical domain is the globin fold, found in the first protein for which the 3-D structure was determined, the oxygen storage protein, myoglobin (Figure 4.9b), from the muscle of the sperm whale. The four-globin chains of haemoglobin have a very similar 3-D structure to myoglobin. It later became apparent that oxygen-binding haemoproteins from invertebrate insect larvae (*Chironomus thummi*) and from the lamprey, a blood-sucking eel-like cyclostome (a subclass of jawless parasitic fish), also have the globin fold. However, the pair-wise arrangements of the bundle of eight α-helices in the globin fold is

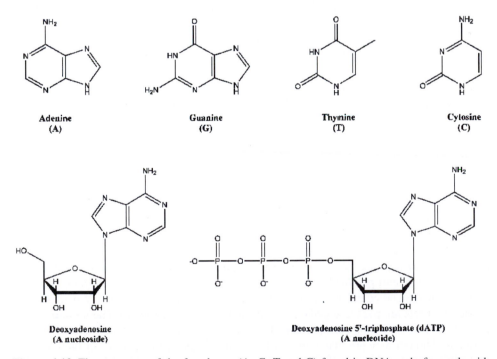

Figure 4.12 The structures of the four bases (A, G, T and C) found in DNA and of a nucleoside (deoxyadenosine) and a nucleotide (deoxyadenosine-5′-triphosphate).

quite different from that found in the four-helix-bundle domains. The helices, which vary in length from 7 (helix C) to 28 (helix H) residues, wrap around the core of the molecule, with its haem-binding pocket, such that most adjacent helices in the 3-D structure are not adjacent in the amino acid sequence (an exception are the last two helices, G and H).

The structural building blocks of nucleic acids

Nucleic acids are made up of three components—nitrogen-rich bases of the pyrimidine and purine families, illustrated in Figure 4.12 by the DNA bases adenine and guanine, and thymine and cytosine, respectively, which are linked in an *N*-glycosidic bond to a sugar (either ribose or deoxyribose in RNA and DNA, respectively) and phosphate groups, which link the sugar residues. The combination of a nucleobase with a sugar generates a nucleoside, hence adenine becomes deoxyadenosine when bound to deoxyribose; while a nucleobase plus a sugar plus one or more phosphate residues (usually on the 5′-hydroxyl of the

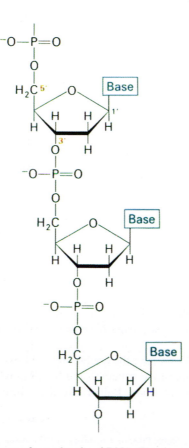

Figure 4.13 The structure of part of a molecule of DNA. The deoxyribose residues are linked by phosphodiester bonds between the 3′OH of one nucleoside and the 5′OH of the next.

ribose or deoxyribose) constitute a nucleotide (in the example given, deoxyadenosine 5′-triphosphate).

In RNA, the base T found in DNA is replaced by uracil, which is similar in structure to T, but lacks the methyl group. The nucleotides in nucleic acids are linked by phosphodiester bonds between the 3′-hydroxyl of one nucleoside and the 5′-hydroxyl of the sugar of its neighbour in the sequence, as was first shown by Alexander Todd[3] in 1952 (Figure 4.13).

SECONDARY AND TERTIARY STRUCTURES OF NUCLEIC ACIDS

An enormous kick-start to modern molecular biology was given by the seminal 1953 Nature paper of Francis Crick and Jim Watson on the double-helical structure of DNA. It was based on two important observations. Firstly, the determination of the base com-position of DNA from a number of sources had shown that while the base composition varied widely, there was always the same amount of A as of T, and the same amount of G as of C. Secondly, high quality X-ray photographs of DNA fibres were consistent with a helical structure composed of either two or three polynucleotide chains. The major physi-ological form of the DNA double helix is the B-DNA described by the X-ray pictures of Rosalind Franklin, with 10 base pairs per turn (Figure 4.14), each pair of bases separated by 0.34 nm and by a helical twist of 36° per base pair resulting in extensive interactions between the bases (often referred to as stacking). The purine and pyrimidine bases project into the interior of the helix, forming hydrogen bonds between A and T and between G and C (Figure 4.15), such that the helix has a solid core. The deoxyribose-phosphate backbone winds about the outside of the molecule, with the two sugar-phosphate chains running anti-parallel to one another. As will become important later, when we consider regulation of genetic expression, this sugar-phosphate backbone constitutes the potential-binding sites for proteins, which through their binding to DNA will influence the expression of its genetic material. We can clearly distinguish (Figure 4.14) a much broader side of the DNA double helix, the so-called major groove, from the less-accessible minor groove.

The DNA double helix is stabilized by both hydrophobic interactions between the bases (base stacking) and hydrogen bonds between the A–T and G–C base pairs, and it can be reversibly dissociated into the individual strands by heating. This process is termed melting, and the melting temperature (T_m) is defined as the temperature at which half the helical structure is lost. The importance of hydrogen bonding in stabilizing the double-helical structure is underlined by the observation that G–C-rich DNA has a much higher T_m than A–T-rich DNA (G–C base pairs have three hydrogen bonds whereas A–T pairs have two).

While RNA molecules do not have the double stranded structure usually found in DNA, in many RNA molecules stem-loop structures are found in which the anti-parallel strands are connected by a 5–7 residue loop. Rather like the β-turn in proteins, this allows the

---

[3] Sir Alexander Todd, who won the Nobel Prize for Chemistry in 1957, was a former pupil of Allan Glen's school in Glasgow, where I also got my secondary education. He not only established the chemical structure of nucleic acids, but we owe to him our knowledge of the structures of FAD, ADP and ATP.

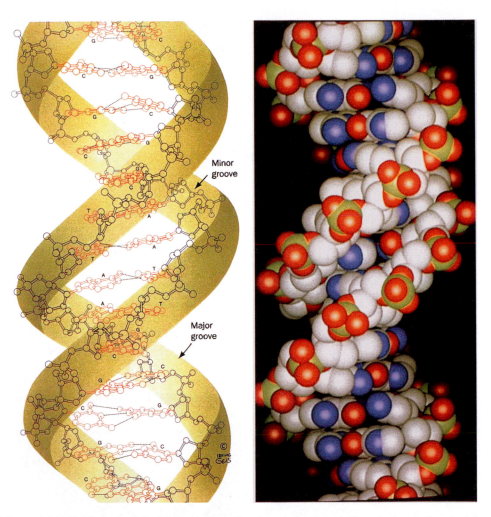

Figure 4.14 The B-form of the DNA double-helix viewed along the helix axis, in a ball-and-stick representation (left) and a space-filling representation (right). (From Voet and Voet, 2004. Reproduced with permission from John Wiley & Sons., Inc.)

polynucleotide chain to change direction by 180°. However, in addition to the classic base pairs (with U replacing T) a number of non-Watson–Crick base pairs are also found. This is particularly well illustrated by the structure of the first RNA molecule to have its 3-D structure determined—the transfer RNA encoding Phe (tRNA[Phe]), which is presented in Figure 4.16. The striking feature of this structure is the optimization of hydrogen-bonding interactions, many of them non-classical Watson-Crick (often called Hoogsteen base pairing, such as that between A5 and U67), which ensures that the molecule attains the maximum degree of hydrogen bonding between bases, utilizing hairpin loops, somewhat like the 3_{10} helix in proteins, to change the direction of the polynucleotide chain by 180°.

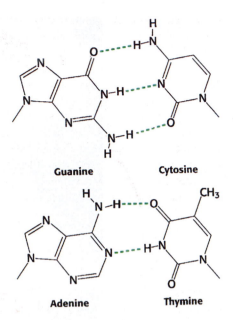

Figure 4.15 The classic 'Watson-Crick' base-pairing between A and T, and between G and C in DNA.

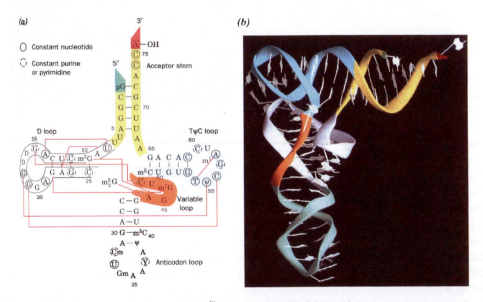

Figure 4.16 The structure of yeast tRNA^Phe (a) the cloverleaf form of the base sequence: tertiary base-pairing interactions are represented by thin red lines connecting the participating bases. Bases that are conserved in all tRNAs are circled by solid and dashed lines, respectively. The different parts of the structure are colour-coded. (b) The X-ray structure showing how the different base-paired stems are arranged to form an L-shaped molecule. The sugar-phosphate backbone is represented as a ribbon with the same colour scheme as in (a). (From Voet and Voet, 2004. Reproduced with permission from John Wiley & Sons., Inc.)

It is interesting to note that the anticodon (G_mAA), which interacts with the mRNA to correctly position Phe for incorporation into the appropriate protein, is situated in the 3-D structure 80 Å away from the Phe residue, bound to the 3′-OH of the tRNA (highlighted in red in the figure).

Whereas DNA is mostly located in the nucleus of cells in higher organisms (with some also in mitochondria and in plant chloroplasts), RNA comes in three major and distinct forms, each of which plays a crucial role in protein biosynthesis in the cytoplasm. These are, respectively, ribosomal RNA (rRNA), which represents two-thirds of the mass of the ribosome, messenger RNA (mRNA), which encodes the information for the sequence of proteins, and transfer RNAs (tRNAs) which serve as adaptor molecules, allowing the 4-letter code of nucleic acids to be translated into the 20-letter code of proteins. These latter molecules contain a substantial number of modified bases, which are introduced enzymatically.

Carbohydrates

Carbohydrates, so-called because they have the empirical formula $C_nH_{2n}O_n$, make up an important class of biomolecules. They are extensively used in the form of simple sugars (monosaccharides) in intermediary metabolism (see Chapter 5), in polymer form (polysaccharides) as storage forms of both carbon and energy (such as polymers of glucose in the starch of plants and the glycogen of animals) and as structural elements (polymers of glucose in the cellulose of plant cell walls, and of *N*-acetylglucosamine in the chitin of invertebrate shells), or even attached to proteins and lipids in highly specific structure involved in biological recognition (such as the blood group determinants).

We begin by describing briefly the main classes of monosaccharides, together with some of the stereochemical properties that give them their particularity with regard to molecular recognition. If we stick with the definition of monosaccharides as $(CH_2O)_n$, it is clear that we need n to be at least three (three carbon 'oses', or sugars called trioses), and we must have one of the carbons at the level of oxidation of an aldehyde or ketone—as illustrated (Figure 4.17) by the structure of the two trioses, glyceraldehyde (an aldose) and dihydroxyacetone (a ketose). These two molecules are *isomers* of one another, and can be inter-converted by enzymes called isomerases. In addition, we note an important point that the hydroxyl group on the central carbon of glyceraldehyde can be either on the right or the left side of the central carbon chain. In fact most sugars (with the exception of the sugar acid ascorbic acid, better known as vitamin C) have the hydroxyl furthest away from their potential reducing function (the aldehyde or ketone group) to the right, and are therefore called *D-sugars*. We will not fatigue the reader with the niceties of the structures of the different aldo- and keto-tetroses, pentoses, hexoses, etc., but simply illustrate a few more, all of which play an important role in metabolism, often in a phosphorylated form—D-ribose, D-galactose and D-glucose (aldoses), and D-fructose (ketose). We illustrate our next important point with glucose (an aldohexose), which can exist in a cyclic form by ring closure between the aldehyde function and the hydroxyl function on carbon 5—this forms a hemiacetal, and when the same thing happens with a ketone, as in fructose, we form a hemiketal (Figure 4.17). The name is less important than the consequence—the product now has an asymmetric carbon atom, such that carbon 1 of glucose in the cyclic form

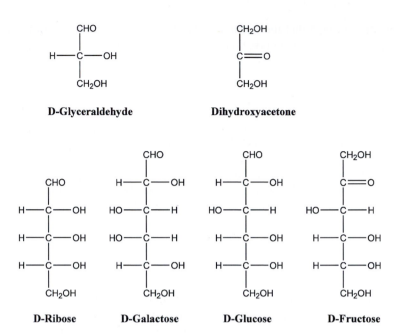

Figure 4.17 The trioses D-glyceraldehyde (aldose) and dihydroxyacetone (ketose), the pentose D-ribose, the hexoses D-galactose and D-glucose (aldoses) and the ketohexose D-fructose in their open chain forms. The configuration of the asymmetrical hydroxyl group on the carbon, the furthest away from the aldehyde or ketone group, determines the assignment of D- or L-configuration.

can have its hydroxyl group either below the plane of the ring (α-D-glucose) or above it (β-D-glucose). These two forms are known as *anomers*, and carbon 1 is the anomeric carbon. This has profound effects, as we will see shortly, when we form links between two sugar residues (contrast the vastly different physical properties of starch and cellulose). Of the sugars represented in Figure 4.18, ribose, fructose and galactose can, and will like glucose, form cyclic forms that are the most stable structures in aqueous solution. Finally, we must distinguish between glucose and galactose, which have a different configuration of the hydroxyl group on carbon 4—these are *epimers*. Many of the monosaccharides described here will be found in metabolic pathways, often phosphorylated. Other biologically important sugar derivatives include deoxy sugars, such as β-D-2-deoxyribose in DNA, and amino sugars, such as N-acetylglucosamine, where the hydroxyl on carbon 2 is replaced by an acetylated amino group, found in chitin.

When two sugars are linked together, a glycosidic bond is formed between them (Figure 4.19), which is illustrated by the Belgian disaccharide[4], maltose, derived from

---

[4] A map of France dating from the late 18 hundreds, which I bought a few years ago, had a small part of Belgium extending at the northern extremities, described as 'kingdom of beer drinkers'. The major substrate for beer production is the disaccharide maltose, produced by partial hydrolysis of starch in the preparation of the malt, prior to addition of the yeast, which then ferments the sugars to ethanol. Since Belgian beers are a major cultural and economic heritage (and in my humble opinion among some of the best in the world), maltose is a good candidate for the national sugar.

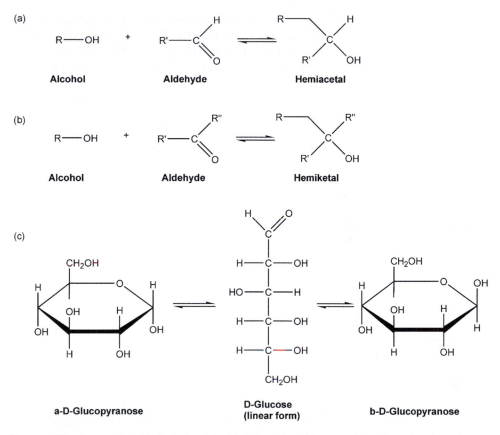

Figure 4.18 The reaction of alcohols with aldehydes (a) and ketones (b) to form hemiacetals and hemiketals. (c) The reaction between the alcohol on carbon 5 and the aldehyde of glucose forms two hemiacetals, α-D-glucopyranose and β-D-glucopyranose (pyranose by comparison with pyran the simplest compound containing this six-membered ring).

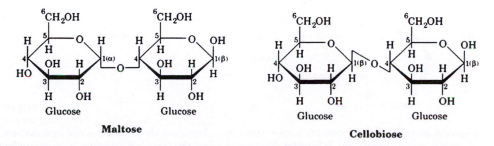

Figure 4.19 The structures of the disaccharides maltose and cellobiose, derived from the hydrolysis of starch and cellulose, respectively.

starch hydrolysis. Here, the two glucose molecules are linked by an α-glycosidic linkage (α-1-4) between the aldehyde group of one glucose molecule and the hydroxyl of carbon 4 of the other, in its α-configuration. Contrast this with cellobiose, the principal disaccharide derived from hydrolysis of cellulose, which has a β-1-4 linkage between the two glucose molecules (Figure 4.19).

There are two main classes of polysaccharides—those that are used as stores of energy and carbon, such as glycogen and starch, and structural polysaccharides such as cellulose. Starch, the principal storage polysaccharide of plants, is a mixture of the linear α(1-4)-linked polyglucose α-amylose, and amylopectin, which although having mostly amylose-like α(1-4)-linkages, also has α(1-6)-branches every 24–30 glucose residues. Glycogen, the storage polysaccharide of animals, is found predominantly as hydrated cytoplasmic granules in tissues such as liver and muscle, which contain up to 120,000 glucose units. It has a structure similar to amylopectin, but with α(1-6)-ramifications every 8–14 residues (Figure 4.20), which allows its rapid degradation to simultaneously release glucose units from the end of each branch point.

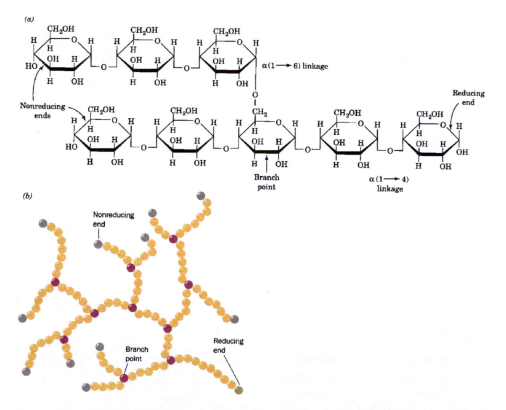

Figure 4.20 (a) Molecular structure of glycogen (the polyglucose chains in the actual molecule are, of course, much longer). (b) Schematic diagram showing the branched structure: note that while there is only one reducing end, there are multiple non-reducing ends from which glucose units can be released. (From Voet and Voet, 2004. Reproduced with permission from John Wiley & Sons., Inc.)

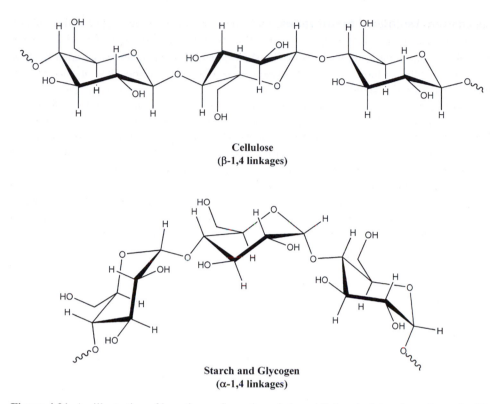

Cellulose
(β-1,4 linkages)

Starch and Glycogen
(α-1,4 linkages)

Figure 4.21 An illustration of how the configuration of glycosidic bonds determine polysaccharide structure and function. The β-1-4 linkages in cellulose favourize straight chains, optimal for structural purposes, whereas the α-1-4 linkages are favourable to bent structures, better adapted to storage in a hydrated form.

Cellulose represents the major structural component of plant cell walls—around 10^{15} kg of cellulose, half of the carbon in the biosphere, are synthesized (and degraded) annually. In contrast to the storage polysaccharides, it is a glucose polymer linked exclusively by β(1-4) glycosidic bonds, typically with up to 15,000 glucose residues. In contrast (Figure 4.21) to the extensively α(1-4) structure of starch and glycogen, both of which have bent structures accessible to hydration (as in the glycogen storage granules mentioned earlier), the β-1-4 linkages of cellulose favour long straight chains. These can form fibrils of parallel chains, which interact with one another through a hydrogen-bonding network. So, a simple change in the configuration of a glycosidic bond can produce spectacular differences between a hydrated granular store of energy and a major component of vegetable cell walls, which, for example in trees, must ensure a considerable role in load bearing. When we turn to the shell of crustaceans such as the lobster, which have a polymer composed of *N*-acetylglucosamine in a β-1-4 linkage, the change in the properties are remarkable.

Lipids and biological membranes

Lipids[5] are fat-soluble molecules, unlike the other biological macromolecules that are all water soluble, which we have described up to now; in addition, they are *not* macromolecules, but of relatively low molecular weight. With the exception of a class of lipids (for reasons of Commissions of Nomenclature[6] known as simple lipids, despite the organic complexity of their structures—they include steroids, isoprenoids, etc.), they are complex lipids. By definition they are saponifiable; in other words, upon treatment with an alkali, they produce soap, containing an alcohol, classically glycerol, and other products, usually fatty acids. Fatty acids typically have an even number of carbon atoms (reflecting their synthesis from acetyl CoA, see Chapter 5) often 16 or 18 and may be saturated, such as stearic acid (mp 69.6 °C) or have one (or more) double bonds, such as oleic acid (mp 13.4 °C) (Figure 4.22a). However, the important message to get across here is that the properties of the lipid will be determined by the nature of the fatty acids that it contains. A simple example from everyday life is the difference between lard (animal fat) and olive oil. Lard, at 20 °C is solid, yet olive oil at the same temperature is liquid—why? Both are triglycerides (Figure 4.22b), but the former consists of glycerol esterified to three molecules

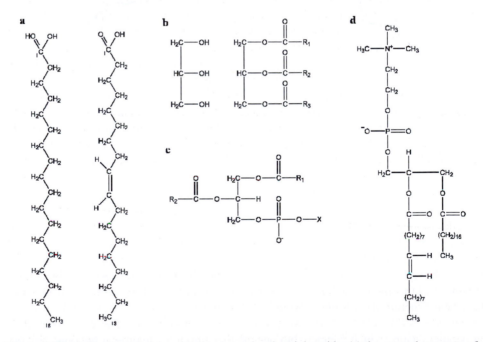

Figure 4.22 (a) Stearic and oleic acid; (b) glycerol and a triglyceride; (c) the general structure of a glycerophospholipid; and (d) the glycerophospholipid 1-stearoyl-2-oleoyl-3-phosphatidylcholine.

---

[5] From Greek *lipos*—fat.
[6] There are two classes of sins—those of omission and of commission. When it comes to scientific nomenclature, these are clearly sins of commission!

of saturated fatty acids (typically stearic acid), in the latter, glycerol, is esterified to three molecules of oleic acid, with a single double bond. Triglycerides are a very efficient reservoir of energy resources, not only because they can produce twice as much energy per gram, than carbohydrates or proteins, but also because they can be stored in an anhydrous form. This is in marked contrast to glycogen, which binds about twice its weight in water. Animals have cells that are specialized in the synthesis and storage of triglycerides, adipocytes, with the fat globules occupying almost the entire cell.

Having established that the simplest 'complex' lipid is a triglyceride, we pass to the biological properties of phospholipids that combine the fat-soluble properties of the triglyceride with the addition of a polar, charged group on one of the glycerol hydroxyl functions (Figure 4.22c). Several alcohols are found in phospholipids, but here we consider the glycerophospholipids. The polar head group is made up of a phosphate group, attached to a diglyceride (this constitutes a phosphatidic acid) to which a polar alcohol is esterified. The alcohol may be, ethanolamine, choline (as in Figure 4.22d), serine, inositol, etc. and the resulting families of phospholipids are called phosphatidyl ethanolamine, phosphatidyl choline, etc. What we have is an *amphiphilic* molecule, which will bury its hydrophobic component in a non-aqueous environment, and maintain its polar head group in contact with water. Such molecules will spontaneously form monomolecular layers on the surface of water, and will readily form bimolecular layers in aqueous solution. An important feature of lipid bilayers is that they have an inherent capacity to close on themselves, leaving no hydrocarbon chains exposed, forming compartments and are self-sealing, because a hole in the lipid bilayer would be energetically very unfavourable. This is the basis for biological membranes in which the amphiphilic phospholipids form a double bilayer of hydrophobic fatty acid side chains, excluding water within the bilayer, yet exposing the polar head groups on either side of the bilayer to the aqueous milieu. The driving force for the self-assembly of lipid bilayers is what we often call the 'hydrophobic effect', i.e. predominantly non-polar parts of biological molecules seek environments in which they are close to other similar molecules, forming a hydrophobic core. Hydrophobic interactions, as we have seen, also play a dominant role in the folding of proteins, creating their hydrophobic cores, and in the stacking of bases in nucleic acids.

However, lipid bilayers are impermeable to ions and most polar molecules, with the exception of water (on account of its high concentration), so they cannot on their own confer the multiple dynamic processes, which we see in the function of biological membranes. All this comes from proteins, inserted into the essentially inert backbone of the phospholipid bilayer (Figure 4.23), which mediate the multiple functions that we associate with biological membranes, such as molecular recognition by receptors, transport via pumps and channels, energy transduction, enzymes and many more. Biomembranes are non-covalent assemblies of proteins and lipids, which can best be described as a fluid matrix, in which lipid (and protein molecules) can diffuse rapidly in the plane of the membrane, but not across it. Biological membranes can therefore be considered as 2-D solutions of oriented lipids into which proteins are selectively inserted, either as integral membrane proteins, which traverse the entire bilipid layer, or as peripheral proteins, associated with one face of the membrane, where they may be bound covalently by a glycolipid linkage. Biological membranes are asymmetric: in the case of the plasma membrane it will often have carbohydrates in the form of oligosaccharides or glycolipids exposed at the outer side

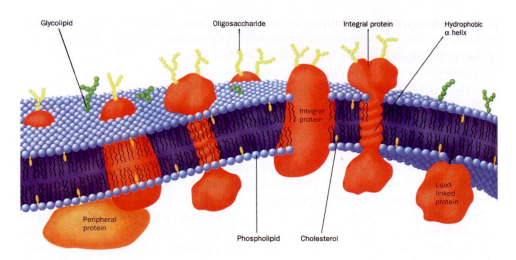

Glycolipid   Oligosaccharide   Integral protein   Hydrophobic α helix

Integral protein

Lipid-linked protein

Peripheral protein

Phospholipid   Cholesterol

Figure 4.23 Schematic diagram of a plasma membrane. (From Voet and Voet, 2004. Reproduced with permission from John Wiley & Sons., Inc.)

of the membrane. Finally, we should point out that, in animal cells, the part of the protein that traverses the hydrophobic membrane phospholipid is frequently a hydrophobic α-helix, some 20–30 residues long.

A brief overview of molecular biology

For reasons that are historical, the part of biochemistry that deals with the 2-D transfer of genetic information, from the 4-letter code of DNA and RNA to the 20-letter code of proteins, has often been referred to as molecular biology[7]. There is little doubt that the recent advances in genome sequencing and the extensive use of gene arrays to detect changes in gene expression have heightened the popular impression that molecular biology in this narrow definition can solve all the problems. It cannot and will not, without the much more time- and labour-consuming tasks of purifying and characterizing the proteins that the genome encodes. Unfortunately, the information contained in the DNA sequence of a gene does not allow us to predict its structure (unless it happens to be homologous to another protein of known structure), far less its biological function. Nonetheless, the tools of molecular biology are sufficiently important for the practice of modern biochemistry that we need to know about them. We begin by a brief description of what has often been termed the central dogma of molecular biology—DNA (which by its *replication* forms two identical daughter molecules) whose information, in the process of *transcription*, is incorporated into RNA and then, by *translation*, is converted into the amino acid sequence of the corresponding protein.

---

[7] A brief look at the contents page of any recent issue of the Journal of Molecular Biology (founded by John Kendrew, protein crystallographer and winner of the Nobel Prize for Chemistry together with Max Perutz for the 3-D structures of myoglobin and haemoglobin) will clearly establish that this is not so.

Replication and transcription

At the end of the classic Watson and Crick (1953) paper the authors write 'It has not escaped our notice that the specific pairing we have postulated immediately suggests a possible copying mechanism for the genetic material'. This may seem, with hindsight, fairly obvious. There is little doubt that DNA polymerases, the enzymes which synthesize DNA, do place the correct base in the daughter strand of newly synthesized DNA (Figure 4.24), by classic Watson-Crick base-pairing. DNA polymerases are in fact metal-loenzymes (Figure 4.25), and typically require two metal ions (usually Mg^{2+}). One metal

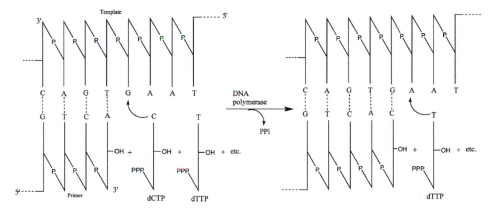

Figure 4.24 The global reaction catalysed by DNA polymerase.

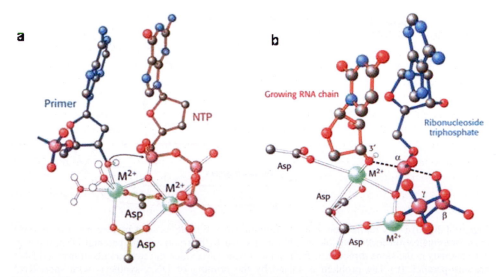

Figure 4.25 (a) DNA polymerase-catalysed phosphodiester bond formation typically requires two metal ions, usually Mg^{2+}. (b) A model of the transition state for phosphodiester-bond formation in RNA polymerase. (From Berg et al., 2002. Reproduced with permission from W.H. Freeman and Co.)

ion coordinates the 3′-hydroxyl group of the primer, while both metal ions coordinate the α-phosphoryl group of the nucleoside triphosphate (NTP). The 3′-hydroxyl group then attacks the α-phosphoryl group to form a new phosphodiester bond, releasing the two terminal phosphates as PP_i (pyrophosphate). The process of DNA synthesis begins with the multi-subunit complex of proteins, which will constitute the replisome binding to an initiation site. Once the two strands of the DNA double helix have been separated locally, by a number of proteins including helicases and single-strand binding proteins, replication can begin.

What makes things a lot more complicated is that the two strands of the DNA molecule are *anti-parallel*, yet only one site of replication is visible by low resolution techniques, such as electron microscopy during the replication of double-stranded DNA, at what is known as the replication fork (Figure 4.26a). There is no way that DNA polymerase can synthesize one strand of DNA in the way that is illustrated in Figures 4.24 and 4.25 and simultaneously synthesize the other in the opposite direction (which logically would have a 5′-phosphate at its terminus rather than what the enzyme actually requires, namely a 3′-hydroxyl). It turns out that this problem is solved by the semi-discontinuous replication of DNA, in which both daughter strands are synthesized in the required 5′ → 3′ direction. However, this requires the introduction on the strand, which lacks a 3′-hydroxyl, of short RNA primers (synthesized by yet another protein, primase), whose 3′-hydroxyl can then be used to synthesize DNA in the opposite direction to that of DNA replication on the other, so-called 'leading strand' that always has a 3′-hydroxyl available for the polymerase, so that DNA is synthesized continuously (Figure 4.26b). On the 'lagging strand',

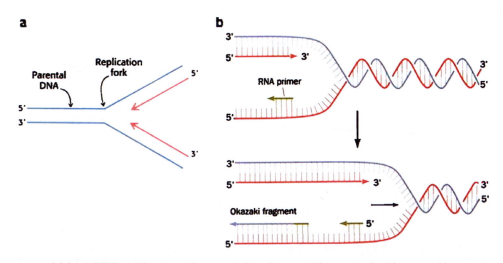

Figure 4.26 (a) DNA replication at low resolution (for example as seen by electron microscopy). Only one replication fork is visible and it appears that both strands of the parental DNA replicate continuously in the same direction, which cannot be the case, since the two strands of parental DNA are anti-parallel. (b) The problem is solved by the priming of DNA synthesis with short RNA primers, whose 3′-hydroxyl can be used by DNA polymerase, producing Okazaki fragments, while on the other strand, DNA synthesis is continuous. (From Voet and Voet, 2004. Reproduced with permission from John Wiley & Sons., Inc.)

DNA synthesis on the 3′-hydroxyl of the short RNA primer produces short DNA fragments, each starting from an RNA primer, called Okazaki[8] fragments (in eukaryotes these consist of only 100–200 nucleotides, while in prokaryotes they are much longer). The primer RNA is then excised to be replaced by DNA by a second DNA polymerase, which can remove RNA from one Okazaki fragment while replacing it by DNA on a neighbouring fragment. Once their RNA has been replaced by DNA, the Okazaki fragments on the lagging strand are joined together by a DNA ligase.

Whereas, in the process of replication, DNA polymerase replicates the entire DNA molecule, in the process of transcription, RNA polymerases synthesize an RNA molecule, which has a sequence complementary to a small fragment of one of the two strands of DNA (the coding strand). The reaction that they catalyse, illustrated by a model of the transition state for RNA polymerase (Figure 4.25b), is similar to that catalysed by DNA polymerase. However, unlike DNA polymerase, which has but one site on the entire DNA molecule to initiate DNA synthesis, RNA polymerases search the DNA for initiation sites (called *promoters*)—the 4.8×10^6 base-pair genome of *E. coli* has ~2000 promoters. RNA polymerases rely upon protein subunits called σ (sigma) factors to recognize promoter sites by gliding rapidly along the DNA duplex until they find a promoter, binding to it and initiating RNA synthesis. The σ factor is then released, allowing the core polymerase to continue synthesizing until it encounters a termination signal, whereupon it releases the newly synthesized RNA, binds to a σ factor, and goes off in search of a new promoter.

Often RNA molecules undergo 'maturation' after their synthesis by the polymerase, referred to as post-transcriptional processing, involving both excision of nucleotides and chemical modification. For example, tRNAs undergo extensive processing (Figure 4.27). In the case of the tRNA^Tyr of yeast, this includes excision of a 19-residue 5′-terminal sequence and a 14-nucleotide intervening sequence from the primary transcript, addition of a –CCA to the 3′-terminus, and extensive modification of a number of bases to form the mature tRNA. This includes transformation of uridines into pseudouridine (ψ), dihydrouridine (D) and thymine. Yet another series of post-transcriptional modifications, of great importance in eukaryote mRNAs (Figure 4.28), include addition of a 5′ cap, which defines the start site for translation of the mRNA (see next section), addition of a 3′ poly(A) tail and splicing. Genes of eukaryotes, unlike prokaryotes, have their coding sequences (*exons*) interspersed with non-coding intervening sequences (*introns*). The lengths of the introns typically represent 4–10 times that of the exons. Thus, the primary transcripts of eukaryotic mRNAs must have not only caps and tails attached, but the non-coding introns must also be excised and the exons joined up to form the mature mRNA, corresponding to the amino acid sequence of the protein. This process is known as gene *splicing*, and must be carried out with absolute precision—a one-nucleotide error in splicing would shift the reading frame of the mRNA and lead to an entirely different amino acid sequence. Splicing of eukaryotic mRNAs requires the cooperation of a number of small nuclear RNAs and proteins that form a large complex called a *spliceosome*.

---

[8] Reiji Okazaki showed that when replicating *E. coli* are labelled for short periods of time with precursors of DNA, some of the newly synthesized DNA is recovered as small fragments of 100–200 nucleotides long, which are subsequently incorporated into double-stranded DNA.

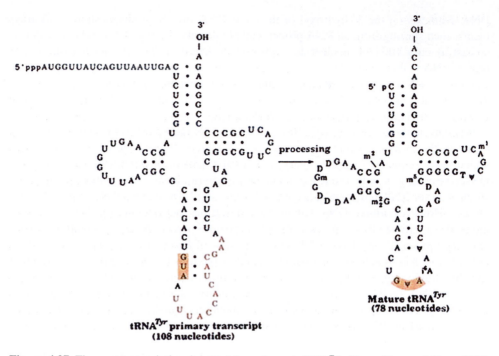

Figure 4.27 The post-transcriptional processing of yeast tRNA^{Tyr}. (From Voet and Voet, 2004. Reproduced with permission from John Wiley & Sons., Inc.)

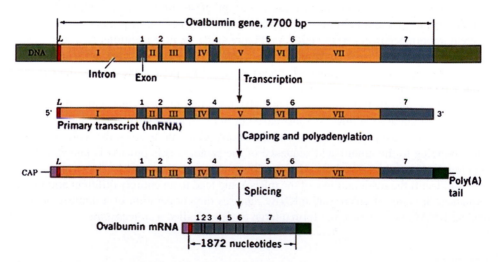

Figure 4.28 The steps involved in the maturation of eukaryotic mRNA, illustrated for the chicken ovalbumin gene. (From Voet and Voet, 2004. Reproduced with permission from John Wiley & Sons., Inc.)

Translation

Translation, the biosynthesis of proteins, involves passing from the 4-letter code of mRNA into the 20-letter code of proteins. It is clear that with 20 different amino acids in proteins, for translation to occur, we require a triplet code with three nucleic acid bases (codons) encoding for one amino acid. This means that the genetic code, as it is called, consists of 64 codons. It is therefore degenerate, i.e. most amino acids have more than one codon. The genetic code is read, not by specific recognition of codons by individual amino acids themselves, but rather by the selective binding of amino acids to adaptor molecules, as first postulated by Francis Crick in 1955 (Figure 4.29). These adaptors, tRNAs, then recognize the corresponding codon on the mRNA by base-pairing between the codon and a three-base anticodon on the tRNA molecule. The key step in determining the specificity of protein biosynthesis is the loading of the amino acids onto their corresponding tRNA by enzymes called aminoacyl-tRNA synthetases. This is a two-step process involving formation of an enzyme-bound aminoacyl-adenylate between the amino acid and a molecule of ATP, followed by transfer of the amino acid to the terminal 2'- or 3'-hydroxyl of its tRNA to form the aminoacyl-tRNA (Figure 4.30). The importance of this reaction is underlined by the classic experiment (Chapeville et al., 1962) in which cysteine, loaded on tRNA^{Cys}, was reductively converted to alanine using Raney nickel; and it was shown that Cys residues in rabbit haemoglobin when synthesized using this Ala-tRNA^{Cys} were systematically

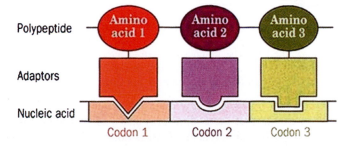

Figure 4.29 The adaptor hypothesis. (From Voet and Voet, 2004. Reproduced with permission from John Wiley & Sons., Inc.)

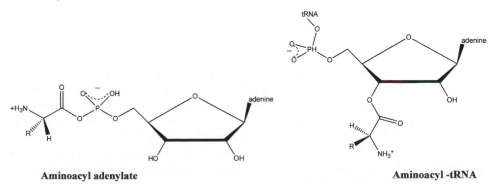

Figure 4.30 Structures of an aminoacyl-adenylate and of an aminoacyl-tRNA.

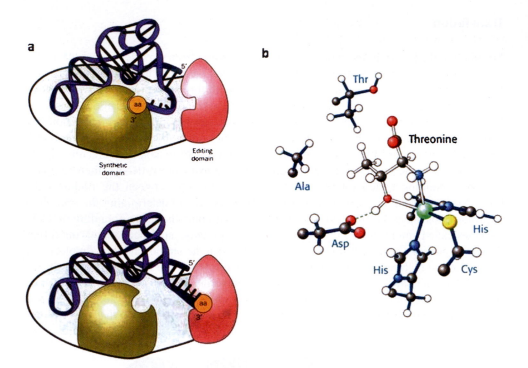

Figure 4.31 (a) A cartoon comparing the positions of the 3′-end of an aminoacyl-tRNA in its synthetic (top) and editing (bottom) modes. (From Voet and Voet, 2004. Reproduced with permission from John Wiley & Sons., Inc.). (b) The amino acid-binding site of threonyl-tRNA synthetase showing the amino acid bound to a zinc atom through its amino and hydroxyl groups. (From Berg et al., 2002. Reproduced with permission from W.H. Freeman and Co.)

replaced by Ala, confirming that the coding properties of this hybrid tRNA are determined by the tRNA, not by the amino acid which is bound.

An aminoacyl-tRNA synthetase exists for each of the 20 amino acids, and is highly specific for its amino acid—the wrong amino acid is introduced into a protein on average only once in every 10^4–10^5 reactions. This is, in part, due to the presence of two physically distinct domains within many synthetases (Figure 4.31a). The catalytic domain (acylation site) carries out the initial recognition of the amino acid, and transfers it to its cognate tRNA; whereas the editing domain, eliminates the wrong amino acid by hydrolysing either the aminoacyl-adenylate or the aminoacyl-tRNA. We can illustrate this by the case of threonyl-tRNA synthetase, which must distinguish between Thr, Val (with a methyl group in place of a hydroxyl) and Ser (with a hydroxyl group but lacking the methyl group). The catalytic site avoids Val by using a zinc ion, bound to the enzyme by two His and one Cys residue. The zinc ion can also coordinate Thr through its side-chain hydroxyl group and its amino group (Figure 4.31b): the hydroxyl group also forms hydrogen bonds to an adjacent Asp residue. Val cannot bind in this way through its methyl group and is therefore not adenylated and transferred to tRNA^{Thr}.

Ser can however be coupled to tRNA^{Thr}, albeit at a 10^{-2}–10^{-3} rate less than Thr. What enables the error rate to be reduced further is that when the Ser-tRNA^{Thr} is transferred to the editing site, more than 20 Å from the acylation site, it is hydrolysed releasing Ser and free tRNA, whereas Thr-tRNA[Thr] is not. This double sieve of acylation and editing sites ensures the observed high fidelity of charging of tRNAs, with the former typically discriminating against amino acids that are larger, and the hydrolytic editing site cleaving activated complexes that contain an amino acid smaller than the correct one.

How, in turn, does the synthetase recognize its specific tRNA? From extensive mutagenesis studies, it appears that the aminoacyl-tRNA synthetases recognize particular regions of the tRNA molecule, most often in their anticodon loops and/or in their acceptor stems.

Once the amino acid has been bound to its tRNA, it can pass to the next phase of protein synthesis, involving its interaction with mRNA, which takes place on the ribosome, a molecular machine of enormous complexity. The ribosome of *E. coli* is a ribonucleoprotein assembly of molecular weight 2700 kDa, and sedimentation constant of 70S[9]. It is made up of roughly two-thirds RNA and one-third protein, and can be separated into a small (30S) and a large (50S) subunit. The 30S subunit contains 21 proteins and one 16S RNA molecule, while the large subunit has 34 different proteins and two RNA molecules, one 23S and one 5S. Despite its size and complexity, the structure of both ribosomal subunits has been determined to atomic resolution (Figure 4.32), and very recently the atomic structure of the 70S ribosome has been determined at 2.8 Å resolution (Selmer et al., 2006).

We know that there are three tRNA-binding sites that bridge the small and large subunits, two of them bound to the mRNA by anticodon–codon base pairs. These sites are called the A (aminoacyl) and the P (peptidyl) sites. The third site, as we will see later, binds

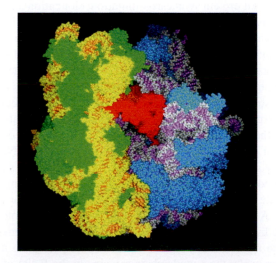

Figure 4.32 A space-filling model of the 70S ribosome: the three RNA molecules—5S, 16S and 23S—are in white, yellow and purple, respectively; ribosomal proteins of the large and small subunit are in blue and green, respectively; the tRNA in the A-site, with its 3′-end extending into the peptidyl-transferase cavity is in red and the P-site tRNA is in yellow. (From Moore and Steitz, 2005. Copyright (2005) with permission from Elsevier.)

[9] Sedimentation coefficients are expressed in Svedbergs (S), after the Swedish biochemist The Svedberg who developed the ultracentrifuge in the 1920s. While S values are indicative of molecular weight, they are not additive-the 70s ribosome is made up of one 50S and one 30S subunit.

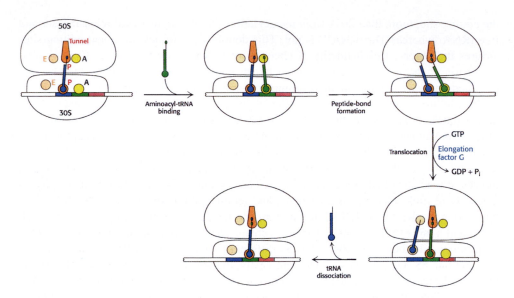

Figure 4.33 Mechanism of protein synthesis. (From Berg et al., 2002. Reproduced with permission from W.H. Freeman and Co.)

an empty tRNA and is designated E (exit). A number of protein factors are involved in the three stages of ribosomal protein synthesis—initiation, elongation and termination. In bacteria, protein synthesis starts with the binding of the 30S subunit to the mRNA. This is followed by binding of the initiator tRNA (charged with *N*-formylMet) to the start codon AUG on the mRNA and completed by binding of the 50S subunit to form the 70S ribosome. With the initiator tRNA (or subsequently a peptidyl-tRNA) in the P site, the elongation cycle can begin with the binding of the corresponding aminoacyl-tRNA to its codon on the mRNA in the A site of both subunits (Figure 4.33). The next step is peptide-bond formation, which does not require energy. The amino group of the aminoacyl-tRNA attacks the carbonyl of the ester linkage of the peptidyl-tRNA to form a tetrahedral intermediate, which collapses with the formation of peptide bond and release of the deacylated tRNA (Figure 4.34). The peptide chain is now attached to the tRNA in the A site of the 30S subunit, whereas both tRNA and its attached peptide are now in the P site of the 50S subunit. In order for translation to proceed, the mRNA must be moved so that the codon for the next aminoacyl-tRNA can enter the A site. This translocation step, involving the protein elongation factor G, is driven by the hydrolysis of GTP. The mRNA advances by one codon, placing the peptidyl-tRNA once again entirely in the P site. It is important to note that throughout the entire elongation cycle, the peptide chain remains in the P site of the 50S subunit, at the entrance to a tunnel[10] that communicates with the exterior of the 50S subunit. The empty tRNA is now in the exit site, and can dissociate from it to complete the cycle.

---

[10] This exit tunnel through the 50S subunit was first revealed by 3-D image reconstruction 20 years ago (Yonath et al., 1987) by two giants of ribosome research, Ada Yonath and the late Heinz-Günther Wittmann (in whose laboratory in Berlin I spent a very fruitful stay 1970-1973).

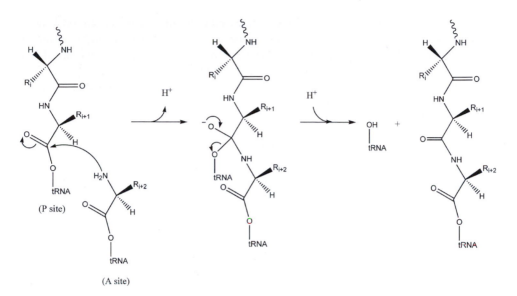

Figure 4.34 Peptide-bond formation (for more details see Rodnina et al., 2007; Yonath, 2005).

Elongation continues until the A site is occupied by one of the three stop codons. Since there are no tRNAs with corresponding anticodons, chain termination occurs, with release of the completed polypeptide.

The peptidyl transferase centre of the ribosome is located in the 50S subunit, in a protein-free environment (there is no protein within 15 Å of the active site), supporting biochemical evidence that the ribosomal RNA, rather than the ribosomal proteins, plays a key role in the catalysis of peptide bond formation. This confirms that the ribosome is the largest known RNA catalyst (ribozyme) and, to date, the only one with synthetic activity. Adjacent to the peptidyl transferase centre is the entrance to the protein exit tunnel, through which the growing polypeptide chain moves out of the ribosome.

Protein biosynthesis by eukaryotic ribosomes, which are larger and more complex than those of bacteria, is very similar in its basic outline to that of bacteria. The major difference is in the initiation of translation, which involves recognition of the 5′-cap on the mature mRNA, mentioned earlier, by the small ribosomal subunit.

Postscript

It is clear that in this brief overview of molecular biology, we have not covered a number of important areas that have an important impact on the study of metalloproteins. These include molecular cloning and recombinant DNA technology, which allow proteins to be over-expressed and individual amino acids to be mutated to any other of the 19 protein amino acids: genome and proteome analysis that enables the sequences of all the genes of the entire organisms to be determined, and the quantification, localization, interactions and, where possible, activities and identification of all of the proteins in an organism, respectively: DNA repair: control of transcription and translation: post-translational modification of proteins: regulation of eukaryotic gene expression: enhancers and silencers of gene expression. Fuller information can be found in one of the excellent textbooks of biochemistry included in the bibliography (Berg et al., 2002; Campbell et al., 2005; Voet and Voet, 2004). But, since this is a textbook about metals in biology and not about biochemistry and molecular biology, brevity was required.

REFERENCES

Anfinsen, C.B. (1973) Principles that govern the folding of protein chains, *Science*, **181**, 223–230.

Berg, J.M., Tymoczko, J.L. and Stryer, L. (2002) *Biochemistry*, 5th edition, W.H. Freeman and Co., New York, 974 pp.

Branden, C. and Tooze, J. (1991) *Introduction to Protein Structure*, Garland Publishing, Inc., New York and London, 302 pp.

Campbell, P.N., Smith, A.D. and Peters, T.J. (2005) *Biochemistry Illustrated Biochemistry and Molecular Biology in the Post-Genomic Era*, 5th edition, Elsevier, London and Oxford, 242 pp.

Chapeville, F., Lipmann, F., von Ehrenstein, G., Weisblum, B., Ray, W.J. Jr. and Benzer, S. (1962) On the role of soluble ribonucleic acid in coding for amino acids, *Proc. Natl. Acad. Sci. U.S.A.*, **48**, 1086–1092.

Creighton, T.E. (1993) *Proteins Structures and Molecular Properties*, 2nd edition, W.H. Freeman and Co., New York, 507 pp.

Eisenberg, D. (2003) The discovery of the α-helix and the β-sheet, the principal structural features of proteins, *Proc. Natl. Acad. Sci. U.S.A.*, **100**, 11207–11210.

Fersht, A. (1999) *Structure and Mechanism in Protein Science: A guide to Enzyme Catalysis and Protein Folding*, W.H. Freeman and Co., New York, 631 pp.

Moore, P.B. and Steitz, T.A. (2005) The ribosome revealed, *TIBS*, **30**, 281–283.

Rodnina, M.V., Beringer, M. and Wintermeyer, W. (2007) How ribosomes make peptide bonds, *TIBS*, **32**, 20–26.

Selmer, M., Dunham, C.M., Murphy F.V., Weixlbaumer, A., Petry, S., Kelley, A.C., Weir, J.R. and Ramakrishnan, V. (2006) Structure of the 70S ribosome complexed with mRNA and tRNA, *Science*, **313**, 1935–1943.

Voet, D. and Voet, J.G. (2004) *Biochemistry*, 3rd edition, Wiley, Hoboken, NJ, 1591 pp.

Watson, J.D. and Crick, F.H.C. (1953) Genetical implications of the structure of deoxyribonucleic acid, *Nature*, **171**, 964–967.

Yonath, A. (2005) Ribosomal Crystallography: Peptide Bond Formation, Chaperone Assistance and Antibiotics Activity. *Mol. Cells*, **20**, 1–16.

Yonath, A., Leonard, K.R. and Wittmann, H.G. (1987) A tunnel in the large ribosomal subunit revealed by three-dimensional image reconstruction, *Science*, **236**, 813–816.

– 5 –

An Overview of Intermediary Metabolism and Bioenergetics

INTRODUCTION

We may ask, what is intermediary metabolism? It is in fact the sum of all of the reactions (Figure 5.1), which are involved in the transformation of the substances that are assimilated by an organism from its environment, their transformation, on the one hand, into energy and their use, on the other hand, to ensure the biosynthesis of molecules necessary for the function of the organism, such as the proteins, nucleic acids, membrane, oligo- and polysaccharides and storage and membrane lipids that we have described in Chapter 4. The former process is often referred to as *catabolism*—essentially the degradation of more complex molecules into simpler and more oxidized products accompanied by the generation of ATP and reducing power in the form of NADPH. A good example is the transformation of glucose (represented here by its empirical formula) to carbon dioxide and water by the combination of glycolysis and the tricarboxylic acid cycle:

$$C_6H_{12}O_6 + 6O_2 \rightarrow 6CO_2 + 6H_2O \qquad (5.1)$$

In contrast, *anabolism*, often referred to as biosynthesis, consumes energy, rather than producing it, typically taking more oxidized molecules and transforming them into more complex and highly reduced end products. The reverse process to that, described in Equation (5.1), is carried out by many photosynthetic organisms and involves the fixation of atmospheric CO_2 to form glucose, catalysed by the enzymes that constitute the Calvin[1] cycle:

$$6CO_2 + 6H_2O \rightarrow C_6H_{12}O_6 + 6O_2 \qquad (5.2)$$

---

[1] After the Californian biochemist Melvin Calvin, who received the Nobel Prize for his discovery that the first product of CO_2 fixation was phosphoglycerate, and went on to establish the cycle.

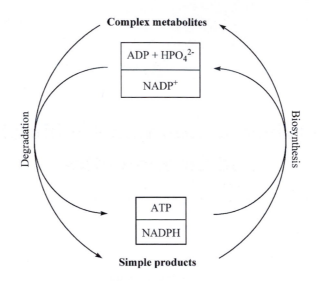

Figure 5.1 Energy (as ATP) and reducing power (in the form of NADPH) for biosynthesis (*anabolism*) are derived from degradation (*catabolism*) of complex metabolites.

Two important implications of the reactions described in Equations (5.1) and (5.2) are: (i) that redox reactions play an important role in metabolic transformations, with the cofactors nicotinamide adenine dinucleotide (NAD$^+$) acting as electron acceptor in catabolic pathways and nicotinamide adenine dinucleotide phosphate (NADPH) as electron donor in anabolism, and (ii) that energy must be produced by catabolism and used in biosyntheses (almost always in the form of adenosine triphosphate, ATP).

The ways in which energy in the form of ATP is produced and utilized constitute bioenergetics, and will be discussed in greater detail at the end of this chapter. However, before turning to a selection of metabolic pathways, we outline some fundamental notions concerning redox reactions followed by a brief description of the central role of ATP in metabolism as an acceptor and donor of phosphoryl groups, and finally a summary of the types of reactions that we will encounter as we wend our way along a sample of some of the pathways of intermediary metabolism.

REDOX REACTIONS IN METABOLISM

Since many of the transformations undergone by metabolites involve changes in oxidation state, it is understandable that cofactors have been developed to act as electron acceptors/ donors. One of the most important is that based on NAD/NADP. NAD$^+$ can accept what is essentially two electrons and a proton (a hydride ion) from a substrate such as ethanol in a reaction catalysed by alcohol dehydrogenase, to give the oxidized product, acetaldehyde and the reduced cofactor NADH plus a proton (Figure 5.2). Whereas redox reactions on metal centres usually involve only electron transfers, many oxidation/reduction reactions in intermediary metabolism, as in the case above, involve not only electron transfer but

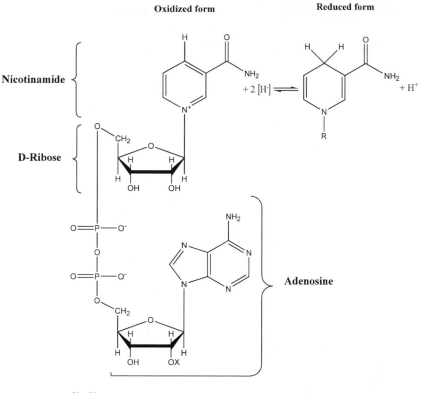

Oxidized form   **Reduced form**

X = H	Nicotinamide adenine dinucleotide (NAD$^+$)
X = PO$_3{}^{2-}$	Nicotinamide adenine dinucleotide phosphate (NADP$^+$)

Figure 5.2 The structure of NAD$^+$ and NADP$^+$ and the role of the nicotinamide moiety as an electron acceptor.

hydrogen transfer as well—hence the frequently used denomination 'dehydrogenase'. Note that most of these dehydrogenase reactions are reversible. In addition to NAD$^+$, which intervenes in redox reactions involving oxygen functions, other cofactors such as riboflavin (in the form of flavin mononucleotide, FMN and flavin adenine dinucleotide, FAD, FMN bound to AMP) (Figure 5.3) participate in the conversion of [-CH$_2$-CH$_2$- to -CH=CH-], as well as in electron transfer chains, lipoate, in α-ketoacid dehydrogenases, and ubiquinone and its derivatives, in electron transfer chains.

THE CENTRAL ROLE OF ATP IN METABOLISM

It might be best to situate the importance of ATP in intermediary metabolism by some anecdotal information. The average ATP molecule is hydrolysed within minutes of its synthesis (in other words, its turnover is very rapid); at rest the average human consumes

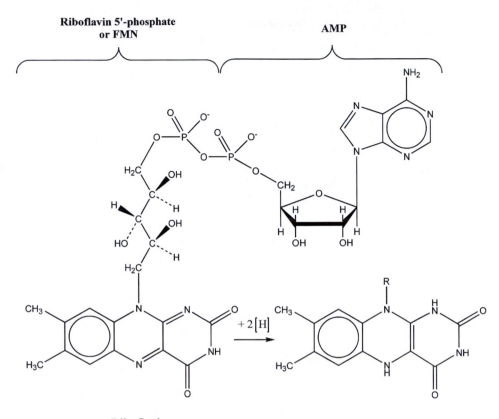

Figure 5.3 The flavin coenzymes FAD and FMN. Note that in contrast to NAD$^+$, flavins can be half-reduced to the stable radical FADH˙ or fully reduced to the dihydroflavin shown.

around 40 kg of ATP per day, while during vigorous exercise, this may rise to around 0.5 kg/min. The hydrolysis of ATP (Figure 5.4) to ADP and P$_i$ is accompanied by a relatively large free-energy change[2] (~50 kJ/mole), as is that of ADP to AMP and P$_i$. In contrast, the hydrolysis of AMP to adenosine and phosphate generates very little free-energy change. In biochemical terms, the importance of ATP as the energetic currency of the cell depends on the capacity of the couple ATP/ADP to accept phosphoryl groups from high-energy donors and to donate phosphoryl groups to low-energy acceptors. Thus, to give an example from the glycolysis pathway described below, ATP can donate its phosphoryl group to glucose to generate glucose-6-phosphate and ADP (the reaction, catalysed by hexokinase, is described in Chapter 10). We can consider this as the sum of two reactions, one

---

[2] Where possible we have used the free energies ΔG calculated from *in vivo* concentrations of metabolites rather than the standard free energies $\Delta G°$, which do not take account of local concentrations of reactants and products.

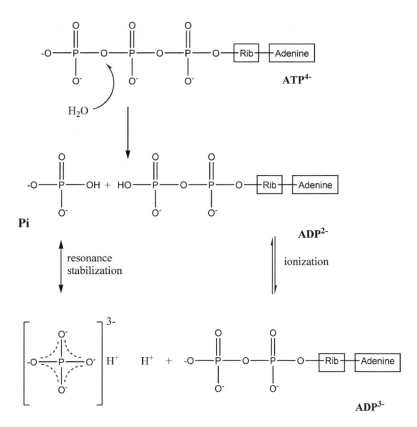

Figure 5.4 The chemical basis of the large free-energy change associated with ATP hydrolysis. Hydrolysis is accompanied by relief of the electrostatic repulsion between the negative charges on ATP by charge separation; the resulting phosphate anion is stabilized by resonance, while the other product, ADP^{2+} releases a proton into a medium where [H$^+$] is very low (~10^{-7} M).

energetically unfavourable, the other extremely favourable (Figure 5.5). This can be readily understood when one considers that the free-energy change for hydrolysis of glucose-6-phosphate is significantly lower than that of ATP. In contrast, the phosphoenolpyruvate molecule can readily transfer its phosphoryl group to ADP, thereby generating ATP. This is the essence of the central role of ATP as the energetic currency of the cell, accepting phosphoryl groups from potential donors, such as phosphoenolpyruvate, and donating them to potential acceptors, such as glucose. Another important feature of ATP is that the hydrolysis of ATP is in a coupled reaction, for example instead of

$$A + ATP + H_2O \Leftrightarrow B + ADP + P_i + H^+$$

the coupled reaction will change the equilibrium ratio of products to reactants by around 10^8, which will make very unfavourable reactions become, energetically, extremely

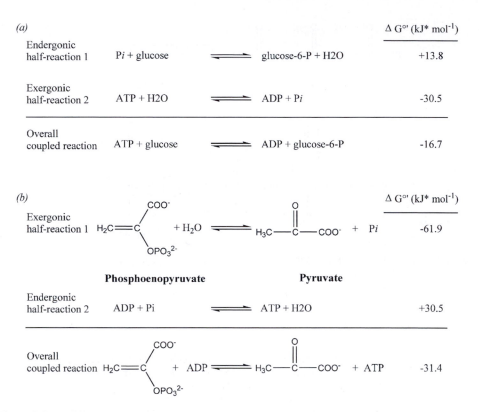

Figure 5.5 The coupled reaction in which ATP supplies the phosphoryl group for glucose-6-phosphate synthesis: in contrast, phosphoenolpyruvate has a phosphoryl-transfer potential sufficiently elevated to enable it to donate its phosphoryl group to ADP, generating ATP.

favourable (indeed coupling one or more reactions to the hydrolysis of nATP molecules will increase the equilibrium ratio by a factor of 10^{8n}).

THE TYPES OF REACTION CATALYSED BY ENZYMES OF INTERMEDIARY METABOLISM

While intermediary metabolism encompasses a vast number of transformations, in reality there are only a few types of reactions that are used. The first class are the redox reactions described above. A second class are nucleophilic displacements (Figure 5.6), often referred to as group-transfer reactions: the most commonly transferred groups are glycosyl groups (A), acyl groups (B) and phosphoryl groups (C). Other examples of group-transfer reactions (Table 5.1), which we have already encountered, include phosphoryl transfer, using ATP or other nucleoside di- or tri-phosphates and electron transfer (described above for nicotinamide and riboflavin derivatives). Other examples include transfer of acyl and aldehyde groups, CO_2, one-carbon units, sugars and phosphatidate.

Nucleophilic Displacement

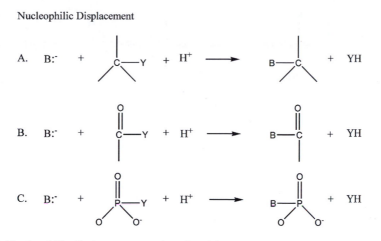

Figure 5.6 Nucleophilic displacement reactions involving glycosyl (A), acyl (B) or phosphoryl (C) group transfers.

Table 5.1

Types of group-transfer reaction involved in intermediary metabolism with the donor of the group to be transferred (left) and the type of group transferred (right)

• ATP	• Phosphoryl
• NADH and NADPH	• Electrons
• $FADH_2$ and $FMNH_2$	• Electrons
• Coenzyme A	• Acyl
• Lipoamide	• Acyl
• Thiamine pyrophosphate	• Aldehyde
• Biotin	• CO_2
• Tetrahydrofolate	• One-carbon units
• S-Adenosylmethionine	• Methyl
• UDP glucose	• Glucose
• CDP-diacylglycerol	• Phosphatidate

Elimination reactions (Figure 5.7) often result in the formation of carbon–carbon double bonds, isomerizations involve intramolecular shifts of hydrogen atoms to change the position of a double bond, as in the aldose–ketose isomerization involving an enediolate anion intermediate, while rearrangements break and reform carbon–carbon bonds, as illustrated for the side-chain displacement involved in the biosynthesis of the branched chain amino acids valine and isoleucine. Finally, we have reactions that involve generation of resonance-stabilized nucleophilic carbanions (enolate anions), followed by their addition to an electrophilic carbon (such as the carbonyl carbon atoms

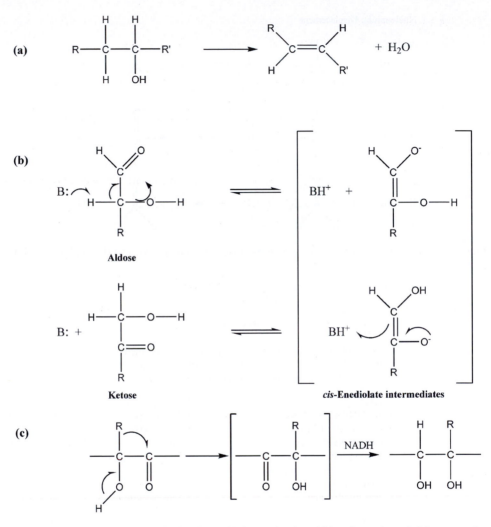

Figure 5.7 Examples of (a) elimination, (b) isomerization (aldose/ketose) and (c) a complex rearrangement of the pinacol–pinacolone type found in the biosynthesis of valine and isoleucine.

of aldehydes, ketones, esters and CO_2), resulting in the formation of carbon–carbon bonds:

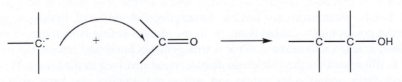

These carbanions can be formed (Figure 5.8) by proton abstraction from ketones resulting in aldol condensations, by proton abstraction from acetyl CoA, leading to Claisen ester

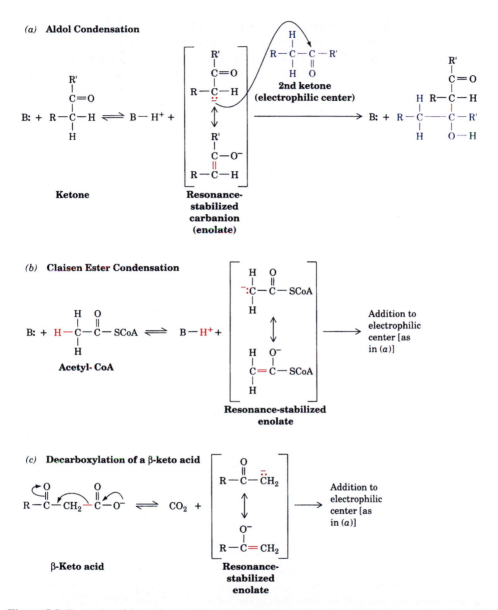

Figure 5.8 Examples of formation and cleavage of carbon–carbon bonds (a) aldol condensation, (b) Claisen ester condensation and (c) decarboxylation of a β-keto acid. (From Voet and Voet, 2004. Reproduced with permission from John Wiley & Sons., Inc.)

condensation and by decarboxylation of β-keto acids leading to a resonance-stabilized enolate, which can likewise add to an electrophilic centre. It should be noted that the reverse of the decarboxylation also leads to formation of a carbon–carbon bond (this is again a group-transfer reaction involving biotin as the carrier of the activated CO_2 to be transferred).

Group-transfer reactions often involve vitamins[3], which humans need to have in their diet, since we are incapable of realizing their synthesis. These include nicotinamide (derived from the vitamin nicotinic acid) and riboflavin (vitamin B_2) derivatives, required for electron transfer reactions, biotin for the transfer of CO_2, pantothenate for acyl group transfer, thiamine (vitamin B_1, as thiamine pyrophosphate) for transfer of aldehyde groups and folic acid (as tetrahydrofolate) for exchange of one-carbon fragments. Lipoic acid (not a vitamin) is both an acyl and an electron carrier. In addition, vitamins such as pyridoxine (vitamin B_6, as pyridoxal phosphate), vitamin B_{12} and vitamin C (ascorbic acid) participate as cofactors in an important number of metabolic reactions.

AN OVERVIEW OF INTERMEDIARY METABOLISM: CATABOLISM

As we pointed out in Figure 5.1, there are two broad strands to intermediary metabolism—those in which energy is produced and reducing power generated (catabolism) and those in which energy is consumed and reducing power utilized (anabolism).

In catabolism, (Figure 5.9) complex macromolecules, such as proteins and storage polysaccharides (glycogen in animals, starch in plants) and fat stores in the form of triglycerides, are first hydrolysed to their basic components—amino acids, monosaccharides, essentially glucose, and glycerol plus fatty acids, respectively. In the second phase, these molecules are transformed into a series of molecules that will feed the final and central core of catabolism, namely, the tricarboxylic acid cycle. In the case of glucose, this is converted into pyruvate by the glycolysis pathway, which is described in greater detail below. Of the 20 amino acids found in proteins, all must first be divested of their amino group, which is usually converted to ammonium ions (which in many organisms must be detoxified by conversion to urea). The resulting carbon skeletons are then transformed either into pyruvate, acetyl CoA or one of the constituents of the tricarboxylic acid cycle. In the case of triglycerides, glycerol can enter the glycolytic pathway, while the long-chain fatty acids are transformed by β-oxidation into acetyl CoA. The oxidation of amino acids and fatty acids is centralized within the mitochondria, where all the enzymes involved in the tricarboxylic acid cycle are also localized.

As was indicated in Equation (5.1), in addition to CO_2, the other final product of glucose oxidation is H_2O, which is produced by the four-electron reduction of dioxygen together with $4H^+$ to give two molecules of water. The reducing equivalents come from the dehydrogenase reactions of catabolism in the form of NADH and $FADH_2$. Their electrons are then transferred through a series of electron acceptors to the terminal oxidase of this so-called respiratory chain, cytochrome c oxidase. As the electrons pass down this electron transport chain, they generate a proton gradient that, as we will see shortly, is used

---

[3] Defined, classically, as things you get ill from when you don't have them! (i.e.—which we are incapable of synthesizing ourselves).

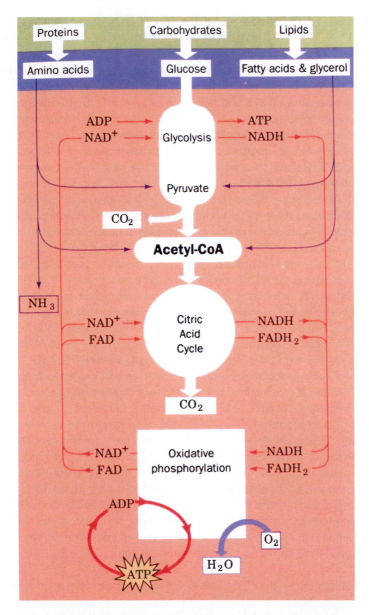

Figure 5.9 An overview of catabolism.

to drive the proton translocating ATP synthase. This aspect of mitochondrial function is usually referred to as oxidative phosphorylation.

We will return to an overview of anabolic pathways shortly, but first we want to examine in more detail the two important catabolic pathways, glycolysis and the tricarboxylic acid cycle.

SELECTED CASE STUDIES: GLYCOLYSIS AND THE TRICARBOXYLIC ACID CYCLE

Glycolysis[4] is an almost universal pathway for glucose catabolism, widely distributed in living organisms, which converts glucose into two molecules of pyruvate with the net production of two molecules of ATP and two molecules of NADH (of which more later). It consists of a sequence of ten reactions (Figure 5.10), the first five of which transform one molecule of glucose into two molecules of triose phosphate, and consume two molecules of ATP. The five subsequent reactions transform the two molecules of triose phosphate, as glyceraldehyde-3-phosphate, into two molecules of pyruvate with production of four molecules of ATP and two of NADH. Of the ten reactions, six are phosphoryl transfers (1,3,6,7,8,10), two are sugar isomerizations (2,5), one is an aldol cleavage (4), one is an NAD$^+$-dependent redox reaction, involving the conversion of a thiohemiacetal to an acyl thioester (6), which is accompanied by a phosphoryl transfer, and one is an elimination reaction (dehydration) resulting in formation of a double bond (9).

The glycolytic pathway is characteristic of many catabolic processes, where you have to invest some energy at the start before getting more energy back from the system. The hexokinase reaction (1), as we have seen in Figure 5.5 uses the energy of ATP hydrolysis to drive phosphorylation of glucose, which not only ensures that the resulting glucose-6-phosphate is not re-exported from the cell (it costs ATP hydrolysis to bring glucose into the cell in the first place), but that glucose is thereby activated for its subsequent degradation. In free-energy terms, this reaction is essentially irreversible (ΔG −27.2 kJ/mol), given the intracellular concentrations of the substrates and products. Following its isomerization (2) to the corresponding ketose (fructose-6-phosphate), a second essentially irreversible phosphoryl transfer (ΔG −25.9 kJ/mol), catalysed by the key regulatory enzyme of the glycolytic pathway, phosphofructokinase, results in formation of fructose-1, 6-bisphosphate. This is subjected to an aldol cleavage[5] generating two triose phosphates, dihydroxyacetone-phosphate (derived from carbons 1–3 of glucose) and glyceraldehyde-3-phosphate (from carbons 4–6). Only one of these products of aldol cleavage, glyceraldehyde-3-phosphate (GAP) proceeds further along the glycolysis pathway. The isomerization of these two triose phosphates, catalysed by triose phosphate isomerase (5), completes the first half of glycolysis. Triose phosphate isomerase has attained catalytic perfection (Knowles, 1991)—the rate of reaction between enzyme and substrate is diffusion controlled, and all encounters between the enzyme and its substrate lead to reaction.

In the second half of the glycolytic pathway, the investment of energy in the first half is repaid by net generation of energy in the form of 4ATP and 2NADH molecules. Glyceraldehyde-3-phosphate dehydrogenase (6) catalyses the NAD$^+$-dependent dehydrogenation of the aldehyde, GAP (which binds to the thiol of an active-site cysteine residue,

---

[4] From Greek *glyko*—sweet and *lysis*—splitting.

[5] This reaction is one of the best examples of the importance of considering ΔG values rather than $\Delta G°$: the latter is +23.9 kJ/mol, whereas in the cell ΔG is −1.3 kJ/mol. This reflects the actual concentrations of the metabolites within the cell.

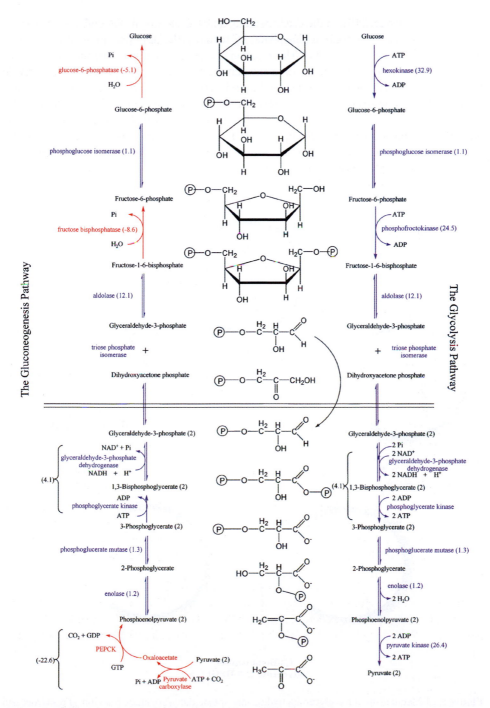

Figure 5.10 The glycolysis and gluconeogenesis pathways.

forming a thiohemiacetal), to the corresponding acid (as an enzyme-bound acyl thioester), which is attacked by inorganic phosphate, P_i, forming the acyl phosphate product, 1,3-bisphosphoglycerate (Figure 5.11). This reaction is a good example of coupling an energetically favourable reaction (oxidation of an thiohemiacetal) with an energetically unfavourable reaction (formation of an acyl phosphate). The product of the reaction, 1,3-bisphosphoglycerate, then transfers its phosphoryl group to ADP (7), recuperating the initial investment of 2ATP molecules in the first phase of glycolysis. The product, 3-phosphoglycerate, then undergoes phosphoryl transfer (8). This involves reaction with a phosphoenzyme to produce a 2,3-bisphosphoglycerate intermediate, which decomposes to form 2-phosphoglycerate and regenerate the phosphoenzyme. Dehydration of 2-phosphoglycerate (9) by the

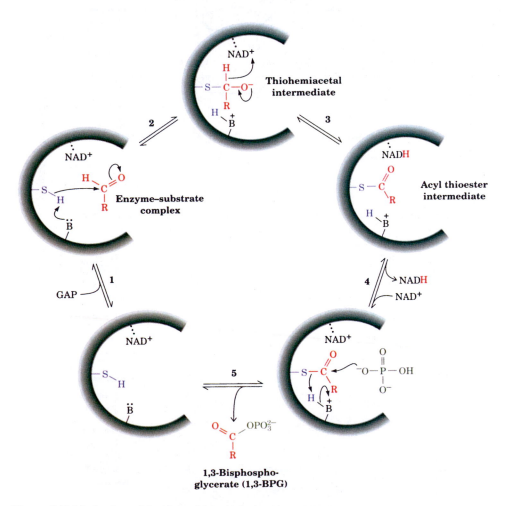

Figure 5.11 Mechanism of the glyceraldehyde-3-phosphate dehydrogenase reaction. (From Voet and Voet, 2004. Reproduced with permission from John Wiley & Sons., Inc.)

Mg^{2+}-dependent enzyme, enolase (described in greater detail in Chapter 10), leads to the 'high-energy' compound, phosphoenolpyruvate. As we have seen earlier (Figure 5.5), it can transfer its phosphoryl group to ADP (10) producing another 2ATP molecules per molecule of glucose oxidized, and represents the third essentially irreversible reaction of the glycolytic pathway (ΔG −13.9 kJ/mol).

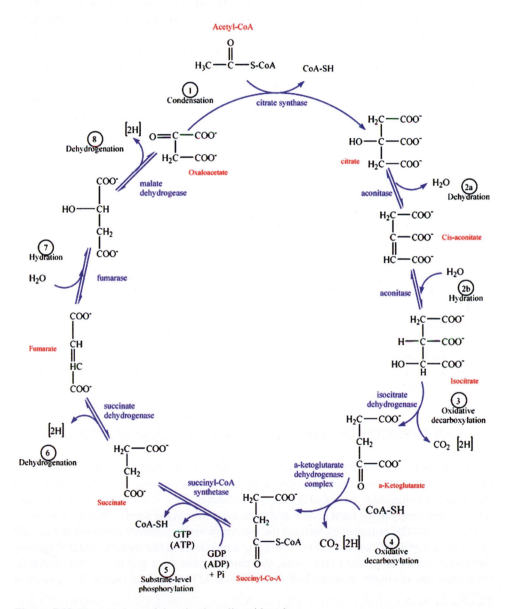

Figure 5.12 The reactions of the tricarboxylic acid cycle.

The second metabolic pathway, which we have chosen to describe, is the tricarboxylic acid cycle—often referred to as the Krebs cycle[6]. This represents the biochemical hub of intermediary metabolism, not only in the oxidative catabolism of carbohydrates, lipids and amino acids in aerobic eukaryotes and prokaryotes, but also as a source of numerous biosynthetic precursors. Pyruvate, formed in the cytosol by glycolysis, is transported into the matrix of the mitochondria where it is converted to acetyl CoA by the multienzyme complex, pyruvate dehydrogenase. Acetyl CoA is also produced by the mitochondrial β-oxidation of fatty acids and by the oxidative metabolism of a number of amino acids. The first reaction of the cycle (Figure 5.12) involves the condensation of acetyl Co and oxaloacetate to form citrate (1), a Claisen ester condensation. Citrate is then converted to the more easily oxidized secondary alcohol, isocitrate (2), by the iron–sulfur centre of the enzyme aconitase (described in Chapter 13). This reaction involves successive dehydration of citrate, producing enzyme-bound *cis*-aconitate, followed by rehydration, to give isocitrate. In this reaction, the enzyme distinguishes between the two external carboxyl groups of the pro-chiral citrate molecule. The hydroxyl group of isocitrate is then oxidized to give an enzyme-bound α-keto acid which readily decarboxylates to α-ketoglutarate and CO_2 (3). In a second oxidative decarboxylation reaction, entirely analogous to the conversion of pyruvate to acetyl CoA, α-ketoglutarate is converted to succinyl CoA, CO_2 and NADH (4). The 'high energy' of the thioester bond in succinyl CoA is conserved (5) by conversion of GDP to GTP (since nucleoside di- and triphosphates are inter-convertible, this is equivalent to conversion of ADP to ATP). The central single bond between the methylene carbons of succinate is then oxidized to a *trans* double bond (6) in a reaction catalysed by the FAD-dependent enzyme, succinate dehydrogenase. Addition of water to fumarate catalysed by fumarase generates malate (7), which, in a final NAD^+-dependent oxidation, is converted to oxaloacetate (8), thus completing the cycle. The sequence of reactions (6–8) is also used in the β-oxidation of fatty acids.

The global outcome of the Krebs cycle is that one molecule of acetyl CoA is converted to three molecules of NADH, two of CO_2, one of GTP and one of $FADH_2$. The reducing equivalents will be used, as we will see later, to generate ATP.

AN OVERVIEW OF INTERMEDIARY METABOLISM: ANABOLISM

As was pointed out in the introduction, the opposite of catabolism, anabolism involves the biosynthesis of more complex and more highly reduced molecules from the simpler and more oxidized molecules generated in the course of catabolism. These biosynthetic pathways require both energy in the form of ATP and reducing power in the form of NADPH. Whereas most of the NADH is funnelled through the mitochondrial respiratory chain for use in ATP synthesis, catabolism can also produce reducing equivalents in the form of NADPH (mostly through a variant of glucose catabolism, the pentose phosphate pathway). Photosynthetic organisms can, of course, generate both NADPH and ATP using light energy. It is also clear

---

[6] Hans Krebs proposed the cycle in 1937 on the basis of experiments on minced pigeon muscle. He received the Nobel Prize for Medicine and Physiology (jointly with Fritz Lipmann) in 1952.

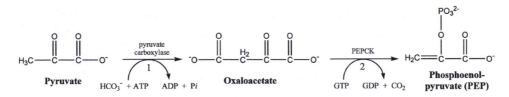

Figure 5.13 Conversion of pyruvate to oxaloacetate and then to phosphoenolpyruvate.

that, for obvious reasons of creating absolute metabolic chaos, biosynthetic pathways cannot use the same enzymic machinery as catabolism. Sometimes, as in the synthesis of glucose from pyruvate (gluconeogenesis), this implies the use of alternative enzymes for only a few specific steps in the pathway, as illustrated in Figure 5.10. Seven of the ten enzymes of the glycolytic pathway are used in gluconeogenesis, and the three that are not, not surprisingly, are those that catalyse essentially irreversible steps. The first two, hexokinase and phospho-fructokinase, which use ATP in the glycolytic pathway, are replaced by hydrolytic reactions catalysed by glucose-6-phosphatase[7] and fructose1,6-bisphosphatase, which remove the phosphoryl groups as inorganic phosphate, P_i. The problem with the conversion of pyruvate to phosphoenolpyruvate is more complex, firstly because the reaction is, energetically, extremely unfavourable, and secondly because pyruvate, available for gluconeogenesis is localized within the mitochondrial matrix—whereas the enzymes of the glycolytic pathway are in the cytosol. The solution (Figure 5.13) involves the energy-dependent carboxylation of pyruvate within the mitochondria to form oxaloacetate. Oxaloacetate (in the form of either malate or aspartate) is then exported to the cytosol (Figure 5.14) where it is converted to phosphoenolpyruvate (Figure 5.13), again in an energy-dependent process (this time GTP is involved), by the enzyme phosphoenolpyruvate carboxykinase (PEPCK). In some species, PEPCK is almost equally distributed between the mitochondria and the cytosol, so that some PEP required for gluconeogenesis can be generated in the mitochondria and exported by a specific transport system. The export of oxaloacetate (Figure 5.14) by the malate shuttle (using malate dehydrogenase) allows reducing equivalents as NADH to be transferred to the cytosol, where they are required for gluconeogenesis.

The tricarboxylic acid cycle not only enables the oxidation of acetyl CoA, but it also supplies a number of molecules that are used in biosynthetic pathways, as is illustrated in Figure 5.15. α-keto acids such as oxaloacetate and α-ketoglutarate can undergo transfer of an amino group with an amino acid (aminotransferase) to give aspartate and glutamate, respectively. Both these amino acids are used extensively in the biosynthesis of other amino acids and in nucleotide biosynthesis. Succinyl CoA is used together with glycine in the synthesis of porphyrins, while citrate is the starting point of both fatty acid and cho-lesterol biosynthesis. We have already mentioned that the enzymes involved in the β-oxidation of fatty acids are located in the mitochondria. The source of two-carbon fragments for the biosynthesis of both fatty acids and isoprenoids such as cholesterol is acetyl CoA, which is generated by oxidative metabolism in the mitochondria. Acetyl CoA

---

[7] Glucose-6-phosphatase is found only in liver and kidney, and allows these tissues to supply glucose to other organs of the body, like the brain, which have little or no reserves of carbohydrates.

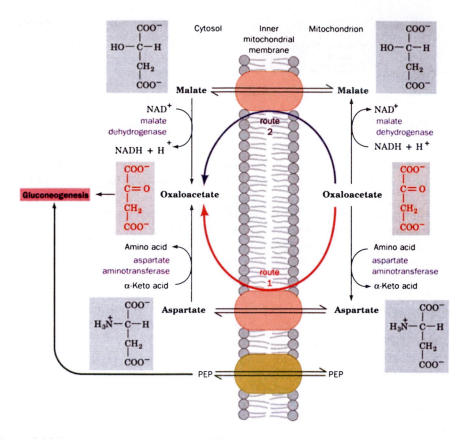

Figure 5.14 Transport of oxaloacetate and PEP from the mitochondria to the cytosol. (From Voet and Voet, 2004. Reproduced with permission from John Wiley & Sons., Inc.)

cannot escape from the mitochondria, but it can be exported to the cytosol as citrate, where it is reconverted to oxaloacetate and acetyl CoA. Fatty acid (and cholesterol) biosynthesis takes place in the cytosol and requires bicarbonate, which is incorporated into acetyl CoA by acetyl CoA carboxylase to form malonyl CoA. The biosynthesis of fatty acids, mostly the C_{16} palmitate (see Chapter 4), requires one molecule of acetyl CoA and seven molecules of malonyl CoA. In animals the seven-enzymatic reactions that are required for fatty acid synthesis are present in a single multi-functional protein complex, known as fatty acid synthase[8]. The synthase also has an acyl carrier protein (ACP), which has a phosphopantothenate side chain attached to a serine side chain, similar to that found in Coenzyme A, which acts as a flexible link transporting the substrate between the different enzymatic domains of the protein.

---

[8] This is not the only example of Nature inventing the assembly line a long time before Henry Ford—both pyruvate dehydrogenase and α-ketoglutarate dehydrogenase, mentioned earlier in the chapter, are also multi-enzyme complexes.

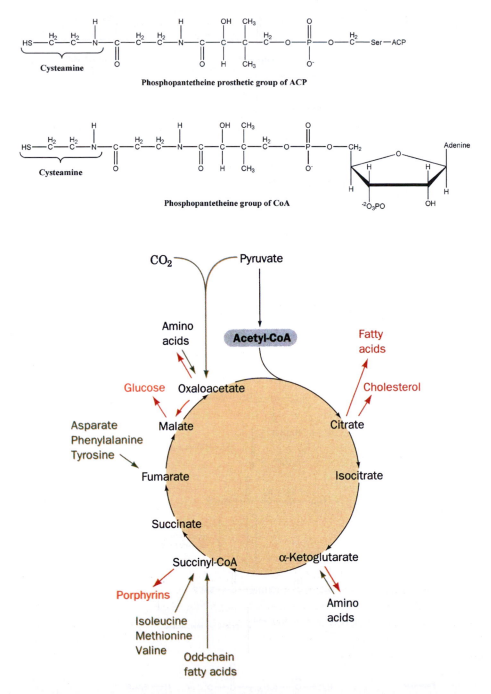

Figure 5.15 The tricarboxylic acid cycle plays a central role in supplying intermediates for biosynthetic pathways as well as receiving intermediates from catabolic pathways. (From Voet and Voet, 2004. Reproduced with permission from John Wiley & Sons., Inc.)

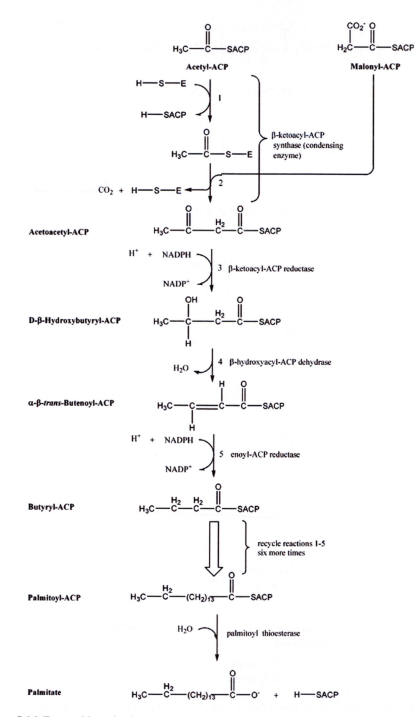

Figure 5.16 Fatty acid synthesis.

In the successive steps of fatty acid synthesis (Figure 5.16), acetyl CoA is transferred to ACP by malonyl/acetyl CoA-ACP transacylase (MAT), and then to the thiol group of β-ketoacyl-ACP synthase (KS, indicated in the figure as E). Malonyl-ACP is formed from malonyl CoA in an analogous fashion to acetyl CoA by the action of MAT. The condensation reaction between the acetyl group and the β-carbon of malonyl-ACP is catalysed by KS accompanied by decarboxylation with the formation of acetoacetyl-ACP and release of the Cys-SH of the active site of KS. The next three steps involving reduction, dehydration and further reduction convert acetoacetyl-ACP to butyryl-ACP, and represent the direct opposite to the β-oxidation that we saw in the Kreb's cycle (succinate to oxaloacetate), which is also found in fatty acid oxidation (oxidation/hydration/oxidation). However, in the biosynthetic pathway, NADPH is the electron donor, whereas in the two-redox steps of β-oxidation the electron acceptors are, respectively, FAD and NAD^+. The condensation reaction, two reduction steps and dehydration are repeated a further six times, resulting in palmitoyl-ACP. At this stage, the thioester bond is hydrolysed by palmitoylthioesterase releasing palmitate, the final product, and regenerating the synthase for another round of biosynthesis.

BIOENERGETICS: GENERATION OF PHOSPHORYL TRANSFER POTENTIAL AT THE EXPENSE OF PROTON GRADIENTS

In our brief discussion of intermediary metabolism, we did not mention that there is an alternative to full-blown respiratory oxidative catabolism, which Louis Pasteur described as '*la vie sans air*'. Many organisms can live without oxygen, albeit producing much less ATP per molecule of substrate oxidized than in respiration. They achieve this by making up, for example in glycolysis, the reduction of NAD^+ to NADH by a compensatory reduction of another organic molecule. Two examples are the reduction of pyruvate to lactate in muscle tissue during exercise, and the reduction of acetaldehyde (derived from the decarboxylation of pyruvate) to ethanol by yeasts. These fermentations, as they are called, can result in a vast number of interesting end products, both culinary (beer, wine, most alcoholic beverages[9] cheese, yoghurt, sauerkraut, etc.) as well as many industrial and medical applications.

However, the overwhelming attraction of respiration is the greatly increased yield of ATP—fermentation of one molecule of glucose to lactate or ethanol yields just 2ATP molecules, whereas full oxidation of glucose to CO_2 and water yields between 36 and 38. So far we have only seen what the biochemist calls 'substrate-level' phosphorylation—ATP production in the course of metabolic processes. To achieve the yields of ATP production we find in respiration, we need to harness the potential energy of the reducing equivalents— NADH and $FADH_2$—by transferring their electrons to an electron acceptor with a much higher redox potential, and in the mitochondrion this is dioxygen.

---

[9] Virtually any source of glucose can undergo alcoholic fermentation—100g of potatoes in an oxygen-free atmosphere at 22°C will give 600 mg of ethanol in 8 days—the product is pretty unpalatable, but distillation can change that.

The standard redox potential E_o' (standard conditions for the biochemist are 1 M oxidant, 1 M reductant, 10^{-7} M [H^+], i.e. pH 7 and 25 °C) for most biological redox couples are known. Remember that in this context E_o' refers to the partial reaction written as:

$$\text{Oxidant} + e^- \rightarrow \text{Reductant}$$

In addition, the standard free-energy change $\Delta G^{o\prime}$ is related to the change in standard redox potential E_o' by

$$\Delta G^{o\prime} = -nF\Delta E_o'$$

where n is the number of electrons transferred, F the *faraday*, a constant equal to 96.48 kJ/mol/V and $\Delta E_o'$ the difference between the two standard redox potentials in volts.

The driving force of oxidative phosphorylation is the difference between the electron transfer potential of NADH or $FADH_2$ relative to that of O_2. For the redox couple

$$NAD^+ \rightarrow NADH + H^+ \quad E_o' \text{ is } -0.315 \text{ V}$$

while for the couple[10]

$$\frac{1}{2} O + 2H^+ + 2e^- \rightarrow H_2O \quad E_o' \text{ is } +0.815 \text{ V}$$

so that for the reaction

$$\frac{1}{2} O + NADH + H^+ \rightarrow H_2O + NAD^+ \quad \Delta E_o' \text{ is } +1.130 \text{ V}$$

we can calculate that $\Delta G^{o\prime} = -220.1$ kJ/mol. For comparison, the $\Delta G^{o\prime}$ for ATP hydrolysis is -31.4 kJ/mol, so we should be able to make a few ATP molecules with this potential bonanza of energy. However, there are two important conditions—first, we cannot simply dissipate all of the potential energy difference in one 'big bang', but pass the electrons through a series of transporters that have progressively increasing redox potentials, and secondly we must use a system coupled to electron transfer, which will allow us to make ATP synthesis turn. This involves generating a proton gradient across the internal mitochondrial membrane.

The first condition is met by having a series of four protein complexes inserted into the mitochondrial inner membrane, each made up of a number of electron (and sometimes proton) acceptors of increasing redox potential. Three of them (Complexes I, III and IV) are presented in cartoon form in Figure 5.17. Complex I, referred to more prosaically as

---

[10] It is clearly absurd to talk about 'half' oxygen molecules given the strength of the O=O double bond. However, for more pedestrian reasons of considering the transfer of two electrons from NADH or $FMNH_2$, through a long series of transporters all the way to the end of the line at molecular oxygen, we would request our more chemically based readers to grant us this small indulgence.

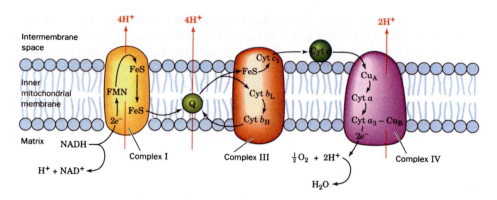

Figure 5.17 The mitochondrial electron-transport chain. (From Voet and Voet, 2004. Reproduced with permission from John Wiley & Sons., Inc.)

NADH-coenzyme Q (CoQ) oxidoreductase, transfers electrons stepwise from NADH, through a flavoprotein (containing FMN as cofactor) to a series of iron–sulfur clusters (which will be discussed in Chapter 13) and ultimately to CoQ, a lipid-soluble quinone, which transfers its electrons to Complex III. $\Delta E_0'$ for the couple NADH/CoQ is 0.36 V, corresponding to a $\Delta G^{\circ\prime}$ of -69.5 kJ/mol; and in the process of electron transfer, protons are exported into the intermembrane space (between the mitochondrial inner and outer membranes).

Complex II (which is not shown in the figure) contains succinate dehydrogenase, the FAD-dependent Krebs cycle enzyme and, like Complex I, transfers its electrons through iron–sulfur centres and a *b*-type cytochrome (more of these haem iron proteins will be discussed in Chapter 13) to CoQ. However, here $\Delta E_0'$ is only 0.085 V, corresponding to $\Delta G^{\circ\prime}$ of -16.4 kJ/mol, which is not sufficient to allow proton pumping.

Complex III (CoQ: cytochrome c oxidoreductase) transfers electrons from CoQ to cytochrome c, through a sequence of cytochrome and iron–sulfur cofactors. Here, $\Delta E_0'$ for the couple CoQ/cytochrome *c* is 0.19 V, corresponding to a $\Delta G^{\circ\prime}$ of -36.7 kJ/mol, again enough to power the synthesis of an ATP molecule and to ensure that protons are pumped across the inner mitochondrial membrane.

Finally, Complex IV, cytochrome c oxidase, takes the electrons coming from four molecules of cytochrome c, a small, water soluble haem protein, moving outside of the membrane in the intermembrane space and carries out the four-electron reduction of a molecule of dioxygen to two molecules of water. The $\Delta E_0'$ for the couple cytochrome c/O_2 is by far the highest of the four complexes, 0.58 V, corresponding to a $\Delta G^{\circ\prime}$ of -112 kJ/mol, and there is certainly proton pumping, which must involve conformational changes, since, in this complex, unlike the three others, there are only one-electron cytochromes and copper atoms, with no obvious proton exchanges possible (more information about cytochrome c oxidase in Chapters 13 and 14).

We now turn our attention to how the gradient of protons pumped by Complexes I, III and IV across the inner mitochondrial membrane into the intermembrane space, together with the associated membrane potential, is used to turn the molecular rotor that ensures

ATP synthesis. Without entering into the detail, we can calculate that the ΔG for pumping a proton from the mitochondrial matrix to the intermembrane space is 21.5 kJ/mol. Since the estimated ΔG (the real *in vivo* free energy) for synthesis of an ATP molecule is between +40 to +50 kJ/mol, we can estimate that at least two protons (most likely three) need to be pumped per ATP generated. From experimental data we know that two electrons descending the respiratory chain from NADH (i.e. via Complexes I, III and IV) to oxygen will produce 3ATP molecules. By comparison, two electrons entering via $FADH_2$ and passing through Complexes II, III and IV to oxygen will result in the formation of only 2ATP molecules.

If, during electron transfer along the respiratory chain, protons are translocated from the matrix to the intermembrane space, how, we may ask, is this proton gradient used to synthesize ATP? Where better to start than the enzyme itself, proton-translocating ATP synthase (Figure 5.18). It is composed of two parts, one called F_0, which is inserted into the inner mitochondrial membrane and contains the proton translocation channel, and the second, F_1, consists of a stalk, which connects with the F_0 component, to which a roughly oval-shaped ball is attached[11]. F_1 consists of five types of subunit (stoichiometry $\alpha_3,\beta_3,\gamma,\delta,\varepsilon$), two of which, the α and β subunits make up the bulk of F_1. Both bind nucleotides, although only the β subunits are directly involved in the catalysis of ATP synthesis. The central stalk consists of the γ and ε subunits, and the γ subunit has a long α-helical-coiled coil, which extends into the centre of the $\alpha_3\beta_3$ hexamer. Crucially for the mechanism, this breaks the symmetry of the $\alpha_3\beta_3$ hexamer, such that each of the catalytic β subunits interacts with a different face of γ.

On the basis of binding studies, Paul Boyer proposed a binding-change mechanism for proton-driven synthesis of ATP by the enzyme (Figure 5.19), which implied that the enzyme (in particular the three β subunits) could exist in three different forms, one which binds ATP with such high affinity (the tight, T form) that it converts bound ADP and P_i to ATP—however, it is incapable of releasing the ATP. At the same time, a second subunit will be in the loose, L, conformation, which can bind ADP and P_i, but cannot release them, while the third subunit is in the open, O conformation. This third subunit can exist in two states—one in which, similar to the T or L forms, a nucleotide is bound, and a second in which it has a more open conformation and releases the bound nucleotide. John Walker and his group were in fact able to crystallize the bovine heart F_1-ATP synthase with each of the three catalytic subunits in one of the three conformational states (Figure 5.20), confirming the predictions of the Boyer mechanism[12]. The structure supports a catalytic mechanism in intact ATP synthase in which the three catalytic subunits are in different states of the catalytic cycle at any instant. As indicated in the insert, the front of the three-β subunits is in the O (here E for empty) form, that to the left has dinucleotide bound (L) and that on the right is in the T form. This convincingly demonstrates that ATP synthase functions by rotational catalysis (Abrahams et al., 1994; Capaldi and Aggeler, 2002).

What drives the inter-conversion of the three states is the rotation of the γ subunit. As the proton flux causes rotation of the γ subunit, say by 120°, the three-β subunits will change

---

[11] I prefer the description of F_1 as a ball on a stick, rather than as sometimes found in textbooks, a lollipop.
[12] Boyer and Walker received the 1997 Nobel Prize for Chemistry together with Jens Skou.

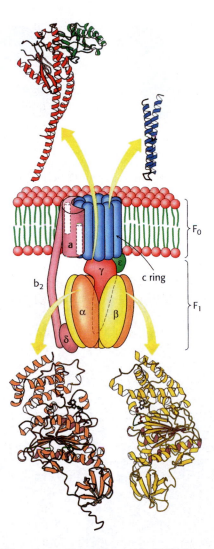

Figure 5. 18 Structure of ATP synthase. (From Berg et al., 2002. Reproduced with permission from W.H. Freeman and Co.)

position and conformational state. So, the subunit that had ATP tightly bound will adopt the open conformation, and ATP will be released. The loosely bound ADP and P_i will find itself in the tight conformation, and its high affinity for ATP will drive ATP synthesis. And finally, the subunit, previously in the open form, will adopt the L form and bind ADP and P_i. The most elegant proof that the ATP synthase is a rotary molecular motor comes from studies in which the $\alpha_3\beta_3$ hexamer was fixed to an Ni-surface (using a short sequence of His residues attached to the end of the protein chain—a His tag), with the γ subunit pointing upwards and attached to a fluorescent-labelled actin filament (Figure 5.21). Addition of ATP (to stimulate the reverse reaction of ATP synthesis) resulted in a rotation of 120° for each equivalent of ATP added.

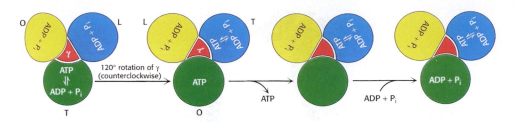

Figure 5.19 Binding-change mechanism for ATP synthase. Rotation of the γ subunit inter-converts the three-β subunits. The subunit in the tight (T) form contains newly synthesized ATP that cannot be released. Rotation by 120° converts it to the open (O) form, from which ATP can be released, allowing it to bind ADP and P_i to begin a new cycle. (From Berg et al., 2001. Reproduced with permission from W.H. Freeman and Co.)

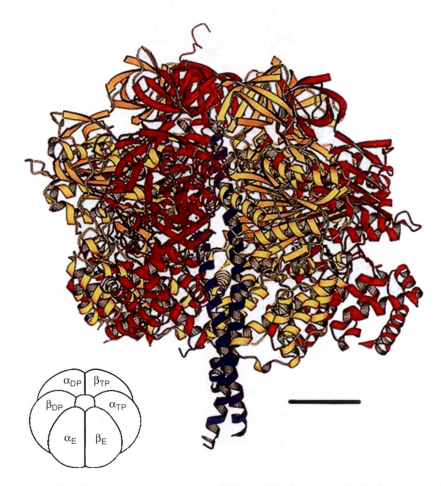

Figure 5.20 The X-ray structure of the F_1-ATP synthase from bovine heart mitochondria. The α, β and γ subunits are in red, yellow and blue, respectively. The inset (bottom left) shows the orientation of the subunits in this view. The bar is 20 Å long. (From Voet and Voet, 2004. Reproduced with permission from John Wiley & Sons., Inc.)

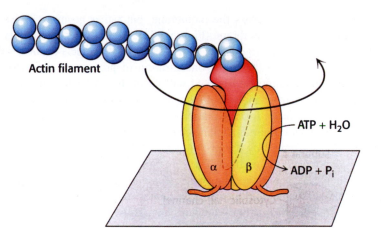

Figure 5.21 Direct observation of ATP-driven rotation in ATP synthase. The $\alpha_3\beta_3$ hexamer is fixed to a surface with the γ subunit pointing upwards and linked to a fluorescent-labelled actin filament. Addition of ATP results in rotation of the γ subunit, which can be observed with a fluorescence microscope. (From Berg et al., 2002. Reproduced with permission from W.H. Freeman and Co.)

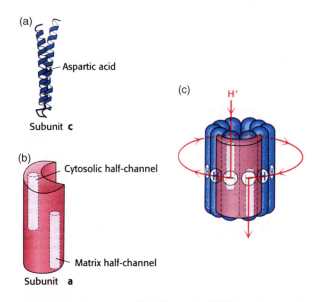

Figure 5.22 Components of the proton-translocating unit of ATP synthase (a, b) and the proton path through the membrane (c). Each proton enters the cytosolic half-channel, follows a complete rotation of the c ring, and exits through the other half channel into the matrix. (From Berg et al., 2002. Reproduced with permission from W.H. Freeman and Co.)

One final conceptual question remains. How does the flow of protons through F_0 drive the rotation of the γ subunit? It is suggested that the c subunit (Figure 5.17), which has an aspartate residue (Asp61) in the middle of a pair of helices that traverse the membrane, plays a key role. There are channels in the a subunit, which surrounds the central ring of *c* subunits, but which do not cross the membrane, but rather go more or less half way across from each side of the membrane (Figure 5.22). Suppose, further, that two residues of Asp61 of two subunits *c* are in contact with the two half channels of the *a* subunit. One has picked up a proton from the high concentration of protons on the cytosolic (inter-membrane space) of the mitochondria, and will be in its protonated (neutral form). The other, coming from the half-channel on the matrix side, which is proton deficient, will be in its charged, non-protonated form. Rotation of the *c* ring by one subunit will now align the protonated Asp61 with the matrix half-channel, whereas the unprotonated Asp 61 of the second will be confronted by the proton-rich cytosolic half-channel (25 times higher $[H^+]$ concentration than in the matrix). The net result is vectorial proton migration across the inner mitochondrial membrane as a consequence of rotation of the *c* ring.

Fuller information on the content of this Chapter can be found in Berg et al., 2002; Campbell et al., 2005; Devlin, 2005; Voet and Voet, 2004.

REFERENCES

Abrahams, K.P., Leslie, A.G., Lutter, R. and Walker, J.E. (1994) Structure at 2.8 Å resolution of F_1-ATPase from bovine heart mitochondria, *Nature*, **370**, 621–628.

Berg, J.M., Tymoczko, J.L. and Stryer, L. (2002) *Biochemistry*, 5th edition, W.H. Freeman and Co., New York, 974 pp.

Campbell, P.N., Smith, A.D. and Peters, T.J. (2005) *Biochemistry Illustrated: Biochemistry and Molecular Biology in the Post-Genomic Era*, 5th edition, Elsevier, London and Oxford, 242 pp.

Capaldi, R. and Aggeler, R. (2002) Mechanism of F_1F_0-type ATP synthase, a biological rotary motor, *TIBS*, **27**, 154–160.

Devlin, T.M. (2005) *Textbook of Biochemistry with Clinical Correlations*, 6th edition, Wiley, Hoboken, NJ, 1208 pp.

Knowles, J.R. (1991) Enzyme catalysis: not different, just better, *Nature*, **350**, 121–124.

Voet, D. and Voet, J.G. (2004) *Biochemistry*, 3rd edition, Wiley, Hoboken, NJ, 1591 pp.

– 6 –

Methods to Study Metals in Biological Systems

INTRODUCTION

The study of metals in biological systems requires techniques, some of them highly specific, some limited to certain aspects of the metal ion in question, some of more general applicability. Thus, Mössbauer spectroscopy in biological systems is restricted to iron-containing systems because the only element available with a Mössbauer nucleus is ^{57}Fe. The EPR spectroscopic techniques will be of application only if the metal centre has an unpaired electron. In contrast, provided that crystals can be obtained, X-ray diffraction allows the determination of the 3-D structure of metalloproteins and their metal centres.

It is not our intention to describe the techniques in detail, but rather to indicate what information can be derived from the application of the method in question (and also what cannot). Two important practical generalizations should be made at the outset. The first is that there is little sense in using sophisticated physico-chemical techniques to analyse impure biological samples, and, reciprocally, that highly purified biological materials should not be subjected to poor analytical techniques. The second, perhaps even more important, is that, in general, the more techniques you can use on a biological sample the better, since there are virtually no situations in which one single method will reply to all of your questions. Table 6.1 summarizes the parameters that can be obtained with each of the techniques and the information that the method should supply (Campbell and Dwek, 1984; Que, 2000).

Spectroscopic techniques have the advantage over protein crystallography in that not only are crystals not required, but also they can allow time-resolved measurements to be made which can detect short-lived intermediates. However, to obtain structural information, the observed spectroscopic data must be fitted to molecular structures. This can be done with reference structures, which model the spectroscopic properties of the metalloprotein site. These could be synthetic low-molecular weight complexes of known molecular structure, or known high-resolution metalloprotein structures obtained by X-ray crystallograhy or high-field NMR. Yet, another promising approach is the use of quantum chemical calculations of spectroscopic properties of metalloproteins and model compounds to elucidate their geometrical and electronic structures (Neese, 2003).

Table 6.1

Spectroscopic methods: an overview

Method	Parameters	Information content
Magnetic susceptibility	Molecular g-value, axial and rhombic zero-field splitting, exchange interaction	Number of unpaired electrons/ground spin state; defines anti-ferromagnetic and ferromagnetic interactions; quantitates ground sub-level splittings
Mössbauer spectroscopy	Quadrupole coupling isomer shift	For ^{57}Fe sites: oxidation and spin state; chemical environment
Electron paramagnetic resonance (EPR)	Quadrupole tensor, nuclear Zeeman splitting, g-values, coupling constants, relaxation times	Usually for odd electron metal sites: probes ground state wavefunction at high resolution
Electron-nuclear double resonance (ENDOR)		Combines sensitivity of EPR and high resolution of NMR to probe ligand super-hyperfine interactions
Nuclear magnetic resonance (NMR)	Chemical shift, nuclear coupling constants, relaxation times	For paramagnetic proteins: enhanced chemical shift resolution, contact and dipolar shifts, spin delocalization, magnetic coupling from temperature dependence of shifts
Vibrational spectroscopy (Raman and IR)	Energies (with isotope perturbation), intensities and polarizations	Identification of ligands coordinated to a metal centre
Electronic absorption spectroscopy (ABS)	Energies, intensities and band shapes	Direct probe of ligand-field and charge-transfer excited states
Magnetic circular dichroism (MCD)	Same as ABS plus circular polarization induced by applied magnetic field and magnetic susceptibility	Greater sensitivity than ABS in observing weak transitions and greater resolution due to differences in circular polarization; complimentary selection rules aiding in assignment of electronic transitions
Circular dichroism (CD)	Same as ABS plus circular polarization due to asymmetric nature of metal site	Allows detection of transitions not readily observable in absorption
Resonance Raman spectroscopy	Intensity profiles, depolarization ratios	Allows study of chromophoric active sites in biological molecules at low concentration; can provide information on metal–ligand bonding
Extended X-ray absorption fine structure (EXAFS)	Energies, intensities and polarizations	Identity of ligand atoms: distance of ligand atoms from metal: number of scattering ligands of a given type
X-ray diffraction	Atomic coordinates at a given resolution	Identity of ligands to metal centre (but distances more precise by EXAFS)

MAGNETIC PROPERTIES

Here, a brief consideration and understanding of the magnetic properties of metal ions will help us enormously in some of the techniques that we will study later. We begin by defining diamagnetic and paramagnetic molecules. The former, with closed shells of electrons, have no inherent magnetic properties, and when weighed in the presence or absence of a magnetic field, will show a small decrease in weight (due to repulsion by the magnetic field). In contrast, paramagnetic molecules show a net attraction to the magnetic field, and a much larger increase in weight. An unpaired electron corresponds to an electric current and, by virtue of its spin and its orbital motion, to a magnetic field. Because transition metals are of great importance in biology, we are particularly interested in the magnetic properties of their unpaired electrons and the information we can derive from studying these properties. Magnetic susceptibilities may be measured by direct methods for small molecule models of metalloprotein cores and from them magnetic moments (expressed in Bohr magnetons) can be readily calculated. The diamagnetic contribution of the rest of the protein molecule and its associated bound water content makes this difficult to apply to metalloproteins. However, the higher sensitivity of superconducting quantum interference device (SQUID) susceptometers and other magnetometers makes direct determination of magnetic properties of metalloproterins possible. Since the magnetic susceptibility of most molecules varies with temperature, the magnetic moment and hence the number of unpaired electrons can be derived from temperature-dependent studies. EPR and NMR can also be used to deduce these parameters.

For paramagnetic molecules, the magnetic moment has two sources: spin and orbital contributions. For transition metal ions (with the exception of Co^{2+} and Co^{3+}), where there is only a small orbital contribution, the magnetic moment reflects the spin-only term, and expected spin-only magnetic moments in octahedral arrangements for d electrons in biologically relevant transition metals[1] are given in Table 6.2. For the first three d electrons, there is no ambiguity concerning their location—they are in the three t_{2g} set of orbitals (Table 6.3). For d^4, two alternative configurations are possible: the fourth electron may remain parallel to the other three and enter the higher energy e_g level or it may pair up with one of the electrons already present in the t_{2g} level (which will produce maximum crystal field stabilization). The first is known as the *high-spin* state, while the arrangement with the paired electrons is the *low-spin* (or strong field) state. Alternative electronic configurations are also possible for d^5, d^6 and d^7 ions in octahedral complexes.

When bridging ligands connect two or more magnetic centres, their electron spins can either cancel each other out or reinforce one another—this constitutes anti-ferromagnetic and ferromagnetic coupling, respectively. This phenomenon is frequently encountered in biological inorganic chemistry, particularly the former, and we will encounter a good example in the case of the iron core stored within the interior of the protein core of the iron storage protein, ferritin (see Chapter 13).

---

[1] Therefore titanium with one d electron is not included.

Table 6.2

Spin-only magnetic moments for octahedral arrangements

Number of d electrons	Magnetic moment (Bohr magnetons)	
	High spin	Low spin
2 (V^{3+})	2.83	
3 (V^{2+}, Cr^{3+})	3.81	
4 (Mn^{3+}, Cr^{2+})	4.93	2.83
5 (Mn^{2+}, Fe^{3+})	5.92	1.73
6 (Fe^{2+}, Co^{3+})	4.90	0.00
7 (Co^{2+})	3.87	1.73
8 (Ni^{2+})	2.83	
9 (Cu^{2+})	1.73	
10 (Cu^{+})	0.00	

Table 6.3

Electronic configurations in octahedral configuration

Number of d electrons	t_{2g}			e_g	
1	↑				
2	↑	↑			
3	↑	↑	↑		
4 high-spin	↑	↑	↑	↑	
4 low-spin	↑↓	↑	↑		
5 high-spin	↑	↑	↑	↑	↑
5 low-spin	↑↓	↑↓	↑		
6 high-spin	↑↓	↑	↑	↑	↑
6 low-spin	↑↓	↑↓	↑↓		
7 high-spin	↑↓	↑↓	↑	↑	↑
7 low-spin	↑↓	↑↓	↑↓	↑	
8	↑↓	↑↓	↑↓	↑	↑
9	↑↓	↑↓	↑↓	↑↓	↑
10	↑↓	↑↓	↑↓	↑↓	↑↓

ELECTRON PARAMAGNETIC RESONANCE (EPR) SPECTROSCOPY

The bioinorganic chemistry of V, Mn, Fe, Co, Ni, Cu, Mo, W, as well as a number of non-biological transition elements, is permeated by their paramagnetism, and EPR spectroscopy is a particularly useful tool for their analysis. It can be used with frozen dilute solutions of metalloproteins, and is quite sensitive (high-spin ferric ions can be detected in the µM range); it has the potential to establish the stoichiometries of complex mixtures of para-magnets. EPR detects unpaired electrons in a sample by their absorption of energy from continuous microwave irradiation (X-band, *ca.* 9–10 Hz), when the sample is placed in a strong magnetic field (around 0.3 T). In standard EPR practice, the EPR absorption is detected by varying the magnetic field at constant microwave frequency, because in order to get the resonance condition, the wavelength of the microwave frequency must be tuned to the dimensions of the resonator cavity. EPR spectra are usually represented as the first derivative of the measured absorption spectrum, and are characterized by the four main parameters: intensity, linewidth, g-value (which defines position) and multiplet structure (Hagen, 2006).

What type of information can we obtain from metalloprotein EPR? Examination of the EPR spectrum should permit (i) the identification of the type of bonding involved, based on its central hyperfine interaction, the oxidation state of the metal ion and, possibly, the type of metallo-ligand centre (for example distinguishing between a 3Fe rather than a 4Fe cluster); (ii) quantification of the concentration of the paramagnet; (iii) structural characterization, which is an extension of (i), involving the identification of ligands based on ligand–hyperfine interaction of atoms in their first-coordination sphere; (iv) functional characterization, such as determining the saturation binding of a metal ion to a specific site on the protein or using EPR spectroscopy to determine the reduction potential of a prosthetic group in the protein. Resonances can be split into multiplet structures by the interaction of the electron spins with nuclear spins: this gives rise to hyperfine interactions. To gain more detailed information on ligand identification and to increase resolving power, it may be necessary to apply advanced EPR techniques such as electron-nuclear double resonance spectroscopy (ENDOR) and electron spin echo envelope modulation (ESEEM).

MÖSSBAUER SPECTROSCOPY

Mössbauer spectroscopy probes high-energy transitions in the atomic nucleus and is based on the phenomenon of Recoil-free γ-ray Resonance Absorption. The effect was discovered by Rudolf Mössbauer in 1957, one year before he received his doctorate from the Technical University of Munich[2]. Under normal conditions, atomic nuclei recoil when they emit or absorb gamma rays, and the wavelength varies with the amount of recoil. Mössbauer found that at a sufficiently low temperature, a significant fraction of the nuclei embedded in a crystal lattice may emit or absorb gamma rays without any recoil. The strictly monochromatic γ-radiation emitted from the excited nucleus of a suitable isotope during a radioactive decay pathway can therefore be absorbed by the same isotope in the sample (biological applications are essentially restricted to ^{57}Fe; the natural abundance of ^{57}Fe is 2%, unless isotope enrichment is used). The interest of the method lies in the fact that, if the energy transitions occur within the nucleus itself, the magnitude depends on the density and arrangement of extranuclear electrons, i.e. on the chemical state of the atoms. The extremely small perturbation (10^{-8} eV) caused by the difference of chemical state between emitter and absorber can be easily offset and measured by Doppler modulation. The γ-ray energy is varied by mechanically moving the source relative to the sample (usually by a few mm/s). This causes a Doppler shift[3] in the frequency of the emitted radiation

---

[2] Mössbauer received the Nobel Prize in Physics in 1962.
[3] The well known effect whereby the pitch of the sound of a moving object (train, plane, etc.) gets higher as it approaches and becomes progressively lower as it recedes.

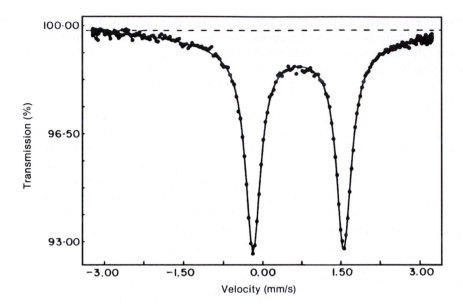

Figure 6.1 Mössbauer spectrum of human deoxymyoglobin.

(higher frequency if the source moves towards the sample, lower if it moves away from it). The counts of γ-rays transmitted at each Doppler velocity are averaged many times to improve the signal-to-noise ratio, and in the resulting spectra, γ-ray intensity is plotted as a function of the source velocity (lower if it moves away from it), and the spectra are collected over a wide range of temperature from 4.2 K to over 300 K The Mössbauer signal is influenced by the nuclear charge, the nature of the surrounding ligands and the symmetry of the ligand field. The observed isomer shift, δ in mm/s, gives information about the metal oxidation and spin states and the nature of the ligands coordinated to the iron. From a structural point of view, the quadrupole splitting, ΔE_Q, is dependent on electric field gradients at the nucleus and reflects the asymmetry of the electric field surrounding the metal centre. Figure 6.1 shows the Mössbauer spectrum of deoxyhaemoglobin crystals at 345 K. The quadrupolar splitting gives rise to two Lorentzian lines which can be well fitted to the experimental data. The isomer shift causes the spectrum to be symmetrical around 0.9 mm/s rather than 0.0 mm/s.

NMR SPECTROSCOPY

Nuclear magnetic resonance (NMR) is a widely utilized technique, which detects the reorientation of nuclear spins in a magnetic field. It can potentially be used to determine the 3-D structure of the protein itself, as well as supplying information on kinetics and dynamics, ligand binding, determination of pK-values of individual amino acid residues, on electronic structure and magnetic properties, to mention only some of the applications. In addition, it can be selectively applied to specific nuclei—^{1}H, ^{13}C, ^{15}N, ^{19}F (often substituted for H as a probe of local structure) and ^{31}P, the latter not only for the study of nucleic acids but also for the study of phosphorylated metabolites within cells. The theory underlying an NMR experiment is very similar to that for EPR, but the setup is very different for technical reasons. In EPR, the promotion of molecules from their ground state to an excited state by microwave radiation is detected by the corresponding *absorption* of energy. In contrast, NMR relies on a *relaxation* process—the radiofrequency radiation raises molecules to their excited state and the experiment then monitors their return (relaxation) to their ground state. The most frequently used method in NMR is to apply a pulse of radiofrequency to the sample and then detect the transient signal as the nuclear spins return to their ground state. The transient signal then undergoes a Fourier transformation to give the NMR spectrum. From NMR experiments four parameters can be derived—the chemical shift (δ), which such as the *g*-value in EPR defines the field position of the NMR signal (in this case with respect to a reference marker added to the sample), the intensity (I), the relaxation times and the coupling constant. Modern NMR instruments can resolve resonances for most of the protons in even large molecules, not just because the magnets themselves now produce very high fields (up to 900 mHz) but also because of the development of multi-dimensional NMR techniques[4]. Because dipolar interactions with neighbouring spins depend on distance, structural information can be deduced. The 2-D NMR techniques, notably of *correlation spectroscopy* (COSY) and *nuclear Overhauser effect spectroscopy* (NOESY), allow the proximity of atoms within a macromolecule to be determined. This is based on the fact that with COSY, protons that are attached to adjacent atoms can be studied, while NOESY spectra can detect two protons that are located closer than ~0.5 nm to each other, because they will perturb each other's spin—even when they are far removed from each other in the amino acid sequence of the protein. The full assignment of the hundreds of different resonances in the NMR spectrum and measurement of a great number of inter-proton distances and torsional angles can allow the complete 3-D structures of medium-sized proteins to be determined in solution[5].

Until a decade ago, metalloproteins containing paramagnetic metal ions were not thought to be suitable for the application of NMR techniques because the presence of paramagnetic centres destroys the resolution of the spectrum. However, the loss of resolution is less severe when the paramagnetic centre exhibits fast electronic relaxation. The application of advanced-pulse techniques and data-handling methods can overcome the limitations that paramagnetism presented previously. The presence of paramagnetism in a protein allows structural and mechanistic information by means of NMR that have no equivalent in the NMR study of diamagnetic proteins. Indeed, the replacement of diamagnetic and NMR-silent metal ions by suitable paramagnetic metal ions can be deliberately introduced into proteins to provide structural and mechanistic information not obtainable otherwise. A good example is the incorporation of lanthanides in Ca^{2+}-binding sites of proteins. (For recent reviews see Arnesano et al., 2005; Bertini et al., 2005; Ubbink et al., 2002).

---

[4] Multidimensional NMR was pioneered by Richard Ernst (Nobel Prize for Chemistry, 1991) and its application to structure determination of biological macromolecules, already heroically undertaken with all the limitations of one-dimensional NMR, was further developed and refined by Klaus Wüthrich (Nobel Prize for Chemistry, 2002).
[5] At typically a concentration of around 10^{-3} M, which for a protein of molecular weight 20 kDa represents a concentration of 20 mg/ml.

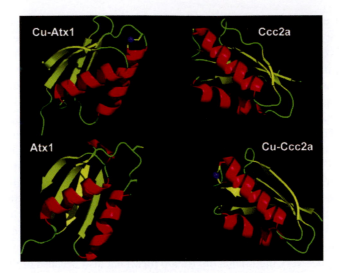

Figure 6.2 Atx1 and Ccc2 exchanging a copper(I) ion. (Reprinted with permission from Fragai et al., 2006. Copyright (2006) American Chemical Society.)

There is growing evidence over several decades that proteins are not rigid objects, but that they can sample a more or less wide range of different conformations. NMR is an appropriate technique to estimate the time scale of these conformational changes, from seconds to picoseconds, providing information both on conformational heterogeneity and on the time scale of motions associated with it. This is illustrated in Figure 6.2 for the transfer of a copper(I) ion from the copper chaperone Atx1 to the soluble domain of the Ccc2 ATPase, where in the adduct between the two proteins, a copper-bridged intermediate is formed (Banci et al., 2006). In Atx1, the two metal-binding cysteines move from buried, in the copper(I)-loaded protein to become solvent-exposed after copper release. In contrast, the structure of Ccc2a remains almost invariant upon binding of copper(I), indicating that the metal-binding site in Ccc2a is more pre-organized than in ApoAtx1.

ELECTRONIC AND VIBRATIONAL SPECTROSCOPIES

Transitions between different electronic states result in absorption of energy in the ultra-violet, visible and, for many transition metal complexes, the near infrared region of the electromagnetic spectrum. Spectroscopic methods that probe these electronic transitions can, in favourable conditions, provide detailed information on the electronic and magnetic properties of both the metal ion and its ligands.

Electronic spectra of metalloproteins find their origins in (i) internal ligand absorption bands, such as $\pi \rightarrow \pi^*$ electronic transitions in porphyrins; (ii) transitions associated entirely with metal orbitals (d–d transitions); (iii) charge-transfer bands between the ligand and the metal, such as the S$\rightarrow$Fe(II) and S$\rightarrow$Cu(II) charge-transfer bands seen in the optical spectra of Fe–S proteins and blue copper proteins, respectively. Figure 6.3a presents the characteristic spectrum of cytochrome c, one of the electron-transport haemoproteins of the mitochondrial

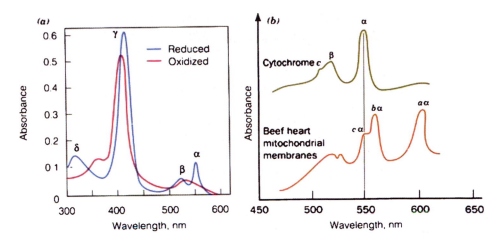

Figure 6.3 (a) Visible absorption spectrum of cytochrome c in its reduced and oxidized states. (b) The three separate α bands in the visible spectrum of beef heart mitochondria (below) indicating the presence of cytochromes a, b and c, with the spectrum of cytochrome c (above) as reference. (From Voet and Voet, 2004. Reproduced with permission from John Wiley & Sons., Inc.)

electron-transport chain. In the reduced form, this consists of four absorption bands, the γ (Soret) band and three others, designated α, β and δ, while in the oxidized form the spectrum is characteristically different. The three α bands of cytochromes a, b and c in beef heart mitochondria can be clearly distinguished (Figure 6.3b).

CIRCULAR DICHROISM AND MAGNETIC CIRCULAR DICHROISM

Because proteins are made up of chiral amino acids (see Chapter 4), they can discriminate between right and left circularly polarized light (rcp and lcp, respectively). The different absorption of lcp and rcp light (reflected by different extinction coefficients) is termed circular dichroism (CD). CD is particularly useful for metals bound within proteins and can often detect and resolve electronic transitions that are not so accessible by classical absorption spectroscopy. This is because some of the *d–d* transitions, which are observed in small molecule inorganic complexes by electronic absorption spectroscopy, cannot be observed in the absorption spectra of metalloproteins. These ligand-field transitions can be intense in the CD spectrum, and potentially give much more information, as illustrated for reduced *Chromatium vinosum* high-potential iron protein in Figure 6.4. In addition, CD is a useful tool for obtaining information about the secondary structure of proteins. Since α-helices, β-sheets and random coils all have characteristic CD spectra, the relative amounts of these different secondary structures can be evaluated. In the presence of a molecular field, even non-chiral molecules exhibit CD spectra, which can be measured by the technique called magnetic circular dichroism (MCD). The intensity developed by the spin-orbit coupling between excited states and between ground states and excited states can be exploited, particularly at low temperature, which generates more intense metal-centred *d–d* transitions in low-temperature MCD relative to absorption spectra. Although the theoretical analysis

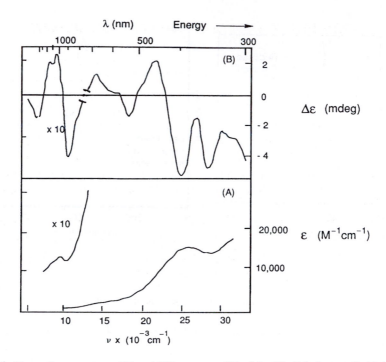

Figure 6.4 Absorption spectrum (A) and CD spectrum (B) of the [Fe$_4$S$_4$] cluster of a high-potential iron protein (HiPIP) from *Chromatium* sp. (From Cowan, 1997. Reproduced with permission from John Wiley & Sons., Inc.)

of MCD spectra is usually complex, it can be a powerful fingerprint for the identification of bound ligands. An example is the use of MCD to identify the presence of a thiolate ligand coordinated to low-spin ferric haem iron in cytochrome P-450s.

RESONANCE RAMAN SPECTROSCOPY

As in Raman spectroscopy, resonance Raman spectroscopy gives information about molecular vibrational frequencies. These frequencies are in the range of 10^{12}–10^{14} Hz, and correspond to radiation in the infrared region of the electromagnetic spectrum. In resonance Raman spectroscopy, the energy of an incoming laser beam is tuned to be near to an electronic transition (in resonance); vibrational modes associated with the particular transition exhibit a greatly increased Raman-scattering intensity, usually overwhelming Raman signals from all other transitions. In haemoproteins, such as haemoglobin, tuning the laser to near the charge-transfer electronic transition of the iron centre gives a spectrum, which only reflects the stretching and bending modes associated with the tetrapyrrole iron. Resonance Raman spectroscopy reduces the complexity of the spectrum, allowing us to look at only a few vibrational modes at a time. Its main advantage over classical Raman spectroscopy is the large increase in the intensity of the peaks (by a factor of as much as 10^6) allowing spectra to be obtained with sample concentrations as low as 10^{-8} M.

EXTENDED X-RAY ABSORPTION FINE STRUCTURE

The availability of easily tunable high-flux X-ray beams from synchrotron radiation has led to the development of new types of X-ray spectroscopy. On technique of particular interest for metalloproteins is extended X-ray absorption fine structure (EXAFS). Here, the absorption of X-rays by a solid or a liquid sample is measured as a function of wavelength at energies just above the absorption transition of a particular metal atom (the absorber). At energies just above the sharp absorption threshold, a pattern of rapid oscillations is observed, which represents an interference effect from the neighbouring atoms of the absorber. A Fourier transform of the oscillations can be analysed to give, in favourable cases, information on the number, types and distances of the neighbouring atoms. This usually requires parallel studies on model compounds of known structure and, for many metal centres in proteins, can give extremely accurate structural information without the requirement for an ordered sample (see the next section on X-ray diffraction). Geometric information derived from fitting EXAFS data to a model structure can be reliable to ± 0.01 Å.

X-RAY DIFFRACTION

Protein crystallography had its beginnings in 1934, when J.D. Bernal and Dorothy Crowfoot (Hodgkin) showed that crystals of pepsin gave an X-ray diffraction pattern, made up of sharp reflections, which showed that the protein had an ordered structure, with most of its 5000 atoms occupying clearly defined positions. Since then, protein crystallography has advanced to become one of the most important techniques for structure determination of macromolecules, with thousands of structures being determined every year. The principal reasons for this explosion of X-ray crystallographic prowess are: (i) more coherent protocols for protein crystallization; (ii) cryo-crystallography; (iii) the use of brighter and tunable synchrotron-generated X-ray beams (which can enable reliable data collection on crystals that was considered too small for study, some years ago); (iv) better data collection facilities; (iv) the quasi generalized use of multiple anomalous dispersion (notably by replacement of methionine residues in the protein by selenomethionine residues) to resolve the 'phase problem'.

Structure determination of proteins requires the availability of an ordered sample in the form of a single crystal. Since protein crystals typically contain large amounts of water, to prevent them from drying, and thereby losing their regular ordered structure, they must be kept moist in presence of the liquid of crystallization during data collection. This is usually achieved by mounting the wet crystals in small glass capillaries, which are placed in a narrow beam of monochromatic X-rays. The crystal is then rotated in order to produce a diffraction photograph. While we can record the intensities and hence the amplitudes of the X-rays diffracted by the crystal, we cannot translate them into atomic structure without the knowledge of both the amplitudes of the scattered beams and their experimentally inaccessible phase constants. The phase problem was first resolved by J. Monteath Robertson[6] for the

---

[6] Gardiner professor of Chemistry in the University of Glasgow, whose lectures I attended as an undergraduate in the early 1960s.

(then) complex organic compound phthalocyanin, by isomorphous replacement in which comparison of the molecule with H, Ni and Cu at its centre allowed him to find phase constants and absolute structures. This approach, multiple isomorphous replacement (MIR) was extended to protein crystals by the preparation of derivatives in which a heavy atom is bound specifically and uniformly to molecules within the crystal. MIR techniques further require that the heavy atom derivatization does not introduce additional changes in the molecular structure or change the crystallographic parameters. MIR has now been replaced by multiple anomalous diffraction (MAD) phasing. MAD exploits the potential of using more than one wavelength along with the known position of anomalous scattering atoms[7] to resolve the phase ambiguity. Most MAD phasing experiments use proteins in which methionine residues have been replaced by selenomethionine. From a practical point of view, such experiments can only be carried out at synchrotron beamlines, and since all the data are collected on the same sample, systematic sources of error are eliminated and resulting phase angles are more accurate (for a review see Ealick, 2000).

REFERENCES

Arnesano, F., Banci, L. and Piccioli, M. (2005) NMR structures of paramagnetic proteins, *Q. Rev. Biophys.*, **38**, 167–219.

Banci, L., Bertini, I., Cantini, F., Felli, I.C., Gonnelli, L., Hadjiliadis, N., Pierattelli, R., Rosato, A. and Voulgaris, P. (2006) The Atx1–Ccc2 complex is a metal-mediated protein–protein interaction, *Nat. Chem. Biol.*, **2**, 367–368.

Bertini, I., Luchinat, C., Parigi, G. and Pierattelli, R. (2005) NMR spectroscopy of paramagnetic metalloproteins, *Chembiochem*, **6**, 1536–1549.

Campbell, I.D. and Dwek, R.A. (1984) *Biological Spectroscopy*, Benjamin/Cummings Publishing Co., Inc., Menlo Park, CA, 404 p.

Cowan, J.A. (1997) *Inorganic Biochemistry An Introduction 2nd edition*, Wiley-VCH, New York, p. 73.

Ealick, S.E. (2000) Advances in multiple wavelength anomalous diffraction crystallography, *Curr. Opin. Chem. Biol.*, **4**, 495–499.

Fragai, M., Luchinat, C. and Parigi, G. (2006) "Four-dimensional" protein structures: examples from metalloproteins, *Acc. Chem. Res.*, **39**, 909–917.

Hagen, W.R. (2006) EPR spectroscopy as a probe of metal centres in biological systems, *Dalton Trans.*, **37**, 4415–4434.

Neese, F. (2003) Quantum chemical calculations of spectroscopic properties of metalloproteins and model compounds: EPR and Mössbauer properties, *Curr. Opin. Chem. Biol.*, **7**, 125–135.

Que, L. Jr. (2000) *Physical Methods in Bioinorganic Chemistry: Spectroscopy and Magnetism*, University Science Books, Sausalito, CA., pp. 59–120.

Ubbink, M., Worrall, J.A.R., Canters, G.W., Groenen, E.J.J. and Huber, R. (2002) Paramagnetic resonance of biological metal centers, *Annu. Rev. Biophys. Biomol. Struct.*, **31**, 393–422.

---

[7] Anomalous scattering occurs when the frequency of the X-rays used falls near the absorption edge of one or more atoms in the sample, e.g. transition metals, often found in metalloproteins, and other atoms such as selenium.

– 7 –

Metal Assimilation Pathways

INTRODUCTION

After outlining the roles of metals in biology, their coordination chemistry, structural and molecular biology, and involvement in metabolism and bioenergetics as well as considering their biological ligands and the plethoric physicochemical array of techniques available for their study in biological systems we complete these introductory chapters by a consideration of how the metal ions essential for living organisms are assimilated from their surroundings. Clearly, this poses three distinct types of problems illustrated by three different kinds of organisms. For single-celled microorganisms, they must acquire the metal ions they require from their immediate environment. If they are motile, and in a liquid milieu they can 'swim around' in search of their food, but find it they must. If they are multicellular organisms, but are rooted to the spot in the soil, such as most members of the plant family are, they must find their source of nutrition wherever they can extend their roots. And finally, if they are multicellular mobile animals who can forage for their food, fish for it, kill it and eat it, or buy it at the supermarket, they ultimately take advantage of pre-processing of their nutritional supply by the organisms from which they have acquired it.

While recognizing that this is a supreme example of reductionist simplification, it nonetheless allows us to situate the three model systems that we will consider here, because the mechanisms of metal assimilation are often significantly different. Finally, once the metal ions have been assimilated, they must be incorporated into the corresponding proteins, we have already presented in Chapter 3 a brief consideration of some of the ways in which metal centres, whether individual metal ions or more complex metal clusters, are engineered into their metalloprotein targets.

METAL ASSIMILATION IN BACTERIA

Iron

The importance of metal assimilation for bacteria is perhaps best illustrated by the correlation between the efficacy of iron uptake systems and virulence in many strains of pathogenic

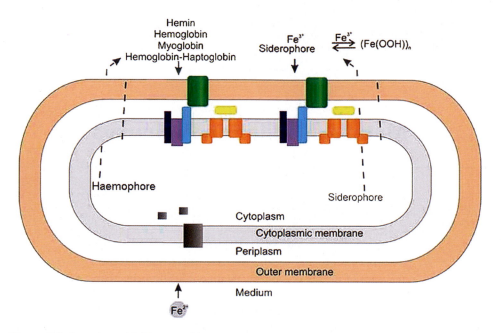

Figure 7.1 Overview of Fe^{3+}- and Fe^{2+}-uptake systems of Gram-negative bacteria. (From Crichton, 2001. Reproduced with permission from John Wiley & Sons., Inc.)

bacteria. We begin this short survey of microbial inventiveness in metal ion accumulation with our favourite metal, iron (unashamedly), not only because we know it well, but also because it represents one of the best studied of all metal ion uptake systems into bacteria.

Some of the mechanisms that are used by Gram-negative bacteria for iron uptake are illustrated in Figure 7.1.

Most nutrients required by Gram-negative bacteria diffuse passively across the outer membrane into the periplasm through transmembrane channels composed of porins. However, scarce metals such as iron and cobalt (the latter in the form vitamin B12) require active transport systems and simply cannot depend on a process of simple diffusion. In the case of iron, this is achieved by synthesizing and secreting siderophores, strong and highly specific Fe^{3+}-complexing compounds, which are taken up by specific transport systems. Some of these bacteria (*E. coli* is an example) also take up Fe^{3+}-loaded siderophores, produced by other bacteria and fungi, which they themselves are unable to synthesize, and in some cases they may also have an uptake system for Fe^{2+}.

Many pathogenic bacteria can also acquire iron from the haem of their mammalian hosts by secreting proteins called haemophores, which release haem from haemoglobin to specific transport proteins in the outer membrane. Yet other pathogens can use iron bound to transferrin and to lactoferrin (not shown in the figure). The list of pathogenic microorganisms which are able to use mammalian host's iron transport systems reads like a role call from Hell's kitchen—*Haemophilus influenzae* (wide range of clinical diseases, but not influenza), *Neisseria meningitidis* (meningitis), *Neisseria gonorrhoeae* (gonorrhea), *Pseudomonas aeruginosa* (opportunistic pathogen, which exploits any break in host defences), *Serratia*

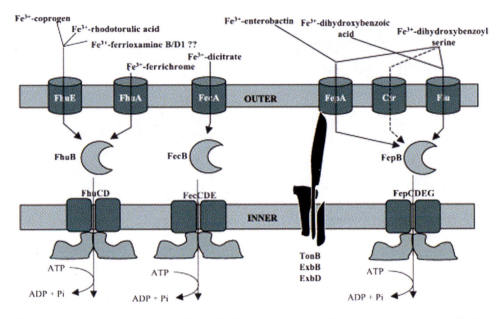

Figure 7.2 Schematic representation of siderophore-mediated iron uptake in *E. coli*. The TonB–ExbB–ExbD complex energizes and interacts with all the outer membrane receptors, not just FepA. (From Andrews et al., 2003. Reproduced with permission from Blackwell Publishing Ltd.)

marcescens (urinary tract and wound infections), *Vibrio cholera* (cholera), *Yersinia pestis* (plague), *Yersinia entercolytica* (gastroenteritis), to mention but a few.

In the best-characterized Gram-negative bacteria, *E. coli*, specific outer membrane receptor proteins (Figure 7.2) bind ferric siderophores, or the iron chelate ferric citrate. Active transport across the outer membrane requires the participation of three cytoplasmic membrane proteins—TonB, ExbB and ExbD (designated as the Ton system), which transmit the energy from the energized cytoplasmic membrane to the non-energized outer membrane. The principal siderophore of *E. coli*, the hexadentate tricatecholate enterobactin (Figure 7.3), is synthesized by the bacteria and excreted into the surroundings, where it complexes ferric iron with a high affinity[1] (pFe = 35.5). When ferric enterobactin binds to its outer membrane receptor FepA, energy transduction from the inner membrane is transmitted by TonB, which recognizes that the iron-charged siderophore has been bound, and opens the periplasmic face of the transmembrane FepA protein. The ferric enterobactin is then transferred to the periplasmic-binding protein, Fep B, which transports it to the plasma membrane. There, an ABC transporter (ATP-binding cassette) ensures its ATP-dependent delivery to the cytosol. The iron is released by reduction to Fe^{2+} by a flavoprotein ferrireductase, which has little regard to the nature of the ferric siderophore. Although *E. coli* produces only enterobactin

---

[1] The stability of Fe(III) siderophore complexes can be evaluated empirically as the pFe (which does not require knowledge of the K_a of the ligand groups nor the denticity of the complex), is defined as the negative logarithm of the free Fe_{aq}^{3+} concentration (pFe = $-\log[Fe_{aq}^{3+}]$ calculated at pH 7.4, with a total ligand concentration of 10 μM and $[Fe_{aq}^{3+}]_{tot}$ of 1 μM. The larger the pFe value, the more stable the metal complex.

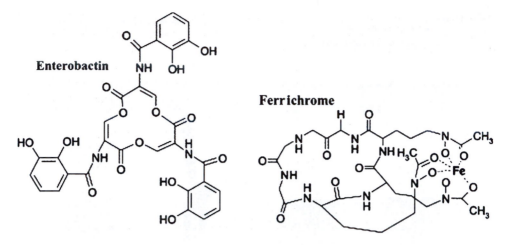

Figure 7.3 Structures of the siderophores enterobactin and ferrichrome. (From Andrews et al., 2003. Reproduced with permission from Blackwell Publishing Ltd.)

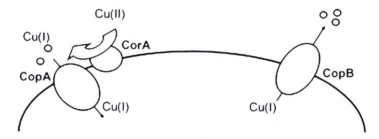

Figure 7.4 Copper uptake and release in *E. hirae*. The extracellular reductase CorA reduces Cu^{2+} to Cu^+ for uptake by CopA. In conditions of copper excess CopB functions as a copper exporter. (From Solioz and Stoyanov, 2003. Reproduced with permission from Blackwell Publishing Ltd.)

(and its precursor and breakdown products, dihydroxybenzoate and dihydroxybenzoylserine) endogenously, it has outer membrane receptors for the uptake of a number of exogenous ferric hydroxamate siderophores, such as ferrichrome and ferrioxamine, which it is itself incapable of synthesizing, as well as the periplasmic transporter FhuB and the inner membrane ABC transporter FhuCD, necessary for their uptake. It also has an uptake system that enables it to acquire iron from ferric citrate, despite the fact that ferric citrate is neither a carbon nor an energy source for *E. coli*.

Copper and zinc

Because of their importance in many enzymes, bacteria have had to develop uptake systems for copper and zinc. Copper uptake (and homeostasis, which is discussed in Chapter 8) has been extensively studied in the Gram-positive bacterium *Enterococcus hirae*. Two of the genes in the *cop* operon, *copA* and *copB*, encode membrane ATPases. An extracellular reductase CorA reduces Cu^{2+} to Cu^+, which is taken up by CopA when copper is limiting.

In contrast, CopB extrudes excess copper (Figure 7.4). Both CopA and CopB belong to the P1 subclass of P-type ATPases, which includes the proteins involved in the human diseases of copper metabolism, Menkes and Wilson's disease (discussed in Chapter 14).

Likewise, for zinc, bacteria have developed active uptake systems (Hantke, 2001). In many bacteria the high-affinity Zn^{2+} uptake system uses an ABC transporter of the cluster 9 family, which mostly transports zinc and manganese and is found in nearly all bacterial species. First identified in cyanobacteria and pathogenic streptococci, but also found in *E. coli*, the system is encoded by three genes ZnuABC and consists of an outer membrane permease ZnuB, a periplasmic-binding protein ZnuA and a cytoplasmic ATPase ZnuC. Low-affinity transporters of the ZIP family, described later in this chapter, such as ZupT, have also been shown to be involved in bacterial zinc uptake.

METAL ASSIMILATION IN PLANTS AND FUNGI

Iron

We have taken the liberty of assimilating fungi with plants not because we think they are more like vegetables than animals, but because many recent exciting findings on transition metal uptake in yeast have important implications for what is going on in plants, particularly since genomic studies have revealed that plants such as *Arabidopsis thaliana* have many of the genes corresponding to those identified as being involved in yeast. Figure 7.5 summarizes the mechanisms that are involved in iron uptake across the plasma membrane in baker's yeast, *Saccharomyces cerevisiae*. The reason why yeast has become the paradigm of fungal iron accumulation is the comparative ease of doing classical and molecular genetic experiments, and that it was the first eukaryote to have its entire genome sequenced.

In common with most prokaryotes many fungi have siderophore-dependent iron uptake systems. The ferrichrome-type siderophores are often employed, although other types of siderophore are also used. Indeed, even if, like the quintessential scavenger baker's yeast (*S. cerevisiae*), they produce no siderophores of their own, they nonetheless have several

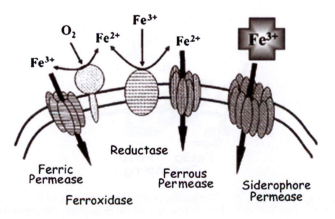

Figure 7.5 Summary of iron uptake across the yeast plasma membrane. (From Kosman, 2003. Reproduced with permission of Blackwell Publishing Ltd.)

distinct facilitators for the uptake of ferric siderophores, including ferric enterobactin, which is produced by many bacteria, but not by fungi. This parasitic behaviour of *S. cerevisiae* is reminiscent of *E. coli*, which does not produce hydroxamate siderophores like ferrichromes, but does have uptake systems that allow it to scavenge fungal siderophores like ferrichrome.

Most fungi produce (or at least encode in their genomes) at least one integral membrane reductase, which is usually relatively unspecific. The most abundant plasma membrane reductase in *S. cerevisiae*, Fre1p, exhibits comparable activity with Fe^{3+} and Cu^{2+} and can use a variety of one-electron acceptors, although the ultimate reducing substrate is most likely NAD(P)H. *S. cerevisiae* has a second reductase Fre2p and both reductases seem to be involved in reduction of both Cu^{2+} and Fe^{3+}. Once in the ferrous form, the iron can be taken up by two possible mechanisms, either involving direct transport of Fe^{2+} or alternatively a second high-affinity pathway, which involves re-oxidation of the Fe^{2+} (the ferroxidase pathway).

In the first, Fe^{2+} is transported directly into the cell through divalent metal ion transporters, which like the reductases are not specific to Fe^{2+} but can also transport Mn^{2+}, Ni^{2+} and Cu^{2+}. One such family of transporters in *S. cerevisiae*, the Smf1 proteins are orthologues of the DCT1 (or Nramp2) divalent cation transporters found in mammalian cells (see below). Transport of Fe^{2+} by both Smf1 and DCT1 is coupled to the transport of H^+, with a stoichiometry of one H^+ for one Fe^{2+}. Another ferrous iron transporter, characterized in *S. cerevisiae* but not in other fungi, is the Fet4 protein, which like the Smf proteins has low substrate specificity, and though of low affinity probably represents the principal iron uptake pathway for baker's yeast at high iron levels in the culture medium.

The second system for ferrous iron uptake in yeast, considered to be the high-affinity uptake pathway, involves the two proteins Fet3 and Ftr1 (Figure 7.6). Fetr3p, a multicopper oxidase, like ceruloplasmin and hephaestin (see below), acts as a ferroxidase to oxidize Fe^{II} to Fe^{III}, whereupon the ferric permease Ftr1p transports the Fe^{III} into the yeast cell (Figure 7.6). These two proteins are probably part of a heterodimeric or higher-order complex in the yeast plasma membrane. It is proposed that Fet3p-produced Fe^{3+} does not equilibrate with the bulk phase, but is transferred directly to Ftr1p for permeation by a classic metabolite channelling mechanism. The transfer of Fe^{3+} from one protein to the other is non-dissociative and probably involves a series of ligand exchange reactions.

Figure 7.6 A cartoon showing a strict channelling mechanism: transfer of Fe^{III} from Fet3p to Ftr1p is non-dissociative and the Fe^{III} does not equilibrate with the bulk phase. (Reprinted with permission from Kwok et al. 2006. Copyright (2006) American Chemical Society.)

Plants also take up iron by mechanisms (Figure 7.7) similar to those described for fungi. However, they can be divided into those which reduce ferric iron to ferrous outside of the root cell (Strategy I) and those which import the iron in the ferric form and reduce it inside the cytosol of the root cell (Strategy II). In Strategy I plants (such as *Arabidopsis*, pea and tomato), iron mobilization is achieved by the combined action of a proton-extruding H^+-ATPase and a ferric chelate reductase, both of which are induced by iron deficiency (Figure 7.7). Sequence similarities with the yeast Frep have allowed the cloning of a plant ferric chelate reductase, designated FRO2 (ferric reductase oxidase) from *Arabidopsis*. An iron transporter of the ZIP family, named IRT1, seems to be the principal transporter of ferrous iron in *Arabidopsis*, and orthologues of IRT1 have also been characterized in tomato and rice.

In grasses (which include barley, maize and rice) high-affinity Fe(III) chelators (phytosiderophores, PS) are synthesized by the plants themselves and excreted into the environment around their roots in order to complex and solubilize the ferric iron in the soil. Transporters specific for the Fe(III)-siderophore complex then take the complex into the cytosol, where the iron is released from the phytosiderophore by an as yet undefined mechanism. The best characterized of these transporters is YS1 (yellow stripe 1) or YSL1 (yellow stripe-like) so named after the phenotypic appearance of a maize mutant deficient in phytosiderophore uptake. However, unlike the bacterial or fungal siderophores, phytosiderophores (Figure 7.8) of the mugeneic acid family are synthesized from L-methionine

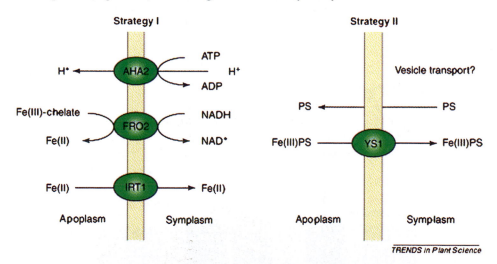

Figure 7.7 Mechanisms of iron uptake by higher plants. (From Schmidt et al., 2003. Copyright 2003, with permission from Elsevier.)

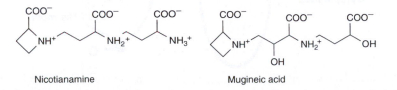

Figure 7.8 Structure of the phytosiderophore mugeneic acid and its precursor nicotianamine.

via nicotianamine. Whereas the Strategy II grasses produce and excrete the mugeneic acid family of siderophores, nicotiniamine is found in both Strategy I and Strategy II plants, where there is much evidence that it is involved in the intercellular transport of iron as the Fe-nicotinamine chelate.

Copper and zinc

Genetic studies in the yeast *S. cerevisiae* identified two proteins involved in high-affinity copper uptake at the plasma membrane, yCtr1 and yCtr3. These proteins transport Cu^+ (Figure 7.9) after the reduction of Cu^{2+} by the same reductases, Fre1p and Fre3p described above for iron. All Ctr family members have three putative transmembrane regions (Figure 7.9), with the N-terminus external and the C-terminus cytoplasmic. The former has two Met-rich regions, which might function in copper binding, while the latter contains a sequence of conserved Cys and His residues. Since the same family of copper transporters are present in plants and animals we will not consider them further. In yeast a third member

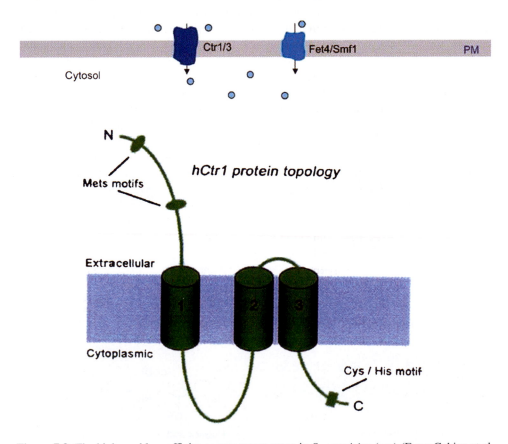

Figure 7.9 The high- and low-affinity copper transporters in *S. cerevisiae* (top) (From Cobine et al., 2006, Copyright 2006, with permission from Elsevier.) and a topological model for the hCtr1 protein (bottom). (From Petris et al., 2004. With kind permission of Springer Science and Business Media.)

of the Ctr family, Ctr2, localizes to the vacuole and is proposed to mobilize copper stored in the vacuolar compartment to the cytoplasm when extracellular copper is limited. There is also a low-affinity copper uptake system through Fet4 and Smf1 permeases.

Zinc uptake in fungi and in plants is carried out to a large extent by the ZIP (Zrt-, Irt-like Protein) family of metal ion transporters. The family name comes from the yeast Zrt1 protein and the *A. thaliana* Irt1 protein. These were the first identified members of the family, which are found at all phylogenic levels, including bacteria, fungi, plants and mammals. The mammalian members of the family are given the systematic designation 'SLC39' (Eide, 2004). Without any known exception the members of the ZIP family transport zinc and/or other metal ions from the extracellular space, or from the lumen of cellular organelles, into the cytoplasm. This, together with their transmembrane topography, distinguishes them from the CDF (cation diffusion facilitator)/Znt family: the mammalian members of the family have been named Znt and given the systematic designation 'SLC30' (Palmiter and Huang, 2004). As we will see when we consider zinc homoeostasis in Chapter 8, this family transports zinc from the cytoplasm into the lumen of extracellular organelles or to the outside of the cell. Thus, CDF proteins work in the opposite direction to ZIP proteins.

Most ZIP proteins have eight predicted transmembrane domains (Figure 7.10) and similar predicted topologies with both the N- and C-termini located on the extracytoplasmic face of the membrane with a His-rich domain frequently in the long cytoplasmic loop between transmembrane domains 3 and 4. In contrast, most CDF transporters have six predicted transmembrane domains and a His-rich domain in the loop between domains 4 and 5, but here the N- and C-termini are on the cytoplasmic side of the membrane.

In the yeast *S. cerevisiae* at least four different transporters are involved in zinc uptake (Figure 7.11). The most important of these is the ZIP family member Zrtr1, which is required for growth under low zinc concentrations. Zrt1 has a high affinity for zinc with an apparent K_m for free Zn^{2+} of 10 nM. A second ZIP protein, Zrt2, with a lower affinity for zinc (apparent $K_m \sim 100$ nM) probably plays a role under less severe zinc limitation. In addition to the two ZIP proteins, two other lower-affinity systems also operate. One is the Fet4 protein, which we saw earlier is also involved in the low-affinity uptake of iron and copper; Fet4 is only found in Ascomycete fungi. It is also likely that the Pho84 high-affinity phosphate transporter is able to transport zinc in addition to phosphate, which could be important for zinc uptake in conditions of low levels of phosphate.

In plants zinc uptake involves members of the ZIP family, some of which are root specific, while others are found in both roots and shoots.

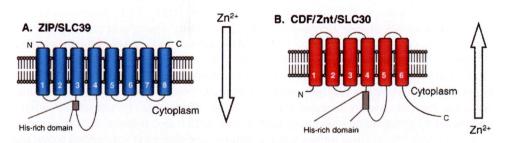

Figure 7.10 Predicted membrane topologies for the ZIP/SLC39 and CDF/Znt/SLC30 families of metal ion transporters. (From Eide, 2006. Copyright 2006, with permission from Elsevier.)

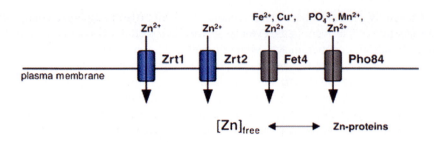

Figure 7.11 Uptake of zinc by the yeast *S. cerevisiae*. (From Eide, 2006. Copyright 2006, with permission from Elsevier.)

METAL ASSIMILATION IN MAMMALS

Since in mammalian species metals first need to be assimilated from dietary sources in the intestinal tract and subsequently transported to the cells of the different organs of the body through the bloodstream, we will restrict ourselves in this section to the transport of metal ions across the enterocytes of the upper part of the small intestine (essentially the duodenum), where essentially all of the uptake of dietary constituents, whether they be metal ions, carbohydrates, fats, amino acids, vitamins, etc., takes place. We will then briefly review the mechanisms by which metal ions are transported across the plasma membrane of mammalian cells and enter the cytoplasm, as we did for bacteria, fungi and plants. The specific molecules involved in extracellular metal ion transport in the circulation will be dealt with in Chapter 8.

Iron

Within the intestinal tract of mammals, dietary iron is in two forms: haem iron and non-haem, essentially ferric, iron. Haem iron, reflecting no doubt the origins of many mammals (including man) as hunters, is generally more readily absorbed than the latter. Non-haem iron is derived from sources such as vegetables, which tend to be a poor source of iron, because of the presence of phosphates, phytates and polyphenols, which decrease absorption by formation of insoluble ferric complexes. Haem iron is taken up by a recently described specific transporter, and Fe^{2+} is released into the intracellular iron pool by haem oxygenase, which degrades haem to Fe^{2+}, porphobilinogen and CO (Figure 7.12). Non-haem dietary iron is taken up in a manner reminiscent of the low-affinity iron uptake pathway in yeast and the Strategy I pathway in plants. The Fe(III) is reduced to Fe(II) by a ferric reductase (Dcytb) at the apical membrane and the Fe(II) is transported into the intestinal cell by DCT1 (also known as DMT1), a proton-coupled divalent cation transporter. Within the intestinal cell iron enters a low-molecular-weight pool: some of it may be stored in ferritin, while the rest can cross to the basolateral membrane. There it can be transferred into the circulation by a transmembrane transporter protein, IREG, also referred to as ferroportin. In the circulation serum iron is transported as diferric transferrin (Tf), described below and in Chapter 8. It has been suggested that iron incorporation into apotransferrin might be

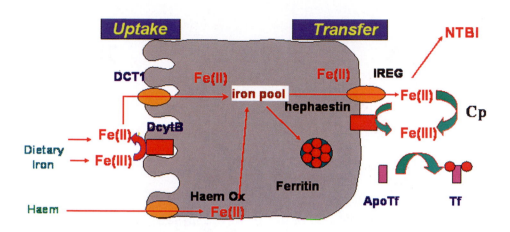

Figure 7.12 Schematic representation of iron absorption in mature intestinal mucosal cells. (From Crichton, 2001. Reproduced with permission from John Wiley & Sons., Inc.)

facilitated by the oxidation of Fe^{2+} to Fe^{3+}. Two candidates for this ferroxidase activity have been proposed (Figure 7.12)—ceruloplasmin (CP), the principal copper-containing protein of serum, or hephaestin, a member of the family of multicopper oxidases (which includes ceruloplasmin) which appears to be bound to the basolateral membrane.

Serum iron is delivered to cells via the transferrin to cell cycle (Figure 7.13). The diferric-transferrin molecule binds to its receptor and the complex is invaginated into clathrin-coated pits, which fuse with the target membranes of endosomes delivering the vesicle contents into the interior of the endosome. The pH of the endosome is reduced to around 5–6 by the action of an ATP-dependent proton pump, and at this pH iron is released from transferrin bound to its receptor as Fe^{3+}, presumably by protonation of the bound carbonate. The divalent cation (metal) transporter DMT1 is thought to assure the transport of iron out of the endosome into the cytoplasm, presumably after reduction of Fe^{3+} to Fe^{2+}. The cytoplasmic iron can then be transferred to the mitochondria for use in haem and iron–sulfur cluster synthesis (see Chapter 3), or stored in ferritin (see Chapter 8). Unlike most other protein ligands taken up by receptor-mediated endocytosis, apotransferrin retains a high affinity for its receptor at acidic pH values and is recycled back to the plasma membrane, where it dissociates from its receptor and goes off into the circulation in search of further iron. This sequence of events constitutes the transferrin to cell cycle, which ensures iron uptake by cells that have transferrin receptors.

Copper and zinc

We will discuss in more detail in Chapter 8 how intracellular copper levels are maintained at extremely low levels by a series of copper chaperone proteins, which sequester newly assimilated copper within the cytoplasm of cells and deliver it in a targeted manner to be incorporated into specific copper-containing proteins. While copper uptake across the gastrointestinal tract is poorly understood—most probably utilising the divalent cation transporter

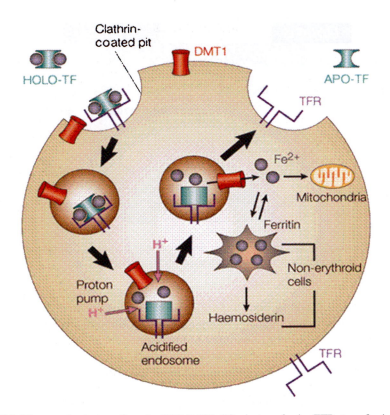

Figure 7.13 The transferrin to cell cycle: HOLO-TF, diferric transferrin; TFR, transferrin receptor; DMT1, divalent metal transporter. (From Andrews, 2000. Reproduced by permission of Nature Reviews Genetics.)

DCT1, at the cellular level Cu is imported across the plasma membrane of mammalian cells by a transport system that requires the product of the *Ctr1* gene. Mammalian Ctr1 mRNA is expressed in all tissues with higher levels in liver and kidney and lower levels in brain and spleen. The role of Ctr1 in mice has been studied using *Ctr1* gene knockout; its importance is underlined by the observation that homozygotes die *in utero*, while heterozygotes are only slightly affected, although they have a severely reduced brain Cu content. An additional human gene, *hCTR2*, similar to *hCTR1*, has been reported but its role in Cu transport remains unknown.

The ZIP family are involved in Zn transport into the cytosol, mostly across the plasma membrane. Although the human genome encodes 14 ZIP-related proteins, only a few (Zip1, Zip2, Zip4) have been shown to be involved in Zn transport. Zip1 is widely distributed in mammalian tissues, while Zip2 is expressed at low levels only in a few tissues. Zip4 is involved in the uptake of dietary Zn into intestinal enterocytes, and mutations in Zip4 have been found in patients with acrodermatitis enteropathica, a recessive disorder of Zn absorption, which results in Zn deficiency. DMT1 is probably also involved in the transport of dietary zinc across the brush border membrane of the intestine.

REFERENCES

Andrews, N.C. (2000) Iron homeostasis: insights from genetics and animal models, *Nat. Rev. Genet.*, **1**, 208–217.

Andrews, S.C., Robinson, A.K. and Rodriguez-Quinones, F. (2003) Bacterial iron homeostasis, *FEMS Microbiol. Rev.*, **27**, 215–237.

Cobine, P.A., Pierrel, F. and Winge, D.R. (2006) Copper trafficking to the mitochondrion and assembly of copper metalloenzymes, *Biochim. Biophys. Acta*, **1763**, 759–772.

Crichton, R.R. (2001) *Inorganic Biochemistry of Iron Metabolism: From Molecular Mechanisms to Clinical Consequences*, Wiley, Chichester, 326 pp.

Eide, D.J. (2004) The SLC39 family of metal ion transporters, *Pflügers Arch. Eur. J. Physiol.*, 447, 796–800.

Eide, D.J. (2006) Zinc transporters and the cellular trafficking of zinc, *Biophys. Biochim. Acta*, 1763, 711–722.

Hantke, K. (2001) Bacterial zinc transporters and regulators, *Biometals*, **14**, 139–249.

Kosman, D.J. (2003) Molecular mechanisms of iron uptake in fungi, *Molec. Microbiol.*, **47**, 1185–1197.

Kwok, E.Y., Severance, S. and Kosman, D.J. (2006) Evidence for iron channeling in the Fet3p-Ftr1p high-affinity iron uptake complex in the yeast plasma membrane, *Biochemistry*, **45**, 6317–6327.

Palmiter, R.D. and Huang, L. (2004) Efflux and compartmentalisation of zinc by members of the SLC30 family of solute carriers, *Pflügers Arch. Eur. J. Physiol.*, **447**, 744–751.

Petris, M.J. (2004) The SLC31 (Ctr) copper transporter family, *Pflügers Arch. Eur. J. Physiol.*, **447**, 796–800.

Schmidt, W. (2003) Iron solutions: acquisition strategies and signalling pathways in plants, *Trends Plant Sci.*, **8**, 188–193.

Solioz, M. and Stoyanov, J.V. (2003) Copper homeostasis in *Enterococcus hirae*, *FEMS Microbiol. Rev.*, **27**, 183–195.

– 8 –

Transport, Storage and Homeostasis of Metal Ions

INTRODUCTION

As in Chapter 7, we successively consider the transport, storage and metal ion homeostasis in prokaryotes, plants and animals. Since the assimilation of metals in unicellular bacteria does not require their transport to other cell types, we confine our discussion only to storage and homeostasis.

METAL STORAGE AND HOMEOSTASIS IN BACTERIA

Iron

Once iron has been assimilated within the bacterial cell, it is made available for intracellular functions. Iron from siderophores, ferric citrate, lactoferrin and transferrin is directly available, whereas haem iron must be released by the action of haem oxygenase. This pool of ferrous iron can be used for intracellular functions; however, in many bacteria, iron can be incorporated into a number of iron-storage proteins (Andrews, 1998), constituting an intracellular reserve that accumulates iron when it is in excess for future use under conditions of iron deficiency. Three types of potential iron-storage proteins are found in bacteria, with a similar molecular architecture of a roughly spherical protein shell surrounding a central cavity within which a mineral core of iron can be deposited. These are the 24-subunit ferritins (Ftns, which are also found in eukaryotes) and haem-containing bacterioferritins, Bfrs, found in eubacteria and the 12-subunit DNA-binding proteins from starved cells (Dps proteins), present only in prokaryotes (Figure 8.1). That Dps proteins should be considered as iron-storage proteins is a matter of debate. Their principal role seems to be to protect bacterial DNA against oxidative stress, notably by H_2O_2, by binding non-specifically and preventing free Fe^{2+} from catalysing Fenton chemistry. Although they form evolutionarily distinct families, they have many structural and functional similarities. They are composed of either 24 (ferritins and bacterioferritins) or 12 (Dps proteins) similar if not identical subunits.

Bfr   **Dps**

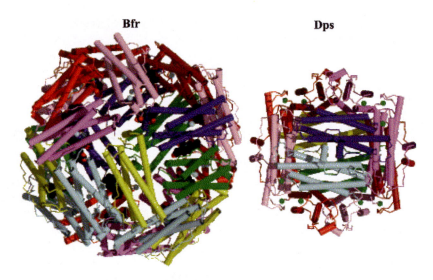

Figure 8.1 Structures of the 24-meric bacterioferritin and the 12-meric Dps protein from *E. coli*, approxoimately to scale. (From Andrews et al., 2003. Reproduced with permission from Blackwell Publishing Ltd.)

Their subunits are folded in a central bundle of four α-helices, which assemble to form a roughly spherical protein shell surrounding a central cavity within which iron is stored (up to 4500 iron atoms per 24mer in ferritins and bacterioferritins, and around 500 in the smaller Dps protein 12mer).

Iron is stored in these proteins in the ferric form, but is taken up as Fe^{2+}, which is oxidized by ferroxidase sites (a more detailed account of iron incorporation into ferritins is given later in this chapter). As we point out in Chapter 13, ferritins are members of the much larger diiron protein family. After oxidation, the Fe^{3+} migrates to the interior cavity of the protein to form an amorphous ferric phosphate core. Whereas the ferritins in bacteria appear to fulfil the classical role of iron-storage proteins, the physiological role of bacterioferritins is less clear. In *E. coli* it seems unlikely that bacterioferritin plays a major role in iron storage.

It was originally thought that iron metabolism in many Gram-negative bacteria was regulated exclusively in response to iron availability by the ferric-uptake regulator (Fur) protein. The Fur protein is a homodimer, which acts as a positive repressor, i.e. repressing transcription when it interacts with its co-repressor, Fe^{2+}, and causing depression in the absence of Fe^{2+} (Figure 8.2). Metal binding (one Fe^{2+}/subunit) increases the affinity of Fur for its DNA-binding sites by ~1000-fold, located between -35 and -10 bases from the initiation site in the promoter region of Fur-repressed genes. The Fur protein controls the iron-dependent expression of more than 90 genes in *E. coli*, and functions as the major iron-responsive regulator in many bacteria. However, in Gram-positive bacteria with a high GC content in their DNA (such as *Mycobacterium* and *Streptomyces*) the diphtheria toxin regulator (DtxR) protein is responsible for global iron regulation. Though Fur and DtxRs have little or no sequence similarity, their α-helical DNA-binding domains are similar, and in both DtxR and Fur, two dimers bind to a single operator that is at least 27 base pairs long.

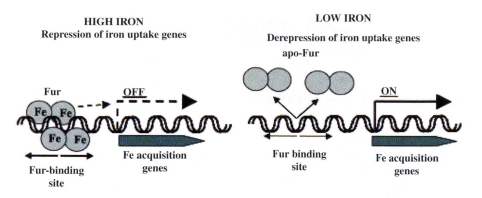

HIGH IRON
Repression of iron uptake genes

LOW IRON
Derepression of iron uptake genes
apo-Fur

Fur   OFF   ON

Fur-binding
site

Fe acquisition
genes

Fur binding
site

Fe acquisition
genes

Figure 8.2 Schematic representation of Fur-mediated gene repression. (From Andrews et al., 2003. Reproduced with permission from Blackwell Publishing Ltd.)

Fur is itself part of the family of gene regulatory proteins throughout many bacterial species. The major subclass is mainly involved, like Fur in *E. coli*, in the control of iron homeostasis, but it can also function in acid tolerance and protection against oxidative stress. Fur also controls the iron-regulated expression of bacterial virulence determinants. One class of the Fur family, Zur, is involved in the regulation of zinc uptake (see below).

The structure of Fur from several bacteria has been determined and Figure 8.3 shows the structure of two *Pseudomonas aeruginosa* Fur dimers docked to a target DNA fragment where the two dimers follow the pseudo-twofold symmetry of the DNA. This is consistent with the observation that whereas binding of a single Fur dimer would not be expected to protect more than ~20 bp against nuclease digestion, Fur has never been shown to protect less than 27–30 bp.

However, a puzzle in iron regulation, until recently, has been that, despite being a negative repressor, Fe^{2+}-Fur is found to be a positive regulator for a number of genes. In *E. coli*, these include the tricarboxylic cycle enzymes aconitase, fumarase and succinate dehydrogenase, both of the ferritins, FtnA and Bfr, and the Fe-superoxide dismutase (SODB). The explanation has come from the discovery of small non-coding RNAs in all organisms that mostly function as regulators of translation and messenger RNA stability. One of these, RyhB, a 90-nucleotide RNA, down-regulates a set of iron-storage and iron-using proteins when iron is limiting. Further, *ryhB* is itself negatively regulated by Fe^{2+}-Fur. RyhB RNA levels are inversely correlated with the mRNA levels of the *sdhCDAB* operon, which codes for succinate dehydrogenase and the other five proteins that had been found to be up-regulated by Fe^{2+}-Fur. When Fur is active, transcription of RyhB RNA is repressed, and in the absence of RyhB, the mRNAs for the proteins found to be up-regulated are no longer degraded. This means (Figure 8. 4) that iron in abundant, active Fur not only switches off iron acquisition genes, but, by switching off *ryhB*, iron-storage genes are switched on. Conversely, in iron penury, inactive Fur allows the genes for iron acquisition to be switched on, and those for iron storage to be switched off. We will see later in the chapter that a similar up- and down-regulation of iron uptake and storage pathways, albeit at the level of translation rather than transcription, operates in animals.

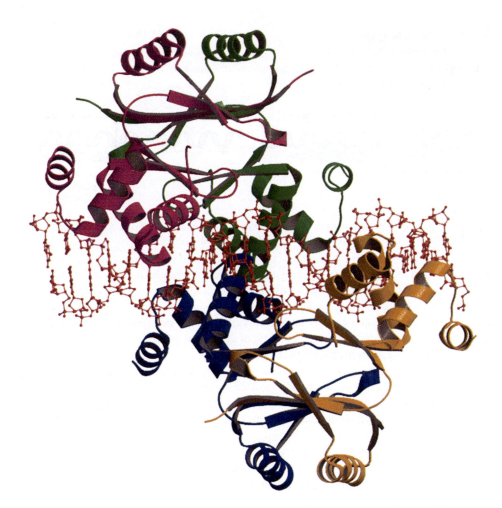

Figure 8.3 Model of two Fur dimers binding to opposite sides of canonical B-DNA. The twofold symmetry between the Fur dimers follows the symmetry of the DNA sequence. (From Pohl et al., 2003. Reproduced with permission of Blackwell Publishing Ltd.)

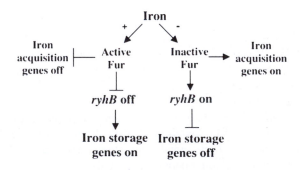

Figure 8.4 Model of Fur and RyhB interaction to regulate iron utilization. (From Massé and Gottesman, 2002. Copyright (1993) National Academy of Sciences, USA.)

Copper and zinc

Both copper and zinc appear to be stored in many bacteria in cysteine-rich proteins, called metallothioneins, which will be discussed from a structural point of view later in the chapter. The expression of these metal sequestering, low-molecular weight, cysteine-rich proteins, is often induced by both monovalent Cu(I) and divalent Zn(II), as well as by the non-biologically necessary, but potentially toxic, Ag(I) and Cd(II).

Because of their importance in the activity of many enzymes, bacteria have had to develop uptake-efficient systems for copper and zinc. However, since both these metals are toxic in excess, their intracellular content must be tightly regulated. Copper-uptake homeostasis has been most extensively studied in the Gram-positive bacterium *Enterococcus hirae*. The four genes involved, *copY*, *copZ*, *copA* and *copB* are arranged in the *cop* operon. As mentioned in Chapter 7, the proteins CopA and CopB are copper-transporting ATPases, copY is a copper responsive repressor and copZ is an intracellular copper chaperone (copper chaperones are discussed in more detail later in the Chapter). The *cop* operon enables *E. hirae* to grow in copper-limiting conditions, as well as in copper concentrations up to 8 mM. Figure 8.5 shows a model for copper homeostasis in *E. hirae*. Copper is reduced to the cuprous form, Cu(I), by an as yet uncharacterized reductase before being imported into the cell by CopA, whence it is carried in the cytoplasm by the specific metallochaperone, CopZ. Copper is then transferred from CopZ to the dimeric Zn-containing repressor CopY, displacing the Zn and releasing the CopY from the promoter. This allows transcription of the four *cop* genes to proceed. Under high-copper conditions, excess CopZ is degraded by a copper-activated protease.

In many bacterial species, zinc storage is apparently not a major mechanism in attaining homeostasis, the exception being cyanobacteria which detoxify and store zinc in a metallothionein. A more common way of ridding the cell of excess zinc is by exporting it. Pathways for iron uptake and efflux in Gram-negative bacteria are represented in Figure 8.6. The highly Zn^{2+} resistant bacterium, isolated from a decantation tank in a zinc factory, *Ralstonia metallidurans*, has a minimal inhibitory Zn^{2+} concentration of 12 mM. Its first

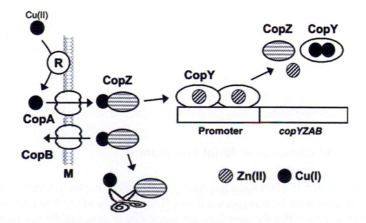

Figure 8.5 Model of copper homeostasis in *E. hirae*. (From Magnani and Solioz, 2005. With kind permission of Springer Science and Business Media.)

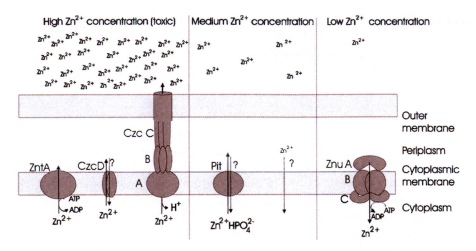

Figure 8.6 Pathways for uptake and efflux of Zn^{2+} in Gram-negative bacteria. (From Hantke, 2001. With kind permission of Springer Science and Business Media.)

line of defence against high Zn^{2+} concentrations seems to be the cation diffusion facilitator CzcD. Exporters of the CzcABC-like RND transporter family also seem to be a very effective way of protecting the cells against toxic Zn^{2+} concentrations. P-type ATPases like ZntA may also play a role in protection against high Zn^{2+} concentrations. Under Zn^{2+}-replete conditions, Pit-type proteins may be involved, while at low-Zn^{2+} concentration the Znu ABC transporters represent the high-affinity uptake pathway. Most of the high-affinity zinc-uptake systems are regulated by Zur proteins, a sub-group of the Fur family.

METAL TRANSPORT, STORAGE AND HOMEOSTASIS IN PLANTS AND FUNGI

While unicellular fungi do not require metal transport systems, multi-cellular fungi and plants most certainly do, and we consider their transport in plants, and then consider how metal ions are sequestered in storage compartments before addressing their homeostasis. Once again, we consider in turn these processes for iron, copper and zinc. Since iron metabolism has been most intensively studied in *S. cerevisiae*, of all the fungi, we will focus our attention on iron homeostatic mechanisms, however, as the reader will see shortly, copper and zinc homeostasis have many similarities.

Iron storage and transport in fungi and plants

The acquisition of iron in plant roots has been described in Chapter 7. Once in the apoplast of the root, the iron must be transported through the roots to the xylem and thence to the leaves. In order to ensure that the iron does not precipitate or generate oxygen radicals during its transport, the iron is bound to an intracellular transporter of iron (both Fe^{2+} and Fe^{3+}),

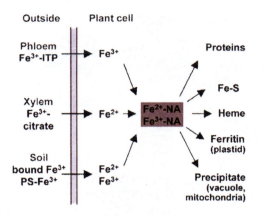

Figure 8.7 Simplified model of nicotiniamine (NA) function in plant cells. Iron is transported across the plasma membrane by the Strategy I or Strategy II uptake systems. Once inside the cell, NA is the default chelator of iron to avoid precipitation and catalysis of radical oxygen species. The iron is then donated to proteins, iron-sulfur clusters and haem, while ferritin and iron precipitation are only present during iron excess. (From Hell and Stephan, 2003. With kind permission of Springer Science and Business Media.)

most often nicotianamine (NA) (Figure 8.7), a non-protein amino acid found in all plants and a precursor of the plant siderophores. Not only does NA keep iron in a bioavailable, non-toxic form, it is also needed for the intracellular transport of iron to proteins responsible for regulating iron-uptake genes. There is also good evidence for participation of citrate, after oxidation of iron, in the transport of iron in the xylem, as the Fe^{3+}–citrate complex. The photoreduction of xylem-transported ferric carboxylates, like citrate, is thought to be an important driving force in the reduction of iron in shoots: thereafter the distribution of iron to the leaves is probably mediated again by the NA–iron complex.

Plants contain phytoferritins, which accumulate in non-green plastids[1] in conditions of iron loading. They are targeted to the plastids by a putative-transit peptide at their N-terminal extremity, and possess the specific residues for ferroxidase activity and iron nucleation, found in mammalian H-type or L-type ferritin subunits.

Iron homeostasis in fungi and plants

As we saw earlier (Chapter 7), *S. cerevisiae* has a variety of genes coding for proteins that are involved in iron acquisition at the cell surface, and many of them are transcriptionally induced in response to low iron. The high-affinity transport system (Fet3, Ftr1) contains a ferroxidase, which requires copper as a cofactor, and as a consequence, genes that are involved in the trafficking and transport of copper to this Fet3 protein (Atx1 and Ccc2, described later) are also regulated at the transcriptional level by iron. It is clear that high-affinity iron uptake is seriously compromised by low levels of extracellular copper.

---

[1] Plastids are members of a family of organelles found in the cytoplasm of eukaryotic cells, all of which contain DNA and are bounded by a double membrane.

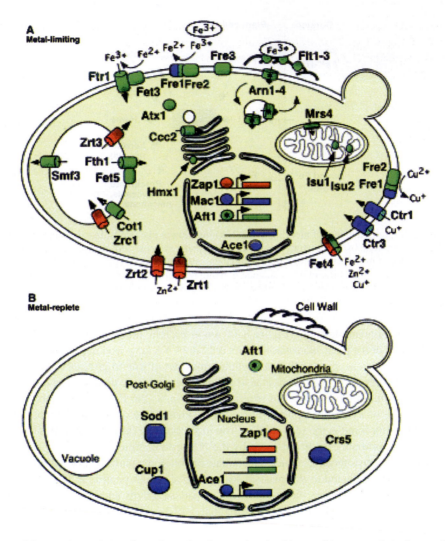

Figure 8.8 Protein products of metal-regulated genes involved in metal homeostasis in *S. cerevisiae.* (From Rutherford and Bird, 2004. With kind permission of Springer Science and Business Media.)

Iron-dependent gene regulation in *S. cerevisiae* is mediated by two transcription factors, Aft1 and Aft2 (Activator of Ferrous Transport). The regulation of gene products by Aft1 under iron-limiting conditions are shown in Figure 8.8 in green, and include the two reductases Fre1and Fre2, the high-affinity (Fet3, Ftr1) and low-affinity (Fet4) systems for free iron uptake, as well as the family of transporters (Arn1–Arn4) which cycle between the cell surface and an endosomal compartment, mediating ferric siderophore uptake. It also regulates the expression of a number of other genes, including other cell surface reductases, a mitochondrial iron transporter (Mrs4) and proteins involved in the biosynthesis of iron–sulfur clusters (Isu1 and Isu2). Aft2 also regulates the expression of some of the same genes as Aft1, and of some other genes as well. The mechanisms by which the transcription factors Aft1 and Aft2 sense iron levels are not fully understood.

In plants, comprehensive surveys of genes whose expression is altered in response to iron-deficiency have yielded a large set of transcripts. For example, in the Strategy I plant *Arabidopsis*, about a quarter of a set of 16,128 genes were found to be differentially expressed after 3 days of growth in Fe-free medium, while in the Strategy II barley plant some 200 genes out of nearly 9000 probed were found to be induced by iron-deficiency stress. While it is difficult to interpret such a large spectrum, some coherent conclusions can be drawn. In the phytosiderophore-dependent barley, the high-methionine requirement for the biosynthesis of mugeneic acids is reflected in the up-regulation of genes involved in methionine biosynthesis.

Much of what we know in strategy I plants comes from the studies of the tomato mutant, *fer*, which is characterized by its inability to induce iron-deficiency responses and to take up iron. FER encodes a transcription factor, and a protein with considerable similarity to FER, FIT1 has been characterized in *Arabidopsis*. The expression of FIT1 in the *fer* mutant of tomato rescues the ability of the plants to induce the iron-deficiency response. Like FER, FIT1 is root-specific; it is however induced under Fe-deficient conditions. In *fit1* mutants, the Fe^{2+} transporter IRT1 is absent, and the mRNAs for the reductase FRO2 and the Fe(III) chelate reductase activity are both absent. Both IRT1 and FRO2 are also regulated post-transcriptionally, most likely by ubiquitination in the presence of iron followed by endocytosis and degradation in the vacuole, in a similar manner to Zrt1 in the presence of zinc.

Phytoferritin transcription is induced by iron excess and repressed by iron deficiency in leaves as well as in roots in several plant species; in the case of maize, there are two ferritin genes which are differentially regulated by two independent-signalling pathways, one involving an oxidative step and one dependent on the plant growth hormone, abscissic acid.

A model for the regulation of iron-deficiency responses in strategy I plants (Figure 8.9) involves a regulatory circuit consisting of two parts: (a) a shoot–root loop and (b) a root–shoot loop. In the root–shoot part of the circuit, differentiation of root epidermal cells is controlled by iron availability, either within the root cell or in the rhizosphere (the immediate external environment of the root). Development of extra numbers of root hairs (which would increase iron uptake) is repressed by iron in the vicinity of the roots. Iron taken up and translocated to the leaves in the xylem regulates the uptake of iron into the leaf cells. When the iron levels in the leaf declines, this is detected by a sensor and a signal molecule is synthesized, which conveys this information to the root via the phloem. The effect of this signal is to increase the transfer of electrons, the net proton excretion and the activity of the Fe^{2+}-transporter, resulting in increased iron uptake by the root cells. It is likely that more than one sensor is involved in this process. The outcome of the interaction between the two signalling cascades is to positively or negatively regulate the uptake of iron.

Copper transport and storage in fungi and plants

We have already described the plasma membrane systems employed in yeast for copper uptake, and here we briefly describe the chaperone proteins involved in the intracellular transport and delivery of copper to target proteins (Figure 8.10), which were first described

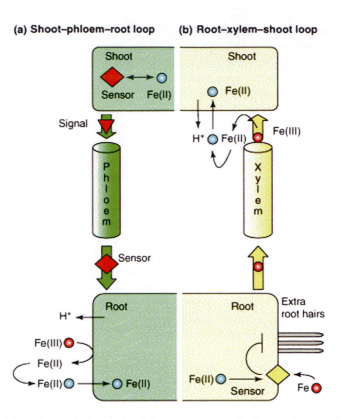

(a) Shoot–phloem–root loop   **(b) Root–xylem–shoot loop**

Figure 8.9 Model for the regulation of iron deficiency responses in Strategy I plants. (From Schmidt, 2003. Copyright 2003, with permission from Elsevier.)

in the *S. cerevisiae*. The first of the copper chaperones to be identified was the yeast Atx1 protein, which delivers copper to Ccc2, a P-type ATPase on the Golgi membrane. It transports copper into the lumen of the secretory pathway where it can be incorporated into newly synthesized cuproproteins like the plasma membrane Fet3 oxidase, involved in the high-affinity iron-uptake system. As we will see a little later in this chapter, Ccc2 corresponds to the human P-type ATPases that are mutated in Mentkes and Wilson's diseases. A second chaperone, Ccs1, provides Cu(I) for the activation of Cu–Zn superoxide dismutase SOD1 in both the yeast cytoplasm and the intermembrane space (IMS), between the inner and outer mitochondrial membranes. It also catalyses formation of an essential disulfide bond in SOD1, and the mechanism by which it functions was described in Chapter 7. Finally, the most complex of the three chaperones, Cox17, present in both the cytosol and IMS, transports copper to the mitochondria for incorporation into the important terminal oxidase of the respiratory pathway, cytochrome c oxidase (Figure 8.11). Cytochrome *c* oxidase has two haem *a* moieties and three copper atoms, and in eukaryotes consists of 12–13 subunits, with the three subunits Cox1–Cox3 that constitute the core enzyme encoded by the mitochondrial DNA (described in more detail in Chapter 13). The mononuclear Cu_B in subunit 1 (Cox1) interacts with the five-coordinate haem a_3 that has

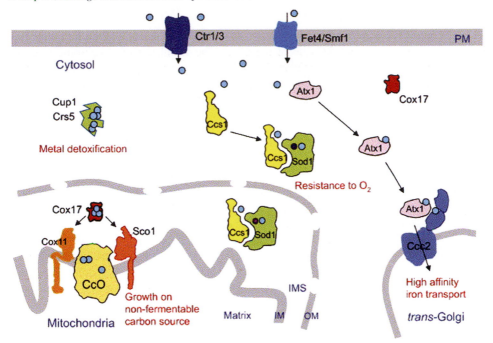

Figure 8.10 Model of copper homeostasis in *S. cerevisiae*. (From Cobine et al., 2006. Copyright 2006, with permission from Elsevier.)

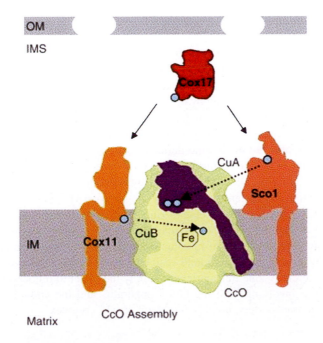

Figure 8.11 Copper delivery to cytochrome c oxidase within the mitochondrial IMS. (From Cobine et al., 2006. Copyright 2006, with permission from Elsevier.)

an open coordination site to which oxygen binds. There is also an additional dinuclear-mixed valence copper site, designated Cu_A located in Cox2. The metallation of cytochrome c oxidase involves a pathway that consists of Cox17 and the co-chaperones, Sco1 and Cox11. Cox17 delivers copper to Cox11 for subsequent insertion into the Cu_B site in Cox1 and to Sco1 for insertion into the Cu_A site in Cox2.

Copper homeostasis in fungi and plants

In fungi, copper homeostasis involves, as in mammals, some degree of post-transcriptional control of copper transporters. However, we consider, here, the transcriptional regulation of genes involved in copper acquisition, mobilization and sequestration. A number of copper-responsive factors have been characterized—in *S. cerevisiae* the two most important are Ace1, which activates gene expression in response to elevated copper, and Mac1, which activates gene expression in response to copper deficiency. Mac1 protects cells from copper deficiency (Figure 8.7) by activating expression of the high affinity copper-uptake systems CTR1 and CTR3 as well as the cell surface reductase FRE1. Mac1 is also required for the post-translational degradation of Ctr1 under conditions of copper excess. However, the principal resistance to copper excess is the Ace1-dependent induction of the *CUP1* gene (Figure 8.7). *CUP1* encodes a small cysteine-rich copper-binding metallothionein, which protects the cells by sequestering copper, thereby preventing its toxicity. Ace1 also regulates the expression of a second metallothionein gene (*CRS5*) and the copper–zinc superoxide dismutase gene (*SOD1*).

Zinc transport and storage in fungi and plants

As we saw in Chapter 7, there are several plasma membrane zinc-uptake transporters in yeast. Within the cell a number of other proteins are involved in zinc transport. *S. cerevisiae* is unusual in that it does not appear to have any plasma membrane zinc efflux transporters. This is to a large extent compensated by the capacity of the vacuole to serve as a major site of zinc sequestration and detoxification, enabling wild type cells to tolerate exogenous zinc concentrations as high as 5 mM. The zinc stored in the vacuole can attain millimolar levels, and can be mobilized under zinc-deficient conditions for use by the cell. Vacuolar zinc uptake is mediated by two members of the CDF family, Zrc1 and Cot1 (Figure 8.12). Zrc1 is a Zn^{2+}/H^+ antiport[2], allowing the zinc accumulation in the vacuole to be driven by the proton concentration gradient generated by the vacuolar H^+-ATPase. The Cot1 protein may function in the same way. The release of zinc to the cytosol is mediated by the Zip family member Zrt3. Within the vacuole zinc may be bound to organic anions. Vesicular storage sites for zinc may also exist in mammalian cells, where they have been designated 'zincosomes', and such membrane-bound vesicles have also been observed in zinc-treated yeast. There are certainly zinc transporters to supply the metal to mitochondria, and in the secretory pathway, involving both the Golgi apparatus and the endoplasmic reticulum; there are also zinc transporters, notably the Msc2–Zrg17 complex.

---

[2] An antiport simultaneously transports two molecules (or in this case, ions) simultaneously in opposite directions.

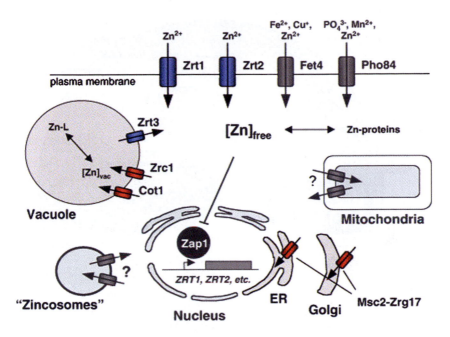

Figure 8.12 An overview of zinc transport and trafficking in the yeast *S. cerevisiae*. Zip family transporters are in blue, CDF family transporters are in red and other transporters are in grey. (From Eide, 2006. Copyright 2006, with permission from Elsevier.)

As we saw in Chapter 7, zinc uptake in plants involves proteins of the ZIP family, some of which are root specific while others are found in both roots and shoots. The transport of zinc from the cytosol in many organisms is often associated with members of the cation diffusion facility (CDF) family. Although there are 12 predicted family members in *Arabidopsis*, only one, MTP1, has been characterized, which seems to function in the transport of Zn into the vacuole. Two members of the heavy metal ATPase (HMA) family, HMA2 and HMA4, have been shown to function in the transport of zinc out of the cells across the plasma membrane.

Zinc homeostasis in fungi and plants

Zinc homeostasis in general parallels copper homeostasis, involving both transcriptional and post-translational regulatory mechanisms. In *S. cerevisiae* the high-affinity zinc-uptake gene *ZRT1* increases in response to zinc limitation, whereas in zinc repletion Zrt1 undergoes zinc-induced endocytosis and is degraded in the vacuole. The transcription factor Zap1 increases the expression of the three-uptake systems encoded by the *ZRT1*, *ZRT2* and *FET4* genes (Figure 8.7). It also stimulates the release of zinc from the vacuole by activation of the *ZRT3* vacuolar efflux system. Finally, and somewhat surprisingly, Zap1 also appears to increase the expression of *ZRC1*, a gene that modulates zinc influx into the vacuole.

A summary of the membrane distribution of Cu, Fe and Zn transporters in plants is presented in Figure 8.13.

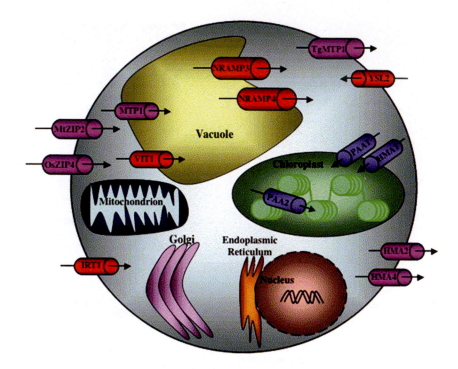

Figure 8.13 Summary of the membrane distribution of Cu (blue), Fe (red) and Zn (purple) transporters localized in plants. (From Grotz and Guerinot, 2006. Copyright 2006, with permission from Elsevier.)

METAL TRANSPORT, STORAGE AND HOMEOSTASIS IN MAMMALS

Finally, we deal with metal storage, transport and homeostasis in mammals. We have already described iron uptake from the gastrointestinal tract and the transferrin to cell cycle, which delivers most of the iron to mammalian cells. We will therefore focus our attention on some interesting aspects of the transport protein transferrin, and on the mechanism of iron uptake into the storage protein, ferritin. The transport and storage of copper and zinc will then be discussed. We then conclude with an account of metal homeostasis, first treating the question at the cellular level and finally addressing the way in which mammals regulate their dietary iron, copper and zinc absorption in order to prevent either metal ion deficiency or excess. As we will see, in man this is a non-trivial problem on account of our severely limited capacity to excrete iron, making the control of dietary-iron absorption a matter of capital importance.

Iron transport and storage in mammals

Under normal physiological conditions, iron is transported in serum by transferrin, an 80 kDa bilobal protein with two almost identical iron-binding sites, one in each half of the molecule.

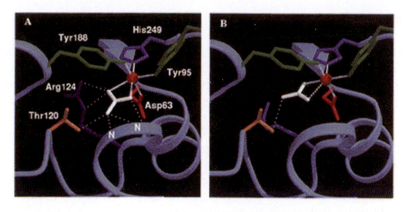

Figure 8.14 Structures of alternative conformations of the iron-binding sites of the recombinant N-lobe of human transferrin. (Reprinted with permission from MacGillivray et al., 1998. Copyright (1998) American Chemical Society.)

The coordination of the iron atom involves four protein ligands and a carbonate anion The N-lobe of human transferrin can be crystallized at acidic pH in two different conformations (Figure 8.14). The A conformation corresponds to the structure of the iron-binding site found in other transferrins, whereas in the B form, the hydrogen bonding of the carbonate has been considerably disrupted, consistent with the idea that protonation of the carbonate might be the first step in iron release. The decreased charge on the bicarbonate would lead to its detachment from the N-terminus of helix 5 and from Arg124, and a change from bidentate to monodentate anion coordination.

There is one known pathway for cellular iron export, involving the export of iron into the plasma from the basolateral membrane of duodenal enterocytes, from macrophages, hepatocytes and a number of other cell types. This involves the protein known as IREG1 or ferroportin described already in Chapter 7. We will discuss ferroportin in more detail in the next section on iron homeostasis, since ferroportin is the target of hepcidin, a recently described iron regulatory peptide.

About a quarter of the total body iron is stored in macrophages and hepatocytes as a reserve, which can be readily mobilized for red blood cell formation (erythropoiesis). This storage iron is mostly in the form of ferritin, like bacterioferritin a 24-subunit protein in the form of a spherical protein shell enclosing a cavity within which up to 4500 atoms of iron can be stored, essentially as the mineral ferrihydrite. Despite the water insolubility of ferrihydrite, it is kept in a solution within the protein shell, such that one can easily prepare mammalian ferritin solutions that contain 1 M ferric iron (i.e. 56 mg/ml). Mammalian ferritins, unlike most bacterial and plant ferritins, have the particularity that they are heteropolymers, made up of two subunit types, H and L. Whereas H-subunits have a ferroxidase activity, catalysing the oxidation of two Fe^{2+} atoms to Fe^{3+}, L-subunits appear to be involved in the nucleation of the mineral iron core: once this has formed an initial critical mass, further iron oxidation and deposition in the biomineral takes place on the surface of the ferrihydrite crystallite itself (see a further discussion in Chapter 19).

Iron homeostasis in mammals

The regulation of cellular iron homeostasis is to a large degree controlled at the level of the translation of mRNAs of proteins involved in cellular iron metabolism. The key players in this post-transcriptional regulation are two iron regulatory proteins (IRP1 and IRP2), which function as cytosolic iron sensors. In conditions of iron deficiency, IRPs bind with high affinity ($K_D \sim 20–100$ pM) to stem loops, known as iron regulatory elements (IREs), in mRNAs encoding the regulated proteins (Figure 8.15). When the IREs are in the 5′-untranslated region of the mRNA, as is the case for ferritin and ferroportin, binding of IRPs prevents initiation of translation, whereas in the case of the transferrin receptor and DMT1, where the IREs are in the 3′-untranslated region, the mRNA is protected against degradation by nucleases. This results in increased iron uptake and blockage of iron storage and export. Conversely, when iron is abundant, the IRPs are no longer active in binding, allowing ferritin and ferroportin mRNAs to be translated and resulting down-regulation of transferrin receptor and DMT1 synthesis as a result of the nuclease catalysed degradation of their mRNAs.

Iron homeostasis in mammals depends on the regulation of dietary-iron absorption by the enterocytes of the duodenum and the recycling of iron by macrophages recovered from the breakdown of the haemoglobin from senescent red blood cells. In hereditary disorders of iron loading, known collectively as haemochromatosis, when iron is deposited in parenchymal cells, transferrin saturation increases, and as the iron load increases, serious damage is done to many tissues, notably liver, endocrine tissues like pancreas, and heart. Hereditary haemochromatosis can be divided into three classes—classical haemochromatosis, juvenile haemochromatosis and ferroportin disease. Classical haemochromatosis is associated with mutations in *HFE*, a gene that encodes a protein of the major histocompatability complex, although in rare cases another gene, the *TFR2* gene that codes for a homologue of the

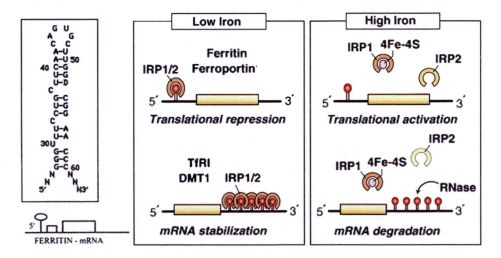

Figure 8.15 IRE in ferritin mRNA from Crichton (2001) and outline of translational regulation of mRNAs of a number of proteins involved in iron metabolism in low and high iron. (From Wallander et al., 2006. Copyright 2006, with permission from Elsevier.)

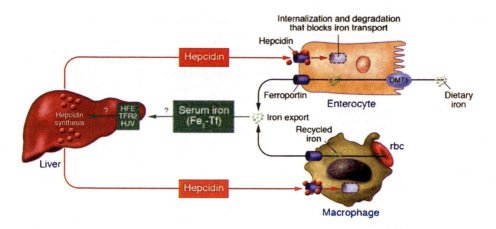

Figure 8.16 Regulation of systemic iron homeostasis. (From Vaumont et al., 2005. Copyright (2005) the American Society for Clinical Investigation.)

major transferrin receptor gene, *TFR1*, can be involved. Juvenile haemochromatosis is a rare form of hereditary haemochromatosis, characterized by early and severe onset of symptoms, particularly cardiac and endocrine defects. Most patients have mutations in the recently cloned *HJV* gene (haemojuvelin), which is expressed in muscle, liver and heart, but whose function is still unknown. A very small subset of juvenile haemochromatosis patients have mutations in the *HAMP* gene that codes for the prepro form[3] of hepcidin, a peptide synthesized by the liver, which is a major regulator of iron metabolism. Hepcidin is positively regulated by iron and negatively regulated by iron deficiency and hypoxia: it limits intestinal iron absorption and iron release by macrophages. Ferroportin disease, the third class of hereditary haemochromatosis is caused by pathogenic mutations in the gene encoding the iron exporter, ferroportin.

Recent results have led to the discovery of how systemic[4] iron homeostasis is regulated (Figure 8.16). The first indice of iron loading, transferrin saturation leads to increased levels of diferric transferrin, Fe_2-Tf, which is detected by the liver via a complex pathway involving HFE, TFR2 and HJV. Hepatocytes respond to this signal by increased expression of *HAMP* resulting in increased secretion of hepcidin. The circulating hepcidin acts to block dietary-iron uptake by the duodenal enterocytes and iron recycling from macrophages, in both cases through internalization of ferroportin, which blocks iron export. The outcome of this is to decrease serum iron levels, leading logically to the feedback response of down-regulating hepcidin synthesis and secretion. This once again allows ferroportin to be displayed on the surface of enterocytes and macrophages, allowing them once again to export iron into the circulation. In classical and juvenile haemochromatosis, mutations in *HFE*, *TFR2* and *HJV* lead to down-regulation of hepcidin synthesis, decreased levels of circulating hepcidin and ferroportin hyperactivity. The latter accounts for the increased iron absorption

---

[3] Prepro proteins are synthesized with a signal peptide, which directs the protein to the endoplasmic reticulum, where the signal peptide is cleaved; thereafter the propeptide is excized and the mature protein or peptide secreted into the circulation.

[4] Systemic-concerning the whole body of an animal or plant rather than an individual part.

and uncontrolled release of iron from macrophages, defects characteristic of hereditary haemochromatosis.

Copper and zinc transport and storage in mammals

Contrary to popular belief, ceruloplasmin[5], the principal copper-containing protein in plasma, ceruloplasmin, is not involved in copper transport. This is clearly underlined by the clinical observation that patients with aceruloplasminaemia (i.e. lacking ceruloplasmin in their blood) have perfectly normal copper metabolism and homeostasis. Copper is transported in plasma mostly by serum albumin with smaller amounts bound to low-molecular weight ligands like histidine. Likewise zinc is mostly transported in plasma bound to proteins (albumin and α_2-macroglobulin).

When animals are fed experimental diets lacking copper or zinc, their copper or zinc status rapidly declines, suggesting that there is not a storage pool of these metals. Thus, while the small, cysteine-rich protein metallothionein (see below) can avidly bind zinc and copper, this may reflect its role in detoxification rather than as a specific storage form. This is reflected by the fact that metallothionein genes are typically expressed at a basal level, but their transcription is strongly induced by heavy metal load.

Copper and zinc homeostasis in mammals

Regulation of total body copper occurs essentially at the level of the large intestine, the major site of dietary-copper absorption. Whereas on low-copper diets (less than 1 mg/day) up to 50% of the copper is absorbed, when copper content is greater than 5 mg/day, less than 20% is absorbed. Total body copper levels are controlled by the liver, which is the major storage site for copper and which also regulates its excretion in the bile. In mammals, post-translational mechanisms, such as intracellular trafficking of copper transporters and the copper-stimulated endocytosis and degradation of proteins involved in copper uptake, play a major role in copper homeostasis. Figure 8.17 presents the major pathways for copper trafficking in mammalian cells (albeit a fictitious cell, which includes all possibilities from intestinal mucosal cells to hepatocytes). It also indicates how these pathways are affected by copper deficiency or overload. As we saw in Chapter 7, copper uptake at the plasma membrane is mediated by Ctr1. When the cell is exposed to high levels of copper, Ctr1 may distribute to intracellular vesicles in certain tissues, where at elevated copper concentrations the Ctr1 protein is degraded. Once inside the cell, copper can interact with metallothioneins (MT) and the tripeptide glutathione (GSH). The latter has been shown to bind Cu(I), to transfer copper to metallothioneins and to play a role in the biliary excretion of copper. As mentioned earlier in this chapter, copper delivery to target apoproteins is achieved by copper chaperones. CCS inserts copper into Cu–Zn superoxide dismutase, and is up-regulated in

---

[5] Ceruloplasmin, akin to Pirandello's Six Characters in Search of an Author, has long been a protein in search of a function. It is certainly involved in tissue iron mobilization, since systemic iron loading is found in the tissues of patients with aceruloplasminaemia and other mutations of the ceruloplasmin gene.

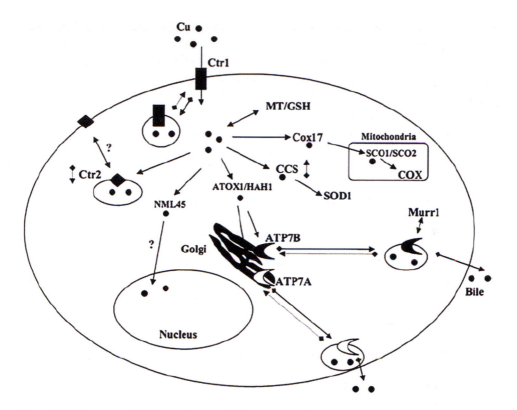

Figure 8.17 Copper-trafficking pathways within a mammalian cell. Regulation by copper deficiency (arrows connected by dotted lines) or overload (arrows connected by solid lines) as currently understood. (From Bertinato and L'Abbé, 2004. Copyright 2004, with permission from Elsevier.)

copper deficiency. As was described for yeast, Cox17 delivers copper to mitochondria where, with the participation of other proteins it supplies copper for cytochrome oxidase (COX). The mammalian homologue of ATX1, ATOX1 (also known as HAH1) supplies copper to the P-type ATPases, ATP7A and ATP7B. These two proteins are mutated in two human diseases: Menkes disease (ATP7A), characterized by copper deficiency and Wilson's disease (ATP7B), by copper overload. When copper levels are normal, both proteins localize to the trans-Golgi network, where they function in the insertion of copper into nascent cuproproteins. In Wilson's disease, this results in greatly reduced activity of ceruloplasmin because of impaired copper incorporation into apoceruloplasmin. In the presence of high levels of copper, both ATP7A and ATP7B are thought to be involved in exporting copper from cells. *MurrI*, the gene mutated in inbred Bedlington terriers, which develop toxic levels of hepatic copper, is known to interact directly with ATP7B, and may also be involved in biliary copper export. NML45, a truncated form of the Menkes protein with a nuclear-targeting insertion, may shuttle copper to the nucleus. Finally, Ctr2 is localized to cytoplasmic vesicles and may function in intracellular copper storage: it is down-regulated in copper deficiency.

In a manner similar to copper homeostasis, zinc homeostasis in mammals involves post-translational mechanisms. Both Zip1 and Zip3 are found predominantly at the plasma membrane in zinc-deficient cells, while in zinc-replete medium, they are largely present in intracellular organelles. In addition, zinc homeostasis is controlled at the transcriptional level by the zinc-responsive transcription factor, MTF1, which protects cells against zinc toxicity by increasing the expression of zinc-binding metallothionein genes, and also acting on a number of other zinc transporters. MTF1 can also be induced by copper, but this requires the presence of zinc-saturated metallothionein.

REFERENCES

Andrews, S.C., Robinson, A.K. and Rodriguez-Quinones, F. (2003) Bacterial iron homeostasis, *FEMS Microbiol. Rev.*, **27**, 215–237.

Andrews, S.C. (1998) Iron storage in bacteria, *Adv. Microb. Physiol.*, **40**, 281–351.

Bertinato, J. and L'Abbé, M.R. (2004) Maintaining copper homeostasis: regulation of copper-trafficking proteins in response to copper deficiency or overload, *J. Nutr. Biochem.*, **15**, 316–322.

Cobine, P.A., Pierrel, F. and Winge, D.R. (2006) Copper trafficking to the mitochondrion and assembly of copper metalloenzymes, *Biochim. Biophys. Acta*, **1763**, 759–772.

Crichton, R.R. (2001) *Inorganic Biochemistry of Iron Metabolism: From Molecular Mechanisms to Clinical Consequences*, Wiley, Chichester, 326 pp.

Eide, D.J. (2006) Zinc transporters and the cellular trafficking of zinc, *Biochim. Biophys. Acta*, **1763**, 711–722.

Grotz, N. and Guerinot, M.L. (2006) Molecular aspects of Cu, Fe and Zn homeostasis in plants, *Biochim. Biophys. Acta*, **1763**, 595–608.

Hantke, K. (2001) Bacterial zinc transporters and regulators, *BioMetals*, **14**, 139–249.

Hantke, K. (2005) Bacterial zinc uptake and regulators, *Curr. Opin. Microbiol.*, **8**, 196–202.

Hell, R. and Stephan, U.W. (2003) Iron uptake, trafficking and homeostasis in plants, *Planta*, **216**, 541–551.

MacGillivray, R.T., Moore, S.A., Chen, J., Anderson, B.F., Baker, H., Luo, Y., Bewley, M., Smith, C.A., Murphy, M.E., Wang, Y., Mason, A.B., Woodworth, R.C., Brayer, G.D. and Baker, E.N. (1998) Two high-resolution crystal structures of the recombinant N-lobe of human transferrin reveal a structural change implicated in iron release, *Biochemistry*, **37**, 7919–7928.

Magnani, D. and Solioz, M. (2005) Copper chaperone cycling and degradation in the regulation of the *cop* operon of *Enterococcus hirae*, *BioMetals*, **18**, 407–412.

Massé, E. and Gottesman, S. (2002) A small RNA regulates the expression of genes involved in iron metabolism in *Escherichia coli*, *Proc. Natl. Acad. Sci., U.S.A.*, **99**, 4620–4625.

Pohl, E., Haller, J.C., Mijovilovich, A., Meyer-Klaukе, W., Garman, E. and Vasil, M.L. (2003) Architecture of a protein central to iron homeostasis: crystal structure and spectroscopic analysis of the ferric uptake regulator, *Molec. Microbiol.*, **47**, 903–915.

Rutherford, J.C. and Bird, A.J. (2004) Metal-responsive transcription factors that regulate iron, zinc, and copper homeostasis in eukaryotic cells, *Eukaryot. Cell*, **3**, 1–13.

Schmidt, W. (2003) Iron solutions: acquisition strategies and signaling pathways in plants, *TIBS*, **8**, 188–193.

Vaulont, S., Lou, D.-Q., Viatte, L. and Kahn, A. (2005) Of mice and men: the iron age, *J. Clin. Invest.*, **115**, 2079–2082.

Wallander, M.L., Leibold, E.A. and Eisenstein, R.S. (2006) Molecular control of vertebrate iron homeostasis by iron regulatory proteins, *Biochim. Biophys. Acta*, **1763**, 668–689.

– 9 –

Sodium and Potassium—Channels and Pumps

INTRODUCTION: —TRANSPORT ACROSS MEMBRANES

Before examining the important roles of the alkali metals sodium and potassium we should briefly review how ions are transported across membranes. As we pointed out in Chapter 3, the phospholipid bilayer of biological membranes is essentially impermeable to polar molecules and to ions such as Na^+ and K^+—their permeabilities (expressed as cms^{-1}) are of the order of 10^{-12}. This can be compared, for example, to H_2O, for which the corresponding value is around 10^{-2}. Transport across membranes is conferred by two classes of membrane proteins, namely channels and pumps. Channels allow ions to flow down a concentration gradient by a process known as passive transport or facilitated diffusion. Of course, channels cannot remain open all of the time, and so they are usually *gated*, which simply means that, like regular garden gates, they usually remain shut, and can only be opened, either by the binding of a ligand (*ligand-gated*), or by changes in the membrane potential (*voltage-gated*). *Ligand-gated* channels, such as the acetylcholine receptors in post-synaptic membranes, are opened by the binding of the neurotransmitter acetylcholine, whereas the *voltage-gated* sodium and potassium channels, which mediate the action potentials in neuronal axons described below, are opened by membrane depolarization.

In contrast, pumps use energy, in the form of ATP or light, to drive the unfavourable uphill transport of ions or molecules against a concentration gradient; in other words, they are involved in active transport. There are two types of ATP-driven pumps, so-called P-type ATPases and ABC (ATP-binding cassette) transporters, both of which use conformational changes induced by ATP binding and its subsequent hydrolysis to transport ions across the membrane. The sodium–potassium (Na^+–K^+)-ATPase described below is one of the P-type ATPases. Another mechanism of active transport, which uses the electrochemical gradient of one ion to drive the countertransport of another will be illustrated by the Na^+/H^+ exchanger, crucial among other things for the control of intracellular pH. Yet another example, discussed in Chapter 11, is the Na^+/Ca^{2+} exchanger, which plays an important role in removing Ca^{2+} from cells.

SODIUM *VERSUS* POTASSIUM

Sodium and potassium are relatively abundant in the earth's crust, although sodium is much more prevalent in seawater. The Na^+ and K^+ content in the average man is ~1.4 and 2.0 g/kg, respectively, making them among the most important of metal ions in terms of concentration. However, their distribution is quite different. Whereas in most mammalian cells, 98% of K^+ is intracellular, for Na^+ the situation is the reverse. This concentration differential ensures a number of major biological processes, such as cellular osmotic balance, signal transduction and neurotransmission. It is maintained by the $Na^+–K^+$ ATPase, which we will discuss below. However, despite the presence of only 2% of total body K^+ outside of cells, this extracellular K^+ concentration plays a crucial role in maintaining the cellular membrane resting potential. Fluxes of these alkali metal ions play a crucial role in the transmission of nervous impulses both within the brain and from the brain to other parts of the body. The opening and closing of gated ion channels (*gated channels*), which are closed in the resting state, and which open in response to changes in membrane potential) generates electrochemical gradients across the plasma membranes of neurons. A nerve impulse is constituted by a wave of transient depolarization/repolarization of membranes which traverses the nerve cell, and is designated an action potential. Hodgkin and Huxley (1952) demonstrated that a microelectrode implanted into an axon (the long process emanating from the body of a nerve cell) would record an action potential (Figure 9.1a). In the first ~0.5 ms the membrane potential increases from around − 60 mV to about +30 mV, followed by a rapid repolarization, which overshoots the resting potential (hyperpolarization) before slowly recovering. The action potential results from a rapid and transient increase in Na^+ permeability followed by a more prolonged increase in K^+ permeability (Figure 9.1b). The opening and closing of these gated Na^+- and K^+-ion channels across the axonal membranes creates the action

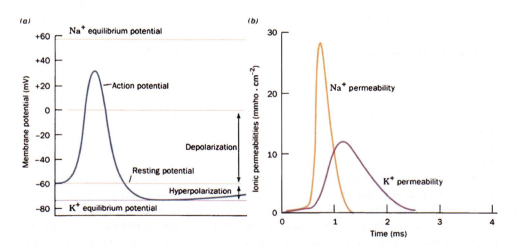

Figure 9.1 Time course of an action potential. (From Voet and Voet, 2004. Reproduced with permission from John Wiley & Sons., Inc.)

potentials (essentially electrochemical gradients) across these membranes, which allows information transfer and also regulates cellular function.

The regulation of the flow of ions across cell membranes is absolutely essential for the functioning of living cells. Because of the hydrophobicity of cellular membranes (as we saw earlier) the energetically driven preference of ionic species such as Na^+, K^+, Cl^-, H^+ and Ca^{2+} to cross, never mind to find themselves preferentially on one side or other of a biological membrane, would be impossible. Without ionic gradients, which maintain high concentrations of K^+ within the cell and low concentrations of Na^+, cells would not be able to carry out their normal metabolic activities. This means, in simplistic terms, that some molecular machines must be able to distinguish between Na^+ and K^+ ions (presumably unhydrated, since the degree of hydration could make for difficulties in discrimination). So, before even beginning a discussion of 'active' transport proteins, whether ion pumps or ion exchangers, we ask the question how do these transporters distinguish between these two closely related cations? (Reviewed in Corry and Chung, 2006; Gouax and MacKinnon, 2005.)

If we had asked this question a few years ago, the answer would have been at best equivocal. However, as was pointed out in an earlier chapter, since the enormous progress in the determination of the structure of membrane proteins, we can, on the basis of the X-ray structures of an increasing number of ion transport proteins, begin to advance hypotheses that have more and more likelihood to be close to reality. In the case of Na^+ channels we are still pretty much in the dark. However, the successful determination of the structure of a number of K^+ channels of both bacterial and mammalian origins represents a great leap forward in our understanding of how these channels function.

Potassium channels

K^+ channels selectively transport K^+ across membranes, hyperpolarize cells, set membrane potentials and control the duration of action potentials, among a myriad of other functions. They use diverse forms of gating, but they all have very similar ion permeabilities. All K^+ channels show a selectivity sequence of $K^+ \sim Rb^+ > Cs^+$, whereas the transport of the smallest alkali metal ions Na^+ and Li^+ is very slow—typically the permeability for K^+ is at least 10^4 that of Na^+. The determination of the X-ray structure of the K^+-ion channel has allowed us to understand how it selectively filters completely dehydrated K^+ ions, but not the smaller Na^+ ions. Not only does this molecular filter select the ions to be transported, but also the electrostatic repulsion between K^+ ions, which pass through this molecular filter in Indian file, provides the force to drive the K^+ ions rapidly through the channel at a rate of 10^7–10^8 per second. (Reviewed in Doyle et al., 1998; MacKinnon, 2004.)

The first voltage-gated potassium channel to be identified was the gene encoding the *Shaker* mutation in the fruit fly *Drosophila*. Currently we have the structures of a pH-dependent bacterial K^+ channel, voltage-gated and calcium-gated K^+ channels from bacteria, and most recently a voltage-gated mammalian K^+ channel. What is most striking is that they all have a similar architecture. They are all tetramers with fourfold symmetry about the central K^+-conducting pore. On the basis of hydrophobicity analysis there are two closely related families of K^+ channels, those containing two membrane-spanning segments per subunit, and those containing six. In the latter case, such as the *Drosophila* and the vertebrate voltage-gated K^+ channels, the last two transmembrane helices, S5 and S6, together with the P-loop that connects them, constitute the pore itself. Several other families of ion channels have similar architectures, including the calcium-activated K^+ channels. The two membrane-spanning families include the inwardly rectifying K^+ channels and some bacterial K^+ channels. They are made up of four subunits, each having only two transmembrane segments. The analogous M1 and M2 segments and pore loop form the complete transmembrane structure of the two transmembrane K^+ channels. Sequence homology is very high between the two families in the channel region, particularly in the pore region itself. K^+ channels allow some other monovalent cations through (but not Na^+), but do not allow the passage of anions and are blocked by divalent cations.

The 4.5 nm long pore contains three distinct regions of variable width. On the cytoplasmic side (Figure 9.2a) it starts with a channel (the internal pore), 0.6 nm wide by 1.8 nm long, which has four negatively charged side chains at its entrance (perhaps a precaution to prevent anions entering the channel). The central cavity is wider, ~1 nm in diameter and contains around 50 water molecules, enabling a hydrated K^+ ion to move through the hydrophobic interior of the membrane. The third part, termed the selectivity filter, believed to be constituted by the P-loops, is only some 0.3 nm wide, such that only an unhydrated K^+ ion can pass through. The selectivity filter, 1.2 nm in length, consists of four K^+-binding sites in a row. Each of these sites can bind a dehydrated K^+ ion, which interacts with eight carbonyl oxygens (main-chain carbonyl or side-chain hydroxyl atoms) of a highly conserved TVGYG amino acid sequence (Figure 9.2b). Mutations in this sequence lead to disruption of the ability of the channel to distinguish between K^+ ions and Na^+ ions. The size of the cavity is a good match for the K^+ ion, with a mean K^+–O distance of 0.284 nm. From the relative electron density of the sites, it is clear that each site is occupied by K^+ only half of the time; in other words, at any given time only two of the binding sites are occupied, with water molecules in the intermediate sites. Thus, two K^+ ions permanently occupy the sites 1 and 3 or 2 and 4, with a water molecule sandwiched between them (Figure 9.3). The X-ray structures also give support to a previously suggested 'knock-on' mechanism whereby K^+ ions can traverse the channel. Additional K^+ ions coordinated by eight water molecules are observed at the extracellular mouth of the channel and in the central cavity. When one of these ions enters either end of the filter, it displaces the equilibrium of the two K^+ ions already present, with the consequence that the column of K^+ ions moves along until one of the ions is ejected, and the new K^+ ion takes its place in the filter.

The 3-D structure of a two-transmembrane bacterial K^+ channel, KcsA, analogous to the inwardly rectifying K^+ channels, reveals an 'inverted teepee' arrangement of the MI and M2 segments in a square array around a central pore, with the narrow outer mouth of the pore formed by the pore loop (Figure 9.4a). Information on how the pore might be opened comes from the structure of a bacterial 2TM calcium-activated K^+ channel, MthK (Figure 9.4b), analysed in its calcium-bound, presumably open form. The M2 helices are bent at a highly conserved glycine residue, and this bend appears to open the intracellular mouth of the pore sufficiently to allow permeation of ions.

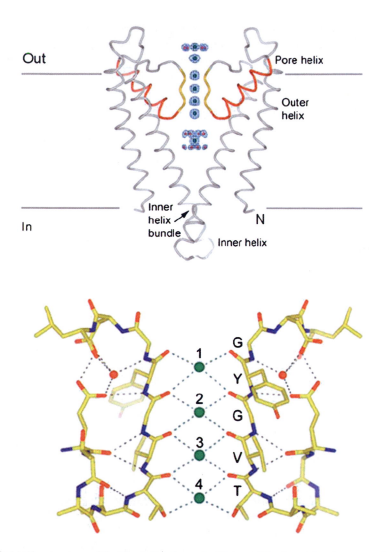

Figure 9.2 (a) The structure of the KcsA K$^+$ channel with two of the four subunits removed: the pore helices are in red, the selectivity filter in yellow; the electron density along the ion pathway is in blue mesh. (b) The K$^+$-selectivity filter (two subunits) with eight carbonyl groups (red) coordinating K$^+$ ions (green spheres) in positions 1–4 from the extracellular side. (From MacKinnon, 2004. Reproduced with permission from John Wiley & Sons., Inc.)

SODIUM CHANNELS

The sodium channels consist of a highly processed α subunit (260 kDa) associated with auxiliary β subunits. The pore-forming α subunit is sufficient for functional expression, but the kinetics and voltage dependence of channel gating are modified by the β subunits. The transmembrane organization is shown in Figure 9.5. The α subunit is organized in four

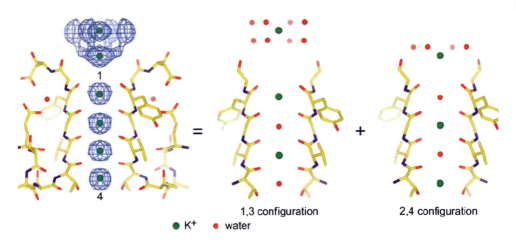

●K+   ● water

Figure 9.3 Two K$^+$ ions in the selectivity filter are hypothesized to exist in the two specific config-urations 1, 3 and 2, 4. (From MacKinnon, 2004. Reproduced with permission from John Wiley & Sons., Inc.)

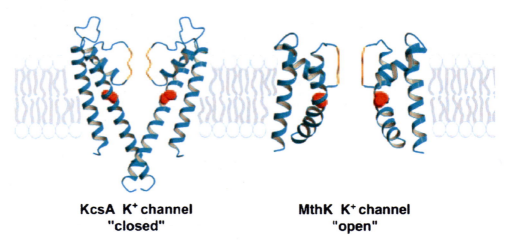

KcsA K$^+$ channel
"closed"

MthK K$^+$channel
"open"

Figure 9.4 Structures of potassium channels in open and closed conformations. The selectivity filter is orange, and the conserved glycine residue is in red. (From Yu et al., 2005. Reproduced with permission of Blackwell Publishing Ltd.)

homologous domains (I–IV) each consisting of six transmembrane α helices (S1–S6) and an additional pore loop located between the S5 and S6 helices, each similar to the individual subunits of the 6TM K$^+$ channels. The pore loops line the outer, narrow entry to the pore, while the S5 and S6 helices line the inner, wider exit from the pore. The S4 segments in each domain contain positively charged amino acids in every third position. These residues serve as gating charges and move across the membrane to initiate channel activation in response to depolarization of the membrane. A short intracellular loop connecting the domains III and IV serves as the inactivation gate, folding into the channel structure and blocking the pore from the inside during sustained depolarization of the membrane. It con-tains an Ile-Phe-Met-Thr (IFMT) motif, which is crucial for inactivation. The circles in the

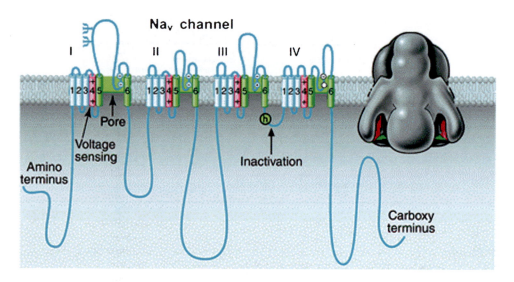

Figure 9.5 Transmembrane organization of sodium channel α subunit. Right 3-D structure of the α subunit at 0.2 nm resolution. (From Yu et al., 2005. Reproduced with permission of Blackwell Publishing Ltd.)

re-entrent loops in each of the four domains represent the amino acids that form the ion selectivity filter: the outer rings have the sequence EEDD and the inner rings DEKA.

The sodium-potassium ATPase

Mammalian cells maintain a lower concentration of Na^+ (around 12 mM) and a higher concentration of K^+ (around 140 mM) than in the surrounding extracellular medium (respectively, 145 and 4 mM). The system $(Na^+–K^+)$-ATPase, which maintains high intracellular K^+ and low intracellular Na^+, is localized in the plasma membrane, and belongs to the family of P-type ATPases. Other members of the family in eukaryotes are the sarcoplasmic reticulum and plasma membrane Ca^{2+} ATPases and, in plants, the H^+-ATPases. The overall reaction catalysed is:

$$3Na^+ \text{ (in)} + 2K^+ \text{ (out)} + ATP + H_2O \Leftrightarrow 3Na^+ \text{ (out)} + 2K^+ \text{ (in)} + ADP + P_i$$

This results in the extrusion of three positive charges for every two that enter the cell, resulting in a transmembrane potential of 50–70 mV, and has enormous physiological significance. More than one-third of the ATP utilized by resting mammalian cells is used to maintain the intracellular $Na^+–K^+$ gradient (in nerve cells this can rise up to 70%), which controls cell volume, allows neurons and muscle cells to be electrically excitable, and also drives the active transport of sugars and amino acids (see later).

The hydrolysis of ATP is used to drive the energetically unfavourable transport of Na^+ and K^+ against a concentration gradient. How this is achieved is still unclear, but a number of experimental observations can be put together to provide a plausible mechanism.

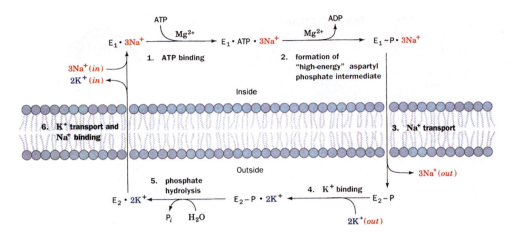

Figure 9.6 A model for the active transport of Na$^+$ and K$^+$ by the (Na$^+$–K$^+$)-ATPase. (From Voet and Voet, 2004. Reproduced with permission from John Wiley & Sons., Inc.)

Key among these is that the ATPase is phosphorylated by ATP in the presence of Na$^+$, and that the resulting aspartyl phosphate residue is only dephosphorylated in the presence of K$^+$. This immediately suggests, as outlined in Figure 9.6, that the enzyme may exist in two distinct conformations, E_1 and E_2, which differ not only in their conformation, but also in their catalytic activity and their ligand specificity. The E_1 form faces towards the inside of the plasma membrane, has a high-affinity site for Na$^+$, and reacts with ATP to form the "high-energy" aspartyl phosphate intermediate $E_1 \sim$ P.3Na$^+$. In relaxing to its "low-energy" conformation E_2–P, the bound Na$^+$ is released. The outward-facing E_2–P, which has a high affinity for K$^+$, binds 2K$^+$, and the aspartyl phosphate group is hydrolysed to give E_2.2K$^+$. This form then changes conformation to the E_1 form, releasing its 2K$^+$ inside the cell, and allowing the cycle to recommence.

The outcome of this is to couple ATP hydrolysis with the vectorial transport of Na$^+$ and K$^+$ across the plasma membrane. The inhibition of the (Na$^+$–K$^+$)-ATPase by cardiac glyco-sides such as digitalis (an extract of foxglove leaves), which blocks the dephosphorylation of the E_2–P form of the enzyme, is the basis for a number of steroid drugs which are commonly prescribed for the treatment of congestive heart failure.

Active transport driven by Na$^+$ gradients

Whereas the (Na$^+$–K$^+$)-ATPase can generate electrochemical potential gradients across membranes, these same gradients can be used to power active transport through the dissipa-tion of the ion gradient. Thus, sugars and amino acids can be transported into cells by Na$^+$-dependent symports. Dietary glucose is concentrated in the epithelial cells of the small intestine by an Na$^+$-dependent symport, and is then transported out of the cells into the cir-culation by a passive glucose uniport situated on the capillary side of the cell (Figure 9.7a). For this system to continue functioning, ATP hydrolysis, which maintains the intracellular Na$^+$ concentration through the (Na$^+$–K$^+$)-ATPase, is absolutely required.

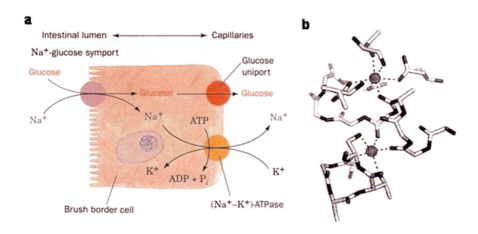

Figure 9.7 (a) The epithelial brush border cells of the small intestine concentrate glucose from the intestinal lumen in symport with Na^+; this is driven by the $(Na^+–K^+)$-ATPase located on the capillary side of the cell. The glucose is then exported by a passive uniport system. (From Voet and Voet, 2004. Reproduced with permission from John Wiley & Sons., Inc.) (b) Two Na^+-binding sites in the LeuT Na^+-dependent pump. (From Gouax and MacKinnon, 2005. Copyright (2005) American Association for the Advancement of Science.)

The LeuT Na^+-dependent pump transports the amino acid leucine and Na^+ across the cell membrane of bacteria, using the energy of the Na^+ gradient to pump leucine into the cell. The atomic structure of LeuT (Figure 9.7b) shows a leucine molecule and two Na^+ ions bound deep inside the protein channel. The Na^+ ions are completely dehydrated and surrounded by five or six oxygen atoms at a mean Na^+–O distance of 0.228 nm. This should be compared with the K^+–O distance of 0.284 nm found in the potassium channel, which, together with the greater number of oxygen atoms (eight), reflects the larger radius of K^+. This confirms studies from the small-molecule chemical literature of the importance of the size of the cavity formed by the binding site. Sites selective for Li^+ (radius 0.060 nm), Na^+ (radius 0.095 nm), K^+ (radius 0.133 nm) and Rb^+ (radius 0.148 nm) were created by simply adjusting the cavity size to match the ion. The data from the LeuT and K^+ channel imply that alkali metal cation specificity in biological systems is also determined by providing an oxygen-lined binding site of the appropriate cavity size.

Sodium/proton exchangers

The carrier-mediated transport of sodium in exchange for protons across membranes is a virtually universal phenomenon in biology, from bacteria to man. It is carried out by a family of Na^+/H^+ exchangers that constitute the solute carrier (SLC) gene family SLC9. They are classified as secondary active transporters, since the driving force is the electrochemical gradient of one of the ions, which drives the counter-transport of the other. However, they play a varied number of functions. Whereas in bacteria such as *E. coli* the inwardly directed H^+ gradient generated by the inner membrane H^+-ATPase is used to export Na^+, in mammals

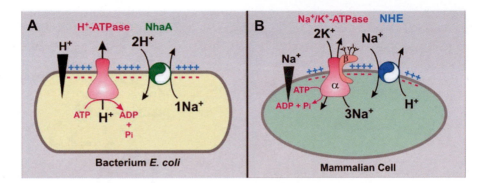

Figure 9.8 The Na$^+$/H$^+$ exchanger of bacteria and mammals. The driving force is provided, respectively, by the H$^+$-ATPase (a) and the Na$^+$/K$^+$-ATPase (b) (From Orlowski and Grinstein 2004. With kind permission of Springer Science and Business Media.)

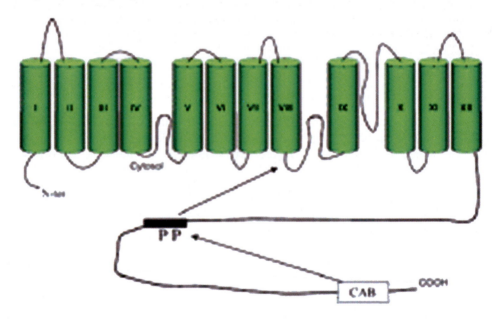

Figure 9.9 Transmembrane organization and regulation of the ubiquitous mammalian plasma membrane Na$^+$/H$^+$ exchanger SLC9A1 by binding of carbonic anhydrase (CAB). (Reprinted with permission from Li et al., 2006. Copyright (2006) American Chemical Society.)

the inward Na$^+$ gradient established by the plasma membrane (Na$^+$–K$^+$)-ATPase drives the counter-transport of H$^+$ (Figure 9.8).

Higher eukaryotic genomes have between seven and nine Na$^+$/H$^+$ exchanger-like proteins. The first to be characterized, SLC9A1 (Figure 9.9), is ubiquitous and localizes to the plasma membrane of most mammalian cells; it is also found at the basolateral membrane of intestinal enterocytes. Based on computer modelling, like all of the other Na$^+$/H$^+$ exchangers, it is predicted to have 12 membrane-spanning segments at the N-terminus (around 500 amino acid residues), with a more hydrophilic C-terminus, which is intracellular and has multiple sites for phosphorylation by protein kinases and binding of other regulatory factors.

SLC9A1 is thought to have two principal functions, namely to regulate cytoplasmic pH homeostasis and to regulate cell volume. The former role, together with bicarbonate transporting systems, is crucial for maintaining cytoplasmic acid–base balance, while the latter provides a major route for Na^+ influx coupled to Cl^- and H_2O uptake, required to restore cell volume to steady-state levels subsequent to cell shrinkage brought about by acute increases in extracellular osmolarity.

Diverse stimuli that activate SLC9A1 are thought to transduce their signals through a common mitogen-activated protein kinase (MAPK) pathway. Carbonic anhydrases produce HCO_3^- and H^+ from the hydration of CO_2, and it was recently demonstrated that carbonic anhydrase II (CAII), the predominant cytoplasmic form of carbonic anhydrase, binds to and modulates the activity of the SLC9A1 isoform of the Na^+/H^+ exchanger. When cells were exposed to weak acid in the form of CO_2, H^+ transport was enhanced when CAII and SLC9A1 were coexpressed. CAII bound to the C-terminal 182 amino acids of the Na^+/H^+ exchanger, and binding was enhanced by phosphorylation of SLC9A1. This suggests that Na^+/H^+ exchanger activity is linked to bicarbonate-based pH regulation through carbonic anhydrase activity (Li et al., 2006).

A number of other members of the SLC9A family are thought to be involved in the reabsorption of Na^+ and HCO_3^- by the epithelia of the kidney and the gastrointestinal tract, while others seem to play a role in regulating the pH homeostasis in intracellular organelles.

Other roles of intracellular K$^+$

Quite a number of enzymes are known to be activated by K^+—a good example is the glycolytic enzyme, pyruvate kinase, where the role of the metal is thought to orient the phosphoenolpyruvate in the substrate-binding pocket. The more active role of Mg^{2+} in this enzyme is discussed in Chapter 10. Since the intracellular concentrations of K^+ and Mg^{2+} are high they dominate metal binding to nucleic acids, with the divalent Mg^{2+} binding more strongly to the polyanionic sugar-phosphate backbone on account of its higher charge. Metal binding reduces electrostatic repulsion between phosphates, stabilizing both base pairing and base stacking; this is underlined by the increase in melting temperature of the DNA in the presence of metal ions. Much of the intracellular K^+ and Mg^{2+} is found bound to ribosomes.

The chromosomes of eukaryotes are linear, and replication of the free ends of these linear DNA molecules presents particular problems. The sequencing of the ends of chromosomes revealed that they consist of telomeres, hundreds of tandem repeats of a hexanucleotide sequence, which in all vertebrates is d(TTAGGG). These G-rich telomeric sequences can fold into a G-quadruplex, a DNA secondary structure consisting of stacked G-tetrad planes, or G-quartets (Figure 9.10), connected by a network of Hoogsteen hydrogen bonds: the cavity in the centre can accommodate a monovalent cation such as Na^+ and K^+, with coordination of the four O-6 oxygens. The quartets can stack upon each other to form a multi-layer structure (Figure 9.11). The repetitive G-rich sequences found at the ends of eukaryotic chromosomes (telomeres) can form several isomeric anti-parallel arrangements where the tetraplex involves intramolecular folding, when one polynucletide supplies two or more strands to the complex.

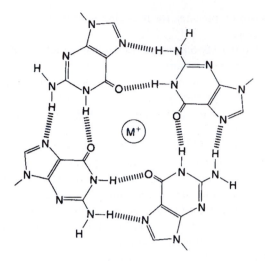

Figure 9.10 The structure of a G-quartet. (From Wozki and Schmidt, 2002. Reproduced with permission from John Wiley & Sons., Inc.)

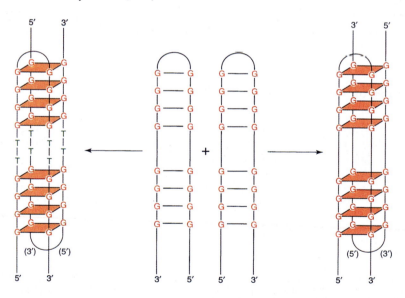

Figure 9.11 G-quadruplex DNA. (From Wozki and Schmidt, 2002. Reproduced with permission from John Wiley & Sons., Inc.)

The formation and stabilization of the DNA G-quadruplex in the human telomeric tandem repeats of the sequence *d*(TTAGGG) inhibits the activity of telomerase, a cancer-specific reverse transcriptase which is activated in 80–90% of tumours, making it an important target for therapeutic intervention. Clearly, knowledge of the intact human telomeric G-quadruplex structure under physiological conditions is a prerequisite for rational, structure-based drug design. The folding structure of the human telomeric sequence in K$^+$ solution has recently

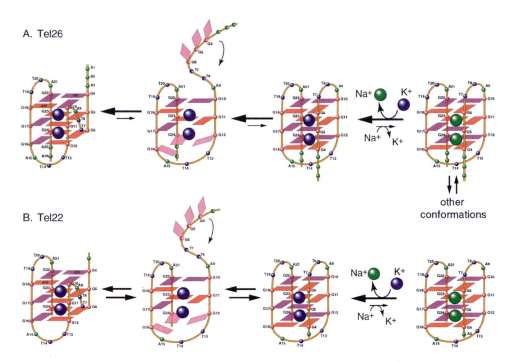

Figure 9.12 Schematic diagram of interconversions between the Na^+ and K^+ forms of telomeric G-quadruplexes. (Reproduced from Ambrus et al., 2006, by permission of Oxford University Press.)

been determined by NMR (Figure 9.12), demonstrating a novel intramolecular G-quadruplex folding topology, which is quite different from that reported previously for the 22-nucleotide Tel22 in Na^+ solution, and is the predominant conformation for the extended 26-nucleotide sequence Tel26 in K^+ solution, whether Na^+ is present or not. Addition of K^+ readily converts the Na^+-form conformation to the K^+-form hybrid-type G-quadruplex. The hybrid-type G-quadruplex topology suggests a straighforward secondary structure folding pathway with effective packing for the extended human telomeric DNA. Furthermore, since this hybrid-type telomeric G-quadruplex is likely to be that which is found under physiological conditions, its distinct folding topology makes it an attractive target for specific targeting by low-molecular-weight drugs which, by stabilizing the telomeric G-quadruplexes, could represent an important cancer therapeutic strategy.

REFERENCES

Ambrus, A., Chen, D., Dai, J., Bialis, T., Jones, R.A. and Yang, D. (2006) Human telomeric sequence forms a hybrid-type intramolecular G-quadruplex structure with mixed parallel/antiparallel strands in potassium solution, *Nucleic Acids Res.*, **34**, 2723–2735.

Corry, B. and Chung, S.-H. (2006) Mechanisms of valence selectivity in biological ion channels, *Cell Mol. Life Sci.*, **63**, 301–315.

Doyle, D.A., Cabral, J.M., Pfuetzner, R.A., Kuo, A., Gulbis, J.M., Cohen, S.L., Chait, B.T. and MacKinnon, R. (1998) The structure of the potassium channel: molecular basis of K^+ conduction and selectivity, *Science*, **280**, 69–77.

Gouax, E. and MacKinnon, R. (2005) Principles of selective ion transport in channels and pumps, *Science*, **310**, 1461–1465.

Hodgkin, A.L. and Huxley, A.F. (1952) A quantitative description of membrane current and its application to conduction and excitation in nerve, *J. Physiol.*, **117**, 500–544.

Li, X., Liu, Y., Alvarez, B.V., Casey, J.R. and Fliegel, L. (2006) A novel carbonic anhydrase II binding site regulates NHE1 activity, *Biochemistry*, **45**, 2414–2424.

MacKinnon, R. (2004) Potassium channels and the atomic basis of selective ion conduction (Nobel lecture), *Angew. Chem. Int. Edn.*, **43**, 4265–4277.

Orlowski, J. and Grinstein, S. (2004) Diversity of the mammalian sodium/proton exchanger SLC9 family, *Pflugers Arch. Eur. J. Physiol.*, **447**, 549–565.

Voet, D. and Voet, J.G. (2004) *Biochemistry*, 3rd edition, Wiley, Hoboken, NJ, 1591 pp.

Wozki, S.A. and Schmidt, F.J. (2002) DNA and RNA: composition and structure, in Devlin, T.M. (ed.) *Textbook of Biochemistry with Clinical Correlations*, 5th edition, pp. 45–92.

Yu, F.H., Yarov-Yarovoy, V., Gutman, G.A. and Catterall, W.A. (2005) Overview of molecular relationships in the voltage-gated ion channel superfamily, *Pharmacol. Rev.*, **57**, 387–395.

– 10 –

Magnesium–Phosphate Metabolism and Photoreceptors

INTRODUCTION

Mg^{2+} is one of the most abundant elements in the earth's crust and in the human body, and is the most abundant divalent cation within cells. Around 50% of total Mg^{2+} resides in bone, the remainder essentially within cells; 50% of cytosolic Mg^{2+} is bound to ATP, and most of the rest, together with K^+, is bound to ribosomes. The intracellular concentration of free Mg^{2+} is around 0.5 mM. Less than 0.5% of the total body Mg^{2+} is in plasma, where its concentration is maintained within fairly strict limits.

Mg^{2+} has properties that make it quite unique among biological cations. Inspection of Table 10.1 reveals that of the four common biological cations, the ionic radius of Mg^{2+} is much smaller than the others, whereas its hydrated radius is the largest of all four. This means that the volume of the hydrated Mg^{2+} cation is 400 times larger than its ionic volume (since the radius enters into the equation to the power three), compared to values around 25 times for Na^+ and Ca^{2+}, and a mere 5 times for K^+.

Other factors that play an important role in determining the biological role of Mg^{2+} are its coordination number and coordination geometry, its solvent exchange rates and its transport number[1]. Like Na^+, Mg^{2+} is invariably hexa-coordinate, whereas both K^+ and Ca^{2+} can adjust easily to 6, 7 or 8 coordination. Thus, Ca^{2+} can accommodate a more flexible geometry, compared to the octahedral geometry of the obligatory hexa-coordinate cations, resulting in deviations from the expected bond angle of 90° by up to 40°, compared with less than half that for Mg^{2+}. Likewise, bond lengths for oxy-ligands can vary by as much as 0.5 Å for Ca^{2+} whereas the corresponding values for Mg^{2+} vary by only 0.2 Å.

In contrast to the other three cations, Mg^{2+} has a much slower exchange rate of water in its hydration sphere (Table 10.1). Mg^{2+} often participates in structures, for example in ATP-binding catalytic pockets of kinases and other phosphoryl-transferase enzymes, where

---

[1] The transport number estimates the average number of solvent molecules associated with a cation sufficiently tightly to migrate with the cation in solution.

Table 10.1

Properties of common biological cations (from Maguire and Cowan, 2002)

Cation	Ionic radius (Å)	Hydrated radius (Å)	Ionic volume (Å³)	Hydrated volume (Å³)	Exchange rate (s⁻¹)	Transport number
Na^+	0.95	2.75	3.6	88.3	8×10^8	$7-13$
K^+	1.38	2.32	11.0	52.5	10^9	$4-6$
Mg^{2+}	0.65	4.76	1.2	453	10^5	$12-14$
Ca^{2+}	0.99	2.95	4.1	108	3×10^8	$8-12$

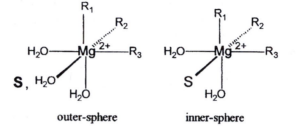

Figure 10.1 Comparison of inner- and outer-sphere modes of activation, where S is the substrate. (From Cowan, 2002. With kind permission of Springer Science and Business Media.)

the metal is bound to four or five ligands from the protein and the ATP. This leaves one or two coordination positions vacant for occupation by water molecules, which can be positioned in a particular geometry by Mg^{2+} to participate in the catalytic mechanism of the enzyme. This capacity is an example of outer-sphere activation of a substrate by a metal ion (Figure 10.1) as distinct from the more usual inner-sphere activation.

In addition, the high charge density on Mg^{2+} ensures that it is an excellent Lewis acid in reactions notably involving phosphoryl transfers and hydrolysis of phosphoesters. Typically, Mg^{2+} functions as a Lewis acid, either by activating a bound nucleophile to a more reactive anionic form (e.g. water to hydroxide anion) or by stabilizing an intermediate.

Unlike the other alkaline earth and transition metal ions, essentially on account of its small ionic radius and consequent high electron density, Mg^{2+} tends to bind the smaller water molecules rather than bulkier ligands in the inner coordination sphere. Many Mg^{2+}-binding sites in proteins have only 3, 4 or even less direct binding contacts to the protein, leaving several sites in the inner coordination sphere occupied by water, or in the phosphoryl trans-ferases, by nucleoside di- or triphosphates.

MAGNESIUM-DEPENDENT ENZYMES

Many enzymes involved in the pathways of intermediary metabolism are Mg^{2+} dependent, as are a great many of the enzymes involved in nucleic acid metabolism. Of the ten enzymes involved in the glycolytic pathway (see Chapter 5), five are Mg^{2+} dependent.

This comes as no surprise since four of the five (hexokinase, phosphofructokinase, phos-phoglycerate kinase and pyruvate kinase) involve phosphoryl transfers. The fifth, enolase, forms a complex with Mg^{2+} before the 2-phosphoglycerate substrate is bound. The inhibition of glycolysis by fluoride results from binding of F^-, in the presence of phosphate, to the catalytic Mg^{2+}, thus blocking substrate binding and inactivating the enzyme. As we could anticipate from its being the most abundant cytosolic divalent cation, Mg^{2+} binds strongly to nucleoside di- and triphosphates such as ATP and ADP, and is therefore directly involved in almost all reactions involving these molecules.

As pointed out above, Mg^{2+} binding to the enzyme can either be directly through protein side chains or peptide carbonyls (inner-sphere) or by indirect interactions through metal-bound water molecules (outer-sphere). Mg^{2+}-dependent enzymes can be divided into two general classes. First there are those in which the enzyme binds the magnesium–substrate complex, and usually the enzyme has little or only weak interaction with Mg^{2+}, its principal binding being to the substrate. Second, there are enzymes to which Mg^{2+} binds directly, altering the structure of the enzyme and/or playing a catalytic role.

Mg^{2+}-binding to enzymes is relatively weak (K_a not more than $10^5 M^{-1}$) such that the enzyme is often isolated in the metal-free form, and Mg^{2+} must be added to the *in vitro* enzyme assay system. As pointed out earlier, the intracellular free Mg^{2+} concentration is $\sim 5 \times 10^{-3}$ M, so that most Mg^{2+}-dependent enzymes have adequate local concentrations of Mg^{2+} for their activity. Two factors that make Mg^{2+}-biochemistry difficult to carry out are that the metal, like Zn^{2+}, is spectroscopically silent, and second, since 1990 the only practically useful isotope ^{28}Mg, with high-energy β and γ emission and a half-life of 21.3 h, has become outrageously expensive ($30,000 per mCi), such that it is no longer used for transport studies. These practical problems may, in part, be resolved by substituting Mn^{2+} for Mg^{2+} to carry out spectroscopic studies, and to use substitute isotopes such as $^{63}Ni^{2+}$ for transport studies.

PHOSPHORYL GROUP TRANSFER: KINASES

Phosphoryl group transfer reactions add or remove phosphoryl groups to or from cellular metabolites and macromolecules, and play a major role in biochemistry. Phosphoryl transfer is the most common enzymatic function coded by the yeast genome and, in addition to its importance in intermediary metabolism (see Chapter 5), the reaction is catalysed by a large number of central regulatory enzymes that are often part of signalling cascades, such as protein kinases, protein phosphatases, ATPases and GTPases.

Kinases are Nature's tools for introducing phosphoryl groups into organic molecules, whether they are metabolites, such as glucose and fructose-6-phosphate in the glycolysis pathway, or proteins that are part of signalling cascades, such as that which activates glycogenolysis and simultaneously inhibit glycogen synthesis via phosphorylation of protein side chains (serine residues in this particular case). The donor of the phosphoryl group is usually Mg^{2+}-ATP.

The resting adult human brain consumes around 80 mg of glucose and 50 ml of O_2 per minute, and once the glucose has been transported across the plasma membrane it is rapidly phosphorylated by hexokinase, the first enzyme of the glycolytic pathway. Hexokinase

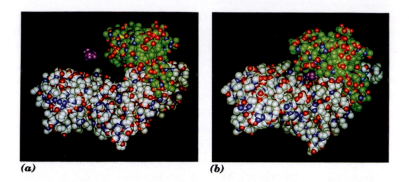

Figure 10.2 (a) Yeast hexokinase; (b) in its complex with glucose. (From Voet and Voet, 2004. Reproduced with permission from John Wiley & Sons., Inc.)

catalyses the transfer of a phosphoryl group from Mg^{2+}-ATP to glucose to form glucose-6-phosphate and Mg^{2+}-ADP. It is a member of a super family of proteins with a common characteristic βββαβαβα-fold, which is repeated in both the N-terminal and the C-terminal domains. The members have a common ATPase domain, and include kinases, which phosphorylate not only sugars but also glycerol, acetate and other carboxylic acids. As illustrated by glucose binding to hexokinase (Figure 10.2), catalysis by these enzymes is known to be accompanied by a large conformational change, which is associated by inter-domain motion. The two lobes of the active site cleft swing together from an open to a closed conformation by ~8 Å. This also has the consequence of excluding water from the active site, which may explain why phosphoryl transfer to glucose is 4×10^4 times faster than to water.

Another characteristic of this kinase family, as has been shown by Jeremy Knowles (1980) using ATP-made chiral in its γ-phosphoryl group, is that phosphoryl group transfer occurs with inversion of configuration. This is taken to be indicative of a direct, in-line transfer of the phosphoryl group from substrate to product by addition of a nucleophile to the phosphorus atom yielding a trigonal bipyramidal intermediate, the apexes of which are occupied by the attacking and leaving groups (Figure 10.3).

Hexokinase forms a ternary complex with glucose and Mg^{2+}-ATP before the reaction takes place, which, as a result of the domain closure, places ATP in close proximity to the C6 hydroxyl group of glucose (Figure 10.4). By complexing the phosphate groups of ATP, Mg^{2+} is thought to shield their negative charges, making the γ-phosphorus atom more accessible to nucleophilic attack by the C6-OH group of the glucose molecule. However, it also seems that, as in many of the other members of the super family, the Mg^{2+} ion binds not only directly to the oxygen atoms of the β- and γ-phosphoryl groups, but also through a water molecule to the carboxylate of a well-conserved Asp residue. This Asp acts as a general base responsible for deprotonating the hydroxyl on the sugar, which will be phosphorylated. This is illustrated for the rhamnulose kinase from *E. coli* (Figure 10.5), which catalyses the transfer of the γ-phosphoryl group from ATP to the 1-hydroxyl group of L-fructose. The γ-phosphoryl group of the ATP can be positioned in such a way that the three oxygen atoms are on the corners of a trigonal bipyramid between O3β of ADP and the O1″ atom of β-L-fructose. Moreover, the required Mg^{2+} can be modelled between the β- and γ-phosphoryl groups and the well-conserved Asp10, as shown in Figure 10.5. The putative

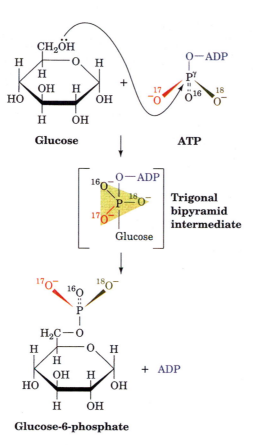

Figure 10.3 In the phosphoryl-transfer reaction catalysed by hexokinase, the γ-phosphoryl group of ATP undergoes inversion of configuration. (From Voet and Voet, 2004. Reproduced with permission from John Wiley & Sons., Inc.)

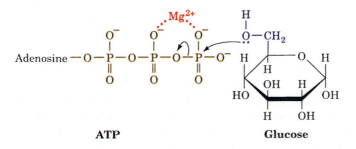

Figure 10.4 Nucleophilic attack of the C6-OH group of glucose on the γ-phosphate of Mg^{2+}-ATP complex. (From Voet and Voet, 2004. Reproduced with permission from John Wiley & Sons., Inc.)

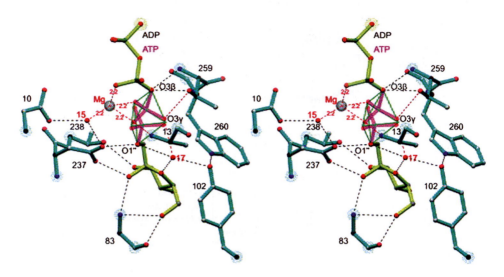

Figure 10.5 Stereoview of the reaction running through a bipyramidal pentavalent phosphorus atom. The γ-phosphoryl group before and after the transfer *is in a transparent mode.* A putative Mg^{2+} was placed at the expected position between Asp10 and the β- and γ-phosphoryl groups. (From Grueninger and Schultz, 2006. Copyright 2006, with permission from Elsevier.)

Mg^{2+} binds directly to the phosphate oxygen atoms and through a water molecule to the carboxylate group. Since Mg^{2+} prefers an octahedral coordination sphere, it requires a water structure different from that observed in holo rhamnulose kinase.

In contrast to the kinases, which phosphorylate metabolites, there are a number of families of protein kinases that phosphorylate Ser, Thr and Tyr residues in specific target proteins, usually part of a signal amplification cascade in response to an extracellular stimulus. Here we consider briefly the family of mitogen-activated protein kinases (MAPKs), which function as mediators of many different cellular signals. MAPKs function as the terminal component of a signalling cascade, which in mammals, following an initial extracellular stimulus, amplifies the signal through at least 14 MAP kinase kinase kinases (MKKKs); these in their turn activate 7 MAP kinase kinases (MKKs), which then activate 12 MAPKs (Figure 10.6). At each step of the cascade, the signal is amplified several fold. The MAPKs then act on other targets, notably transcription factors (that regulate the synthesis of target mRNAs) and other kinases. While it is clearly established that these kinases require Mg^{2+}, in the form of the Mg^{2+}-ATP complex, in a great many cases they also seem to require a second Mg^{2+} ion. It is not clear, in the absence of any *a priori* chemical necessity, what the function of the second magnesium might be. As often happens when biochemists have no clear idea of what is going on, conformational change of the enzyme protein is invoked.

PHOSPHORYL GROUP TRANSFER: PHOSPHATASES

In contrast to kinases, phosphatases catalyse the removal of phosphoryl groups, again, either from phosphorylated metabolites such as glucose-6-phosphate or fructose-1,6-bisphosphate

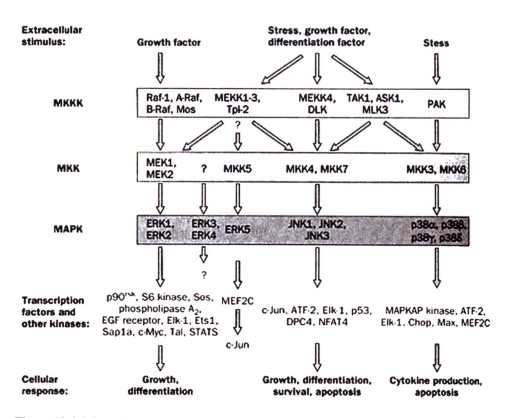

Figure 10.6 Schematic representation of the mitogen-activated protein kinase (MAP) cascades in mammalian cells. (From Voet and Voet, 2004. Reproduced with permission from John Wiley & Sons., Inc.)

in the central metabolic pathways, or from proteins that have been phosphorylated by protein kinases. Unlike the kinases, they catalyse a hydrolytic reaction in which the phosphoryl group is transferred to water. Here we consider, as an illustration, phosphatases and the phosphoglucomutases of the haloacid dehalogenase[2] (HAD) super family of phospho-transferases However, though this still involves nucleophilic catalysis, instead of an in-line mechanism with a penta-coordinate transition state, we have formation of a covalent aspartylphosphate enzyme intermediate (Figure 10.7a). The Mg^{2+}, which is essential for the reaction, binds to both the nucleophilic Asp and to the phosphorylated substrate, providing orientation and charge shielding for nucleophilic attack (Figure 10.7b). The aspartylphosphate has a high energy of hydrolysis, which drives phosphoryl transfer to water, the predominant acceptor of phosphoryl groups. In the phosphatases, hydrolysis of the aspartylphosphate intermediate is facilitated by a general base, which typically contributes a rate enhancement of 10^2–10^4. In contrast, in the ATPases the rate of dephosphorylation of the aspartyl phosphate is greatly reduced by the use of a Thr residue, which contributes only 30-fold to the

---

[2] Although named after a dehalogenase, of more than 3000 sequenced proteins of the HAD family, the majority are phosphoryl transferases.

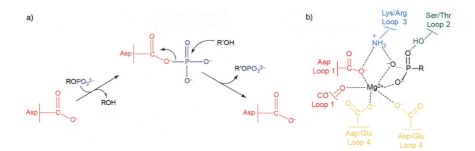

Figure 10.7 (a) In HAD enzymes Asp mediates phosphoryl group transfer to a variety of acceptors, including H₂O; (b) the catalytic scaffold around the essential Mg²⁺ion. (From Allen and Dunaway-Mariano, 2004. Copyright 2004, with permission from Elsevier.)

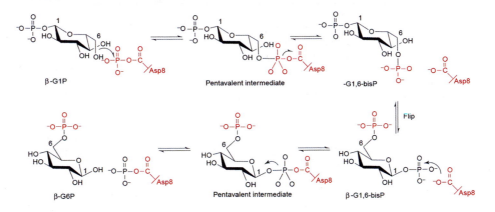

Figure 10.8 The phosphoglucomutase reaction proceeds via two phosphoryl-transfer reactions. (From Allen and Dunaway-Mariano, 2004. Copyright 2004, with permission from Elsevier.)

rate enhancement. And, in the phosphoglucomutase reaction, in which glucose-1-phosphate is converted to glucose-6-phosphate by two phosphoryl-transfer reactions with the formation of a glucose-1,6-bisphosphate intermediate (Figure 10.8), the rate of hydrolysis of the aspartylphosphate is reduced even further. This is probably achieved by the sugar phosphate itself positioning the base catalyst necessary for phosphoryl transfer, whereas a water molecule cannot do so.

The structural and functional role of the divalent cation in the active centre of phosphatases is well illustrated by the case of human phosphoserine phosphatase. When the essential Mg²⁺ is replaced by Ca²⁺, the enzyme is inactivated. Figure 10.9a shows the active site of a bacterial phosphoserine phosphatase with a Mg²⁺ and phosphoserine in the active site, while Figure 10.9b shows human phosphoserine phosphatase with a Ca²⁺ion bound and the modelled substrate in the active site. The hepta-coordinate Ca²⁺ binds to both side-chain oxygen atoms of the catalytic Asp20, unlike the hexa-coordinate Mg²⁺, which ligates only one oxygen atom (Figure 10.9). This prevents the nucleophilic attack by one of the Asp20

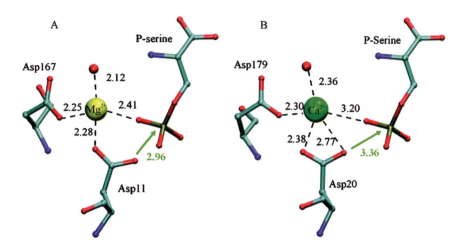

Figure 10.9 Active site of *Methanococcus* phosphoserine phosphatase with Mg^{2+} and phosphoserine in the active site (a) and of human phosphoserine phosphatase with Ca^{2+} bound and the modelled substrate in the active site (b). (From Peeraer et al., 2004. Reproduced with permission of Blackwell Publishing Ltd.)

side-chain oxygens on the phosphorus atom of the substrate, accounting for the inhibition (Peeraer et al., 2004).

STABILIZATION OF ENOLATE ANIONS: THE ENOLASE SUPER FAMILY

Yet another example of a family of Mg^{2+}-dependent enzymes is the enolase super family, which catalyses a series of mechanistically diverse and overall different reactions. However, they all share a partial reaction in which an active site base of the enzyme abstracts the α-proton of a carboxylate substrate to generate an enolate anion intermediate, which is stabilized by coordination to the essential Mg^{2+} ion. This intermediate then is directed to different products in the different active sites (Figure 10.10). The three 'founder' members of this family are the mandelate racemase (MR) and the muconate lactonizing enzyme (MLE) of *Pseudomonas putida* and enolase, the fifth glycolytic enzyme referred to above. The 3-D structures of all three enzymes are remarkably superposable (Figure 10.11). All three have a two-domain structure, with the active sites located at the interface between flexible loops in the capping domain and the C-terminal barrel domain. The capping domain is formed by segments from the N- and C-termini of the polypeptide chain, while the barrel domain is formed from ends of the β-strands of the modified TIM-barrel domain. While the TIM molecule (see Chapter 3, Figure 3.7) has $[(\beta/\alpha)_8\beta]$, the enolase super family instead has $[(\beta/\alpha)_7\beta]$. All members of the super family contain ligands (nearly always Glu or Asp) for the essential Mg^{2+}, located at the ends of the third, fourth and fifth β-strands (illustrated for enolase, Figure 10.12).

In enolase, the substrate, 2-phosphoglycerate (2-PGA), is coordinated to two Mg^{2+}ions, one of which is liganded to the three conserved carboxylate residues (Asp246, Glu295 and Asp320). Currently more than 600 enolase sequences have been identified in the databases,

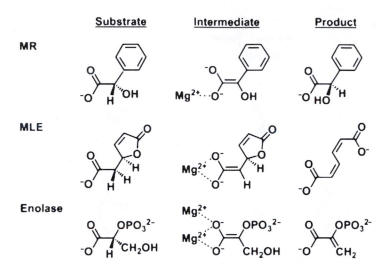

Figure 10.10 The substrates, enolate anion intermediates and products of the MR; MLE and enolase reactions. (From Gerlt et al., 2005. Copyright 2005, with permission from Elsevier.)

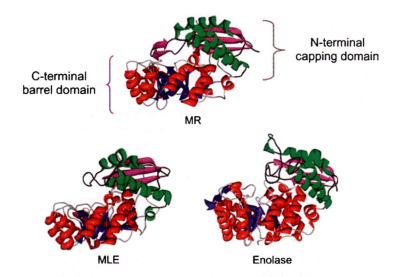

Figure 10.11 Comparison of the structures of MR, MLE and enolase showing the two homologous domains that illustrate divergent evolution. (From Gerlt et al., 2005. Copyright 2005, with permission from Elsevier.)

and all are thought to be isofunctional, catalysing the conversion of 2-PGA to phospho-enolpyruvate. In the MLE subclass of the super family, at least three reactions are known to be catalysed—in addition to the lactonization of muconate, succinylbenzoate synthase and L-Ala-D/L-Glu epimerase reactions are observed within the ~300 members. The MR subclass catalyses at least five reactions, mandelate racemization and four sugar dehydratases. As in the MLE subclass, of the ~400 members identified, only ~50% of these are functionally assigned.

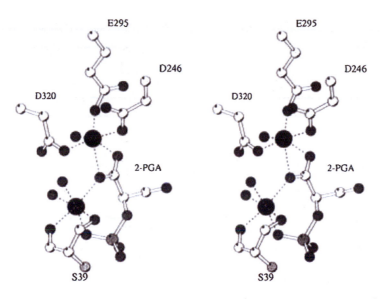

Figure 10.12 Stereoview of the coordination of the two Mg^{2+} ions in the enolase-$(Mg^{2+})_2$-2-PGA complex. (From Larsen et al., 1996. Copyright (1996) American Chemical Society.)

ENZYMES OF NUCLEIC ACID METABOLISM

Clearly, since DNA and RNA molecules are polynucleotides, composed of an invariant sugar-phosphate backbone, it comes as no surprise that many of the enzymes involved in their metabolism require Mg^{2+} ions. Here we do not consider how this large number of enzymes achieve their structure or sequence specificity, although the sensitivity of ligand geometry and electrostatic environment of Mg^{2+} ions is proposed to greatly enhance substrate recognition and catalytic specificity (for a recent review see Yang et al., 2006), but rather consider the role of the metal ions. However, it is salutary to recall a few examples of their specificity. Restriction endonucleases each typically recognize a specific six-base-pair sequence, and more than 3000 type II restriction endonucleases recognize over 200 different sequences. How is this sequence specificity achieved? Endonucleases such as RNase H remove the RNA primer strand from the RNA–DNA hybrid in Okazaki fragments during DNA replication, yet they do not cleave either double-stranded DNA or RNA. DNA and RNA polymerases, even without their elaborate proofreading function, insert the wrong nucleotide only every 10^3–10^4 bases, despite the relatively small free energy difference of only ~2 kcal/mol between Watson–Crick and mismatched base pairs.

Nucleic acid metabolism is dominated by phosphoryl transfer reactions (Figure 10.13). These include the reactions involved in nucleotide polymerization, catalysed by DNA and RNA polymerases, in which a phosphodiester bond is transferred from a (d)NTP to the 3′ hydroxyl of a growing nucleic acid chain (Figure 10.13a), nucleic acid cleavage in which the phosphate being attacked is the backbone of a nucleic acid, and the nucleophile is either a water molecule or a sugar hydroxyl (Figure 10.13b–f). In nuclease reactions, where a water molecule is the nucleophile, the cleavage products are a 5′ phosphate and a

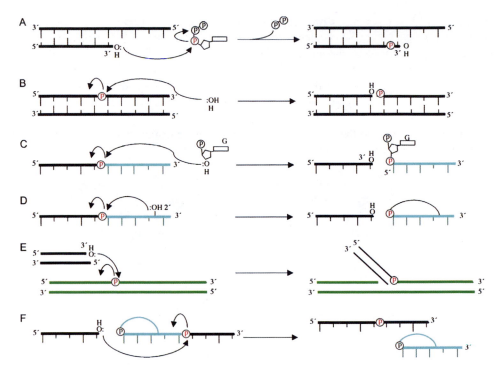

Figure 10.13 Phosphoryl-transfer reactions. The figure shows (a) nucleotide polymerization, (b) nucleic acid hydrolysis, (c) first cleavage of an exon–intron junction by group I ribozyme (d) and by a group II ribozyme, (e) strand transfer during transposition and (f) exon ligation during RNA splicing. (From Yang et al., 2006. Copyright 2006, with permission from Elsevier.)

3′ hydroxyl (Figure 10.13b). When the 2′ or 3′ hydroxyl group of a ribonucleotide is the nucleophile, as in RNA splicing, catalysed by the group I and group II self-splicing ribozymes, the 5′ end product is covalently linked to the ribonucleotide (Figures 10.13c and d). If the nucleophile is the terminal 3′ hydroxyl of a DNA or RNA strand, this can result in strand transfer, as in DNA transposases[3] (Figure 10.13e), or in exon ligation, as in RNA splicing (Figure 10.13f). As in all phosphoryl transfer reactions, the first step is deprotonation and activation of a nucleophile, and the final step is the protonation of the leaving group by a water molecule. All these reactions involve a pentacovalent phosphate intermediate with the inversion of the configuration of phosphorus atom.

The discovery of self-splicing introns showed that RNA could catalyse chemical reactions. Yet, unlike proteins, RNA has no functional groups with pK_a values and chemical properties similar to those considered to be important in protein-based enzymes. Steitz and Steitz (1993) postulated that two metal ions were essential for catalysis by ribozymes using a mechanism similar to DNA cleavage, in which a free 3′ OH is produced. They proposed,

_____

[3] Enzymes required in site-specific DNA recombination, moving transposable elements around in DNA molecules of bacteria and other organisms.

based on the X-ray structures of alkaline phosphatase and of the exonuclease fragment of DNA polymerase, that two divalent cations, almost certainly Mg^{2+}, bound to conserved carboxylate residues were essential for catalysis. The two metal ions are situated roughly in line with the phosphate-sugar backbone on the opposite side from the bases. No protein side chains are directly involved in the reaction; rather, protein residues serve to bind and correctly orient the two Mg^{2+} ions, the 3-terminal and penultimate residue of the DNA structure and the attacking water molecule for catalysis. Not only are they close to one another (~4 Å), but also they more or less bisect the plane of the trigonal core of the pentacovalent intermediate (Figure 10.14). The two metal ions form inner-sphere complexes with the scissile phosphate and water, facilitating formation of the attacking hydroxyl ion and stabilizing the transition state. Metal ion A (Me_A^{++}) decreases the pK_a of a water molecule, assisting nucleophilic attack by formation of a hydroxide anion, which, with the help of Glu357 and Tyr497, is correctly oriented for an in-line attack on the phosphorus of

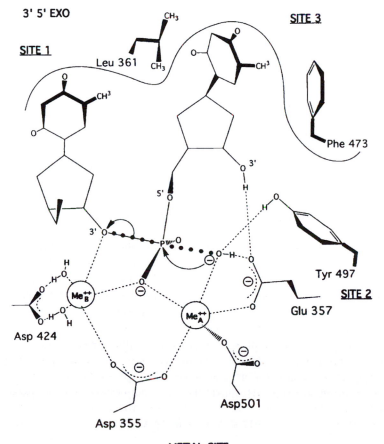

Figure 10.14 The proposed transition state in the mechanism of the 3′–5′ exonuclease activity of DNA polymerase I. (From Steitz and Steitz, 1993. Copyright (1993) National Academy of Sciences, USA.)

the scissile phosphate. Metal ion B (Me_B^{++}) acts as a Lewis acid, facilitating the leaving of 3'-oxyanion and stabilizing the penta-coordinate transition state. Recent high-resolution structural studies (Nowotny and Yang, 2006) on reaction intermediate and product complexes suggest that (i) metal ion A may assist nucleophilic attack by approaching metal ion B, bringing the nucleophile close to the scissile phosphate, (ii) metal ion B seems to move from an irregular coordination in the substrate complex to a more regular geometry in the product complex, (iii) the coordination environment of Mg^{2+} probably destabilizes the enzyme–substrate complex and reduces the energy barrier for product formation and (iv) product release probably requires dissociation of metal ion A. The metal ions here are in close proximity (<4 Å) and are bridged by a common substrate. A large number of nucleases, collectively known as the DEDD family, have their active site Mg^{2+} ions coordinated by four conserved acidic amino acids, three Asp and one Glu residue, as in the DNA polymerase exonuclease active site described above.

Yet another super family, the nucleotidyl-transferase family, also utilizes two-metal-ion-dependent catalysis: the members include transposases, retrovirus integrases and Holliday junction resolvases[4]. Whereas in the nucleases, the Mg^{2+} ions are asymmetrically coordinated, and play distinct roles, in activating the nucleophile and stabilizing the transition state, respectively, in the transposases, they are symmetrically coordinated and exchange roles to alternatively activate a water molecule and a 3'-OH for successive strand cleavage and transfer.

We pass in silence the extensive discussions as to whether some endonucleases, which have a conserved Lys residue in their active site, use the two-metal-ion catalytic mechanism, whether there are situations where the metal ion A alone suffices for catalysis and whether there may not be endonucleases that require three Mg^{2+} ions (for a review see Pingoud et al., 2005).

MAGNESIUM AND PHOTORECEPTION

We terminate this brief overview of the biological chemistry of Mg^{2+} by introducing the green pigment that gives us all joy in springtime, chlorophyll. Not only does it give verdant colour to our trees and garden plants, but it harnesses solar energy to ensure, not only CO_2 fixation but also a plethora of other important metabolic functions. Contrary to popular belief that photosynthesis corresponds to the fixation of CO_2 accompanied by the evolution of molecular O_2, more than 50% of photosynthetic organisms are strict anaerobes. The effective reaction carried out by photosynthetic organism uses the energy of solar photons to oxidize an electron donor, H_2D (which in green plants is H_2O), to supply electrons to an electron acceptor, A (usually $NADP^+$), to generate the oxidized donor, D (in green plants O_2), and the reduced acceptor, AH_2 (NADPH), according to the general equation:

$$H_2D + A \xrightarrow{\text{light}} H_2A + D$$

---

[4] It would simply befuddle the reader to explain what these enzymes do. It is sufficient to know that they cut and paste DNA fragments with exquisite specificity, as we do with our word processors.

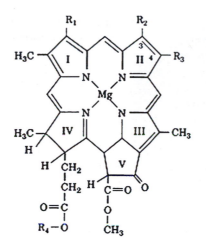

Figure 10.15 The chlorophyll *a* and *b* molecules have vinyl, ethyl and phytyl side chains as R_1, R_3 and R_4, respectively: chlorophyll *a* has a methyl group as R_2, which is replaced by a formyl group in chlorophyll *b*. (From Voet and Voet, 2004. Reproduced with permission from John Wiley & Sons., Inc.)

The principal photosynthetic photoreceptor is chlorophyll, a cyclic tetrapyrrole that is formed, like haem, from protoporphyrin IX. However, it has a cyclopentenone ring (V) fused to the pyrrole ring III, variable modifications of the substituents of rings I and II and esterification of the propionyl side chain of ring IV by a tetraisoprenoid alcohol; has one of its pyrrole rings reduced (ring IV in eukaryotic and oxygen-evolving cyanobacteria, rings II and IV in other photosynthetic bacteria) and, most importantly, has replacement of the central metal ion by Mg^{2+} instead of Fe^{2+} (Figure 10.15). Chlorophylls are very effective photoreceptors, since they contain an extensive network of alternating single and double bonds (they are polyenes). They have very strong absorption bands in the visible region of the spectrum, where the sun's radiation is also maximal. The peak molecular extinction coefficient of the various chlorophylls is in excess of 10^5 M^{-1} cm^{-1} amongst the highest known for organic molecules.

Why, we may ask, does nature use Mg^{2+} as the metal to capture solar energy? Perhaps, as has been suggested by Frausto da Silva and Williams (2001), the reasons are that Mg^{2+} does not have the redox properties of other metal ions such as Mn, Co, Fe, Ni and Cu when inserted into a porphyrin, and that it does not enhance fluorescence as much as the corresponding Zn porphyrin would.

If the light-harvesting system relied only on the special pair of chlorophyll molecules that constitute the reaction centre, it would be inefficient for two reasons. Firstly, chlorophyll molecules absorb only in the blue and red regions of the spectrum, so that light in the middle of the visible region, from 450 to 650 nm, which constitutes the peak of the solar spectrum, would be lost. Secondly, on account of the low density of reaction centres, many of the photons arriving on the photosystem would be unused. Accessory pigments, both additional chlorophyll molecules, and other classes of molecules, such as carotenoids[5] and the linear

---

[5] These are not only responsible for the orange colour of carrots, but for the spectacular autumn colours of deciduous trees as well.

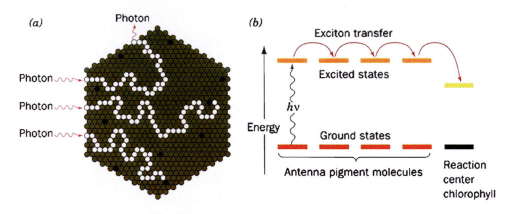

Figure 10.16 Solar energy transfer from accessory pigments to the reaction centre. (a) The photon absorption by a component of the antenna complex transfers to a reaction centre chlorophyll, or, less frequently, is reemitted as fluorescence. (b) The electron ends up on the reaction centre chlorophyll because its lowest excited state has a lower energy than that of the other antenna pigment molecules. (From Voet and Voet, 2004. Reproduced with permission from John Wiley & Sons., Inc.)

tetrapyrrole phycobilins, which together constitute the reaction centre antenna, absorb energy and funnel it to the reaction centres. The phycobilins are particularly important in harvesting the yellow and green light, which reaches the ecological niche of the blue–green (cyanobacteria) and red marine algae. The excited state of the special pair of chlorophyll molecules is lower in energy than those of the other antenna pigment molecules (Figure 10.16).

What happens when the energy of a photon of light absorbed by one of the accessory pigments is transferred to a chlorophyll molecule in the reaction centre? The light energy excites an electron from its ground-state level to an excited level, and can be moved from the initial chlorophyll molecule to a suitable nearby electron acceptor. This results in photo-induced charge separation, a positive charge being formed on the chlorophyll molecule and a negative charge on the acceptor. This is beautifully illustrated in the structure of the bacterial photosynthetic reaction centre from the bacterium *Rhodopseudomonas viridis* (Figure 10.17)[6]. The two bacteriochlorophyll molecules of the reaction centre, which constitute the special pair referred to above, are incorporated into a nearly perfect two-fold symmetry axis of two similar polypeptide chains, L (red) and M (blue). The two chlorophyll molecules are almost parallel, with an Mg^{2+}–Mg^{2+} distance of ~7 Å, and are each in a predominantly hydrophobic region of the protein with a His side chain as fifth ligand to the metal ion. An additional polypeptide chain, designated H (white) and a cytochrome subunit (yellow), which subsequently restores the electron deficit in the special pair, complete the structure.

---

[6] This was the first transmembrane protein to have its structure described in detail by Deisenhofer, Huber and Michel in 1984. They received the Nobel Prize for Chemistry 4 years later.

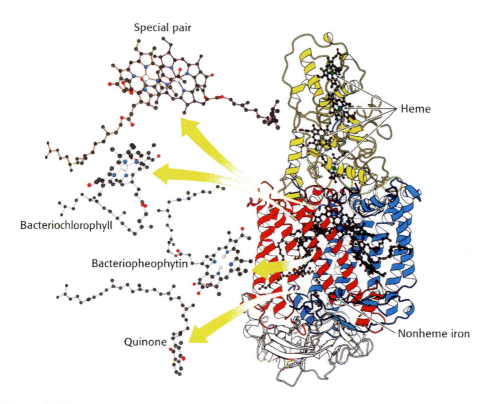

Figure 10.17 The core of the bacterial photosynthetic reaction centre. (From Berg et al., 2001. Reproduced with permission from W.H. Freeman and Co.)

Once the special pair has absorbed a photon of solar energy, the excited electron is rapidly removed from the vicinity of the reaction centre to prevent any back reactions. The path it takes is as follows: within 3 ps (3×10^{-12} s) it has passed to the bacteriopheophytin (a chlorophyll molecule that has two protons instead of Mg^{2+} at its centre), without apparently becoming closely associated with the nearby accessory bacteriochlorophyll molecule. Some 200 ps later it is transferred to the quinone. Within the next 100 μs the special pair has been reduced (by electrons coming from an electron transport chain that terminates with the cytochrome situated just above it), eliminating the positive charge, while the excited electron migrates to a second quinone molecule.

REFERENCES

Allen, K.N. and Dunaway-Mariano, D. (2004) Phosphoryl group transfer: evolution of a catalytic scaffold, *TIBS*, **29**, 495–503.

Berg, J.M., Tymoczko, J.L. and Stryer, L. (2001) *Biochemistry*, 5th edition, Freeman, New York, 974 pp.

Cowan, J.A. (2002) Structural and catalytic chemistry of magnesium-dependent enzymes, *BioMetals*, **15**, 225–235.

Frausto da Silva, J.J.R. and Williams, R.J.P. (2001) *The biological chemistry of the elements 2nd edition,* Oxford University Press, Oxford, p. 270.

Gerlt, J.A., Babbitt, P.C. and Rayment, I. (2005) Divergent evolution in the enolase superfamily: the interplay of mechanism and specificity, *Arch. Biochem. Biophys.*, **433**, 59–70.

Grueninger, D. and Schultz, G.E. (2006) Structure and reaction mechanism of L-rhamnulose kinase from *Eschericia coli*, *J. Mol. Biol.*, **359**, 787–797.

Knowles, J.R. (1980) Enzyme-catalysed phosphoryl transfer reactions, *Annu. Rev. Biochem.*, **49**, 877–919.

Larsen, T.M., Wedeking, J.E., Rayment, I. and Reed, G.H. (1996) A carboxylate oxygen of the substrate bridges the magnesium ions at the active site of enolase: structure of the yeast enzyme complexed with the equilibrium mixture of 2-phosphoglycerate and phosphoenolpyruvate at 1.8 Å resolution, *Biochemistry*, **30**, 4349–4358.

Maguire, M.E. and Cowan, J.A. (2002) Magnesium chemistry and biochemistry, *BioMetals*, **15**, 203–210.

Nowotny, M. and Yang, W. (2006) Stepwise analyses of metal ions in RNase H catalysis from substrate destabilization to product release, *EMBO J.*, **25**, 1924–1933.

Peeraer, Y., Rabijns, A., Collet, J.-F., Van Scaftingen, E. and De Ranter, C. (2004) How calcium inhibits the magnesium-dependent enzyme human phosphoserine phosphatase, *Eur. J. Biochem.*, **271**, 3421–3427.

Pingoud, A., Fuxreiter, M., Pingoud, V. and Wende, W. (2005) Type II restriction endonucleases: structure and mechanism, *Cell.Mol.Life Sci.* **62**, 685–707.

Steitz, T.A. and Steitz, J.A. (1993) A general two-metal-ion mechanism for catalytic RNA, *Proc. Natl. Acad. Sci. U.S.A.*, **90**, 6498–6502.

Voet, D. and Voet, J.G. (2004) *Biochemistry*, 3rd edition, Wiley, New York, Chichester, 1591 pp.

Yang, W., Lee, J.Y. and Nowotny, M. (2006) Making and breaking nucleic acids: two-Mg^{2+}-ion catalysis and substrate specificity, *Mol. Cell*, **22**, 5–13.

– 11 –

Calcium: Cellular Signalling

INTRODUCTION: —COMPARISON OF Ca^{2+} AND Mg^{2+}

How does Nature achieve the high degree of selective binding of Ca^{2+} by biological ligands compared to Mg^{2+}? The differences in structure, thermodynamic stability and reaction rates all stem from the difference in their ionic radii (Mg^{2+}, 0.6 Å; Ca^{2+}, 0.95 Å) measured in an octahedral oxygen donor environment. The Mg^{2+} ion is strictly octahedral with Mg–O distances of around 2.05 Å, whereas the larger Ca^{2+} ion, such as Na^+ and K^+, has an irregular coordination geometry, bond angle, bond distance and coordination number (7–10) with Ca–O distances of 2.3–2.8 Å. In the case of the smaller Mg^{2+} the central field of the cation dominates the coordination sphere, whereas in the larger cations the second and possibly even the third coordination spheres have a more important influence resulting in irregular structures. This also enables Ca^{2+}, unlike Mg^{2+}, ion to bind to a large number of centres at once. Further, the kinetics of Ca^{2+} binding are quite different with water exchange rates close to the collision diffusion limits of 10^{10} s^{-1}, unlike much slower rates of 10^6 s^{-1} for Mg^{2+}. Ca^{2+} can interact with neutral oxygen donors, such as carbonyls and ethers, unlike Mg^{2+} in aqueous media, as well as with anions, which avoids competition with Na^+.

THE DISCOVERY OF A ROLE FOR Ca^{2+} OTHER THAN AS A STRUCTURAL COMPONENT

Calcium, together with sodium, potassium and magnesium, is one of the metals required by living systems in macro-amounts—indeed it represents 1.5–2% of an adult's total body weight. The biominerals that constitute teeth and bones contain the majority of the body's calcium (~99%). Yet the 1% that remains within the cells and tissues has enormous importance in the regulation of a whole series of cellular responses. Like a number of other discoveries, it was made by serendipity, and came far too early for the scientific community to recognize the importance of the discovery. In 1883, the English physiologist Sidney Ringer carried out a rather sloppy experiment, in which he suspended rat hearts in a saline medium made from London tap water (notoriously 'hard' on account of its high calcium content) and observed that they continued beating for a considerable period of time. When he repeated his initial experiments with distilled water, the hearts stopped beating after ~20 min incubation. He subsequently found that the addition of Ca^{2+} to the saline solution of distilled water allowed prolonged cardiac contraction, establishing unequivocally that Ca^{2+} had a role in a tissue that had neither bones nor teeth. However, we had to wait for more than 60 years before the importance of his observation that Ca^{2+} had a real function in cellular biochemistry, totally unrelated to its well-established structural role as a component of bone, became evident (for reviews see Carafoli, 2005; Williams, 2006).

We now recognize Ca^{2+} as an intracellular signalling messenger, which, as Ernesto Carafoli has poetically stated, 'accompanies cells throughout their entire lifespan, from their origin at fertilization, to their eventual demise ... as a conveyor of doom at the moment of cell death'. Yet, like Na^+, the extracellular concentration of Ca^{2+} is much higher (10^4 times) than that in the cytosol of the average mammalian cell (which is around 10^{-7} M). There are, however, at least three intracellular compartments that can accommodate much higher Ca^{2+} concentrations—the mitochondria, the endoplasmic reticulum (ER) and the Golgi apparatus (Figure 11.1).

Since there is such a great difference between the external and internal (or at least cytosolic) Ca^{2+} levels, and since cytosolic Ca^{2+} levels below micromolar must be maintained to allow the signalling role of Ca^{2+} fluxes, it is clear that in order to maintain intracellular Ca^{2+} homeostasis, all cells must have developed mechanisms for regulating both Ca^{2+} uptake and egress, which are described in the next sections.

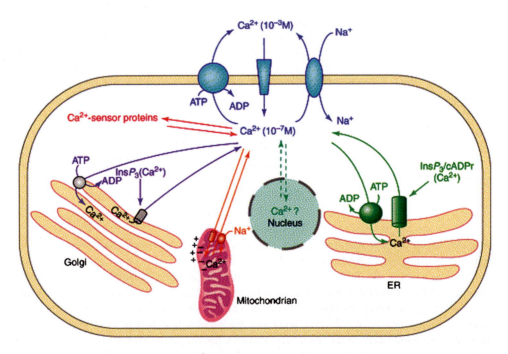

Figure 11.1 The basic concepts of Ca^{2+} cellular homeostasis and signalling. (From Carafoli, 2004. Copyright 2004, with permission from Elsevier.)

PLASMA MEMBRANE UPTAKE PATHWAYS

Ca^{2+} enters cells via a number of plasma membrane channels which belong to three families (Figure 11.2): (i) voltage-gated channels; (ii) ligand-gated channels and (iii) capacitative, or store-operated channels (SOCs).

The voltage-gated Ca^{2+} channels, like the corresponding Na^+ and K^+ channels (see Chapter 9), consist of a channel-forming α_1-subunit made up of four six-transmembrane domain repeats, in each of which the Ca^{2+} pore is formed by the loops which fold within the membrane between transmembrane domains 5 and 6. They also contain a voltage sensor within the transmembrane domain S4 of the α-subunit, rich in polar amino acids (Figure 11.2a), as well as several accessory subunits that influence the properties of the channels. Ten α_1-subunit genes have been cloned, which define four different channel types, three gated by high voltage and one by low voltage.

The ligand-gated Ca^{2+} channels can be activated either directly or indirectly. Among the directly activated ligand-gated Ca^{2+} channels are those that are activated by the neurotransmitter glutamate, like the NMDA receptor (Figure 11.2c), activated by the glutamate agonist *N*-methyl-D-aspartate.

The indirectly activated ligand-gated Ca^{2+} channels act by the interaction of the agonists with G-protein-coupled receptors, which produce inositol-(1,4,5)-trisphosphate (InsP3). This molecule, as we will see below, then acts to release Ca^{2+} rapidly from the ER stores. The resulting depletion of Ca^{2+} within the lumen of the ER serves as the primary trigger for a message that is returned to the plasma membrane, resulting in the relatively slow (10–100 s) activation of the third type of Ca^{2+} channels, the capacitative or SOCs. These mediate Ca^{2+} entry which can then replenish intracellular stores. The mechanism by which the emptying of intracellular ER stores activates the SOC Ca^{2+} entry pathway is still unknown. The SOCs are homologous to the transient receptor potential (TRP) channels, first identified in the retina of *Drosophila*, for which the general membrane topology is presented in Figure 11.2b.

CALCIUM EXPORT FROM CELLS

In order to prevent overloading of the cytosol with Ca^{2+}, cells have two systems to export Ca^{2+}, a high-affinity, low-capacity Ca^{2+}-ATPase, and a low-affinity, but much higher capacity Na^+/Ca^{2+} exchanger. The Ca^{2+}-ATPase, termed the plasma membrane calcium ATPase (PMCA) pump, functions to fine tune intracellular Ca^{2+}, functioning at concentrations below the micromolar level which most Ca^{2+}-dependent regulatory functions seem to prefer (see below). These PMCAs belong to the family of P-type ATPases, characterized by formation of an aspartyl phosphate intermediate during the reaction cycle. Like the Na^+/K^+-ATPase discussed in the previous chapter, the Ca^{2+}-ATPase exists in two conformational states. At the beginning of the transport cycle it is in the E1 state, which, in contrast to its sister protein of the ER, only binds one Ca^{2+} with high affinity on the cytoplasmic side of the plasma membrane, and releases it from the E2 form, which has low affinity for Ca^{2+} at the exterior of the cell. However, one important aspect of the 10 transmembrane domain PMCA is that it has a long C-terminal tail, which contains a binding site for the

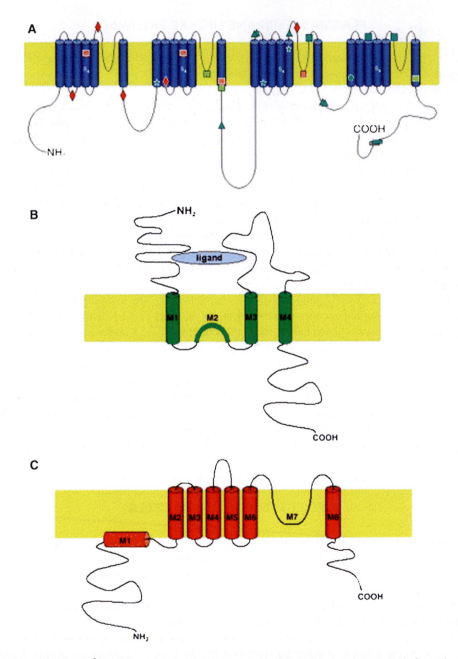

Figure 11.2 The Ca^{2+} influx channels of the plasma membrane. (a) A voltage-gated channel—the membrane topography of the α1-subunit, which forms the channel proper. (b) Membrane topography of TRP channels, homologous to the SOC channels. (c) Membrane topography of the ligand-gated glutamate NMDA receptor. (From Carafoli, 2005. Reproduced with permission from Blackwell Publishing Ltd.)

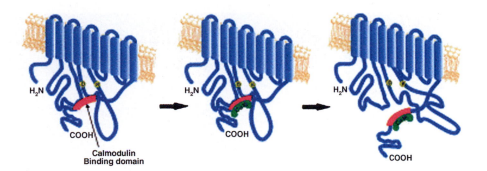

Figure 11.3 The activation of the PMCA pump by calmodulin. (From Carafoli, 2005. Reproduced with permission from Blackwell Publishing Ltd.) Calmodulin is indicated in green.

Ca^{2+}-binding protein, calmodulin. This means that the plasma membrane Ca^{2+}-ATPase is autoregulated by Ca^{2+} itself (a more detailed description of calmodulin and its role in Ca^{2+} signalling follows below). In the absence of Ca^{2+} and calmodulin the pump is inactive, whereas when calcium-saturated calmodulin binds to the C-terminal tail, the pump is activated (Figure 11.3). In the resting state, the calmodulin-binding domain interacts with the first and second main cytosolic protrusions of the pump, blocking access to the active site, indicated in the figure by the ATP-binding Lys (K) and the phosphoenzyme-forming Asp (D). Calcium-saturated calmodulin complexes the binding domain and removes it from its interaction with the pump, thereby freeing access to the active site and activating the pump.

In mammals there are four genes for the PMCA pump, two of which are expressed in all tissues (housekeeping enzymes), whereas the other two, which have a particularly high affinity for calmodulin, are restricted in their cellular distribution, notably to neurons. Additional variants of PMCAs are produced by alternative splicing of their primary RNA transcripts.

The other system for exporting Ca^{2+} from the cytosol comprises the Na^+/Ca^{2+} exchangers, which use the energy of the Na^+ gradient across the plasma membrane, and represents the major Ca^{2+} ejection system in excitable cells such as heart. There are three distinct proteins in the Na^+/Ca^{2+} exchanger gene family (now known as SLC8), NCX1, NCX2 and NCX3, of which NCX1 is expressed ubiquitously whereas both NCX2 and NCX3 are expressed at high levels in neurons and, to a lesser extent, in skeletal muscle. The exchanger operates electrogenically, exchanging three Na^+ for one Ca^{2+}, and its membrane topography consists of nine transmembrane segments (Figure 11.4).

The path for Na^+ and Ca^{2+} across the protein is not known, but it does not appear to involve the large cytosolic loop connecting transmembrane segments 5 and 6. This loop contains a Ca^{2+}-binding site, which is thought to sense intracellular Ca^{2+} changes, a domain termed XIP, which binds calmodulin, as well as a region where alternative splicing occurs.

The importance of the Na^+/Ca^{2+} exchangers is underlined by the observation that in brain ischemia, specific cleavage of NCX3 by Ca^{2+}-activated proteases leads to uncontrolled Ca^{2+} overload and to cell death, and that homozygous NCX1 knockout mice die in the embryo stage.

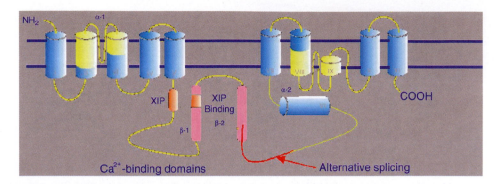

Figure 11.4 Topological model of the Na^+/Ca^{2+} exchanger NCX1. (From Guerini et al., 2005. Copyright 2005, with permission from Elsevier.)

Finally, we should mention the K^+-dependent family of Na^+/Ca^{2+} exchangers (NCKX), the first of which was discovered in retinal photoreceptors, which cotransport Ca^{2+} and K^+ from the cytosol into the extracellular space, in exchange for the entry of Na^+, with a stoichiometry of one K^+ and one Ca^{2+} for four Na^+.

CA²⁺ TRANSPORT ACROSS INTRACELLULAR MEMBRANES

As was mentioned earlier, there are three other intracellular compartments, which are able to store substantial amounts of Ca^{2+}, namely the ER, the Golgi apparatus and the mitochondria. Most of the Ca^{2+} required by cells is liberated from these three stores, rather than by importation from the exterior (Figure 11.1).

The ER takes up Ca^{2+} using a sarco(endo)plasmic reticulum ATPase (SERCA) pump (Figure 11.5), which has the same topology with 10 transmembrane domains and the same mechanism of action, with an aspartyl phosphate intermediate, as the PMCA pump described above.

However, it has two Ca^{2+}-binding sites, and transports two Ca^{2+} per ATP hydrolysed. The structure of the SERCA pump in both the E1 (with two Ca^{2+} bound) and the E2 (Ca^{2+}-free, obtained in the presence of the pump inhibitor thapsigargin) conformations has been determined by the group of Toyoshima. It confirms the predicted transmembrane topology and shows the cytosolic part of the pump to be organized in three domains. These are well separated in the E1 state, allowing access of ATP to phosphorylate Asp351, but take up a much more compact conformation in the phosphorylated E2 state. The transmembrane helices and the loops that connect them also undergo large movements. In the E1 state the Ca^{2+}-ligands, which are well separated in the E2 state, move closer to each other to form two compact Ca^{2+}-binding sites, formed by ligands in helices 4, 5, 6 and 8. The X-ray structure also confirms and extends predictions concerning access and exit routes for Ca^{2+} based on mutagenesis studies (Toyoshima and Nomura, 2002).

The SERCA pump of many types of muscle is regulated by a protein called phospholambin, which binds to SERCA in both its cytosolic and its transmembrane regions, maintaining the pump in an inactivated state when in its non-phosphorylated form, but detaches from

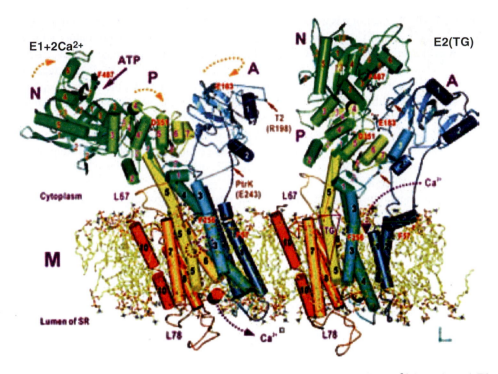

Figure 11.5 Three-dimensional structure of the SERCA pump in the E1 (Ca^{2+}-bound) and E2 (Ca^{2+}-free) state. (From Carafoli, 2005. Reproduced with permission from Blackwell Publishing Ltd.)

the pump upon phosphorylation, presumably due to a conformational change. In this sense the SERCA pump is like the PMCA pump, except that Ca^{2+}-saturated calmodulin activates PMCA, whereas kinase-dependent phosphorylation is involved in SERCA activation (perhaps also dependent on calmodulin).

Once within the endoplasmic (or sarcoplasmic) reticulum, Ca^{2+} is stored complexed to a number of low-affinity Ca^{2+}-binding proteins, the most important of which is called calsequestrin.

Release of Ca^{2+} from the reticulum occurs through two types of specific ligand-gated channels (Figure 11.1), both of which require Ca^{2+} itself in order to be opened. The first, in cells other than striated muscle, requires inositol 1,4,5-trisphosphate (InsP$_3$), which together with diacyl glycerol and Ca^{2+} itself, are the second messengers of the so-called phosphoinositide cascade, described in greater detail later in this chapter. The InsP$_3$-binding site is located in the long cytosolic amino-terminal tail of the protein, whereas the Ca^{2+} channel itself is in the C-terminal part of the molecule, consisting of the characteristic 6-transmembrane domain, with the loop connecting helices 5 and 6 folding back into the membrane to form the walls of the channel, just as is found in plasma membrane voltage-gated channels.

The other ligand-gated channel is commonly referred to as the ryanodine receptor, after the alkaloid ryanodine, which can induce opening or closing as a function of concentration.

Like the InsP$_3$ channel, it has a very large N-terminal cytosolic component, with the Ca^{2+} channel proper in the C-terminal part of the very large protein (<5000 amino acids per subunit of the tetrameric receptor). Again, the classical 6-transmembrane helical structure with the channel formed by the loop between helices 5 and 6 found in voltage-gated plasma membrane channel is present. The ryanodine receptor, like the InsP$_3$ receptor, is activated by Ca^{2+}, although the precise nature of the native messenger is unclear. A prominent candidate is the metabolite of NAD$^+$, cyclic ADP-ribose (cADPr). More recently nicotinic acid adenine dinucleotide phosphate (NAADP) has been found to be an extremely potent calcium-mobilizing messenger. While some evidence suggests that NAADP may act on acidic cellular compartments rather than on the ER, other studies indicate that it may also act on the ryanodine receptor.

The Golgi apparatus also appears to be an important regulator of intracellular Ca^{2+} homeostasis. It has a SERCA pump, as well as an unusual Ca^{2+}-ATPase that can transport Ca^{2+} and Mn^{2+} with high affinity for Ca^{2+} uptake. Ca^{2+} release is mediated by a channel that is modulated by InsP$_3$, similar to that in the ER. However, it does not seem to have ryanodine-type channels.

The third important storage compartment for intracellular Ca^{2+} is the mitochondria (Figure 11.6). The original observations in the 1950s and 1960s (reviewed in Carafoli, 2003) that mitochondria could actively accumulate Ca^{2+} led both to the clarification of the

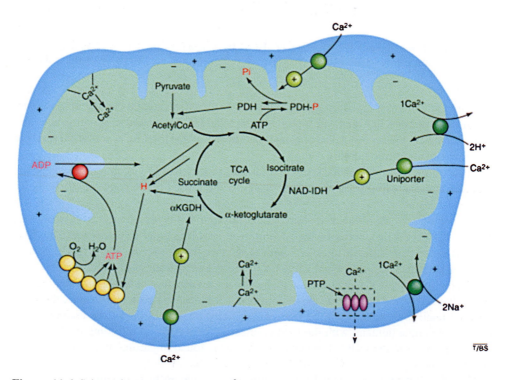

Figure 11.6 Schematic representation of Ca^{2+} transport in and out of mitochondria, showing all the Ca^{2+} transporters and activation of matrix dehydrogenases. PTP—permeability transition pore. (From Carafoli, 2003. Copyright 2003, with permission from Elsevier.)

uptake mechanism and the seemingly predominant role of the mitochondria in cellular Ca^{2+} homeostasis. The Ca^{2+} uptake pathway relies on an as yet unidentified electrophoretic uniporter which gets its energy from the negative membrane potential inside the mitochondrial inner membrane maintained by the respiratory chain (see Chapter 5). Ca^{2+} release is mediated by a Na^+/Ca^{2+} exchanger, which in some types of mitochondria, is replaced by a H^+/Ca^{2+} exchanger. These uptake and egress pathways constitute a mitochondrial energy-dissipating Ca^{2+}-cycle (Carafoli, 1979).

However, it was apparent that the affinity of the mitochondrial uptake system for Ca^{2+} is too low compared to cytosolic Ca^{2+} concentrations (apparent K_m in the 10–15 µM range, compared with sub-micromolar cytoplasmic Ca^{2+} concentrations) to effectively concentrate Ca^{2+} within the mitochondria. Further, the affinity of the ER system Ca^{2+} uptake system is at least one order of magnitude higher than that of the mitochondria (apparent K_m well below 1 µM). Together with the later discovery of the $InsP_3$-gated channels, this seemed to imply that the ER played the principal role in the regulation of cellular Ca^{2+} homeostasis.

The role of the mitochondria in cellular Ca^{2+} homeostasis was saved from oblivion by several important observations. Firstly, it became clear using indicators, notably aequorin, which can sense Ca^{2+} changes within microdomains of the cell rather than in the bulk cytoplasm, that the increase in cytosolic Ca^{2+} induced by the opening of $InsP_3$-gated channels was paralleled by the rapid and reversible increase in mitochondrial Ca^{2+}. This activation of mitochondrial uptake results from the proximity of mitochondria and ER: the release of large amounts of Ca^{2+} by the ER creates local 'hotspots' of Ca^{2+} concentration in the vicinity of the mitochondrial uptake channels attaining levels of 20–30 µM, thereby activating the low-affinity mitochondrial uniporter. This constitutes the so-called 'Ca^{2+} microdomains' concept (Rizutto et al., 1993), illustrated in Figure 11.7. Mitochondria are able to sense localized hotspots of Ca^{2+} concentration, whether penetrating from the plasma membrane or from the ER. The rapid uptake of Ca^{2+} by mitochondria stimulates mitochondrial metabolism, as shown much earlier, by activation of three Ca^{2+}-sensitive matrix dehydrogenases (isocitrate dehydrogenase, α-ketoglutarate dehydrogenase and pyruvate dehydrogenase).

It was also observed in earlier studies that mitochondria not only accumulate Ca^{2+} as an alternative to phosphorylation of ADP (Ca^{2+} uptake uncouples phosphorylation from electron transport), but could also accumulate much larger amounts of Ca^{2+} if phosphate was also taken up, resulting in precipitation of Ca^{2+} within the matrix as insoluble hydroxyapatite, visible as electron-dense granules by EM. An unusual feature of these hydroxyapatite deposits is that they fail to become crystalline and remain amorphous even over protracted periods of time. Their presence in mitochondria in a number of disease conditions underlines the role for mitochondria as a sort of safety device, which can enable the cell to survive, if only for a limited period of time, situations of cytoplasmic Ca^{2+} overload.

Finally, the nucleus may also play a role in Ca^{2+} homeostasis: this matter is controversial; the nuclear envelope, which is a prolongation of the ER, is a genuine Ca^{2+} store. It contains a Ca^{2+} pump identical to that of the ER, and channels sensitive to $InsP_3$ and cADPr.

Ca²⁺ AND CELL SIGNALLING

As we mentioned earlier, Ca^{2+} is a component of a number of intracellular signal-transducing pathways, including the phosphoinositide cascade (Figure 11.8). We also mentioned that

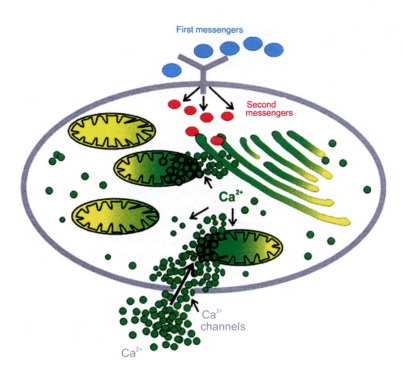

Figure 11.7 The microdomain concept of mitochondrial Ca^{2+} transport. Ca^{2+} penetrating from outside the cell or released from the ER generates local 'hotspots' of high Ca^{2+} concentration in the vicinity of mitochondria sufficient to activate their low-affinity Ca^{2+} uptake system. (From Carafoli, 2002. Copyright (2002) National Academy of Sciences, USA.)

in order to prevent the precipitation of phosphorylated or carboxylated calcium complexes, many of which are insoluble, it is essential to keep the cytosolic levels of Ca^{2+} in unexcited cells extremely low, much lower than that in the extracellular fluid and in intracellular Ca^{2+} stores. This concentration gradient gives cells a superb opportunity to use Ca^{2+} as a trigger—the cytosolic Ca^{2+} concentration can be abruptly increased for signalling purposes by transiently opening Ca^{2+} channels in the plasma membrane or in an intracellular membrane. These increases in intracellular free Ca^{2+} concentration can regulate a wide range of cellular processes, including fertilization, contraction, secretion, learning and memory and ultimately cell death, both apoptotic and necrotic.

Extracellular signals often act by causing a transient rise in cytosolic Ca^{2+} levels, which in turn activates a great variety of enzymes through the action of Ca^{2+}-binding proteins such as calmodulin, as we will discuss in detail later: this triggers such diverse processes as glycogen breakdown, glycolysis and muscle contraction. In the phosphoinositide cascade (Figure 11.8), binding of the agonist to the surface receptor R activates phospholipase C, either, as shown, through a G protein (which uses the energy of GTP hydrolysis to liberate a subunit capable of activating the next partner in the cascade) or alternatively by activating a tryosine kinase. This results in activation of phospholipase C, which hydrolyses phosphatidylinositol-4,5-bisphosphate (PIP_2) to $InsP_3$ (IP_3 in the figure) and diacylglycerol

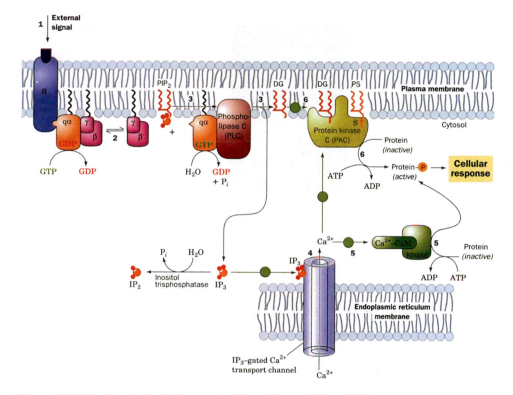

Figure 11.8 The phosphoinositide cascade. (From Voet and Voet, 2004. Reproduced with permission from John Wiley & Sons., Inc.)

(DG). InsP$_3$, as we saw above, stimulates the release of Ca^{2+}, sequestered in the ER, and this in turn activates numerous cellular processes through Ca^{2+}-binding proteins, such as calmodulin. The membrane-associated DG activates protein kinase C to phosphorylate and activate other enzymes, such as glycogen phosphorylase. This step also requires Ca^{2+}.

We now describe the Ca^{2+}-binding proteins and analyse how, in their Ca^{2+}-activated form, they bind to their target sites. Proteins that have been specifically designed to bind Ca^{2+} do so by a number of structural motifs, the best known of which is the helix-loop-helix EF-hand motif (Figure 11.9). This consists of two helices inclined at approximately 90°, flanking a 12-amino-acid loop which coordinates Ca^{2+} via side chain and carbonyl oxygens of five invariant residues. Calmodulin is a well-characterized member of the Ca^{2+}-binding protein family, and consists of two globular domains, each of which has two high-affinity Ca^{2+}-binding sites linked by a seven-turn α-helix. In all EF-hand proteins each calcium is seven-coordinate, with three monodentate Asp or Asn residues, one bidentate Glu residue, one peptide carbonyl group and one bound water molecule (Figure 11.10). Binding of Ca^{2+} to the globular domains of the dumbell-like calmodulin molecule (Figure 11.9) results in a first change in conformation, which does not alter its overall dimensions, but opens up its two Ca^{2+}-binding lobes, exposing hydrophobic residues (essentially Met— unusually calmodulin is rich in Met, having 6% of this residue compared to ~1% in the

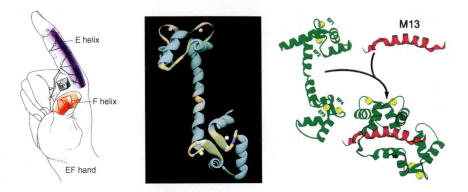

Figure 11.9 (Left) The EF hand helix-loop-helix motif; (centre) rat testes calmodulin. The globular domains each have two Ca^{2+}-binding sites, indicated by white spheres, connected by a seven-turn α-helix; (right) two views of the $(Ca^{2+})_4$ fruit fly calmodulin in complex with its 26-residue target peptide from rabbit skeletal muscle myosin light chain kinase. ((left, centre) From Voet and Voet, 2004. Reproduced with permission from John Wiley & Sons., Inc.; and (right) Carafoli, 2002. Copyright (2002) National Academy of Sciences, USA.)

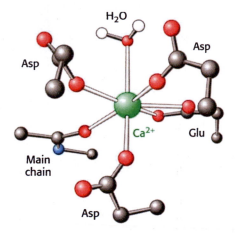

Figure 11.10 The Ca^{2+}-binding site of calmodulin. (From Berg et al., 2002. Reproduced with permission from W.H. Freeman and Co.)

average protein). Then, a second, and much more dramatic conformation change occurs, which collapses the elongated structure of calmodulin protein to a hairpin conformation, which wraps around the binding domain of the target enzyme (Figure 11.9).

Binding of Ca^{2+} to calmodulin transmits changes in intracellular Ca^{2+} to the state of activity of a number of pumps, enzymes, including protein kinases, NAD kinase, phosphodiesterases, and other target proteins. Two targets are particularly interesting, one of which propagates the signal, while the other abrogates it. Calmodulin-dependent protein kinases (CaM kinases) phosphorylate many different proteins, which regulate fuel metabolism, ionic permeability, neurotransmitter synthesis and release. Binding of Ca^{2+}-calmodulin to these CaM kinases activates them and allows them to phosphorylate target proteins

(Figure 11.8). In addition, the activated enzyme phosphorylates itself, and thus remains partly active even after the Ca^{2+} concentration falls and calmodulin is released from the enzyme. In contrast to the CaM kinases, another important target of Ca^{2+}-calmodulin is the plasma membrane Ca^{2+}-ATPase pump, whose activation drives down the Ca^{2+} concentration within the cell, helping to terminate the signal.

One unusual feature of calmodulin is that, unlike other EF-hand proteins, which are usually committed, i.e. only interact with a specific target protein, it interacts with a wide range of targets. Comparisons of amino acid sequences of calmodulin-binding domains of target proteins suggests that calmodulin principally recognizes positively charged amphipathic helices. It is interesting to note that the binding affinity of ~20 segments of these helices bind to Ca^{2+}-calmodulin as tightly as the target proteins themselves. The dramatic conformation change upon binding the target peptide (compare the central and right images of Figure 11.9) shows how the long central helix of uncomplexed calmodulin has unwound and bent to form a globular structure enclosing the helical target polypeptide within a hydrophobic tunnel.

REFERENCES

Berg, J.M., Tymoczko, J.L. and Stryer, L. (2002) *Biochemistry*, 5th edition, Freeman, New York, 974 pp.

Carafoli, E. (1979) The calcium cycle of mitochondria, *FEBS Letts.*, **104**, 1–5.

Carafoli, E. (2002) Calcium signalling: a tale for all seasons, *Proc. Natl. Acad. Sci. U.S.A.*, **99**, 1115–1122.

Carafoli, E. (2003) Historical review: mitochondria and calcium: ups and downs of an unusual relationship, *Trends Biochem. Sci.*, **28**, 175–181.

Carafoli, E. (2004) Calcium-mediated cellular signals: a story of failures, *Trends Biochem. Sci.*, **29**, 371–379.

Carafoli, E. (2005) Calcium—a universal carrier of biological signals, *FEBS J.*, **272**, 1073–1089.

Guerini, D., Coletto, L. and Carafoli, E. (2005) Exporting calcium from cells, *Cell Calcium*, **38**, 281–289.

Rizzuto, R., Brini, M., Murgai, M. and Pozzan, T. (1993) Microdomains with high Ca^{2+} close to IP3-sensitive channels that are sensed by neighboring mitochondria, *Science*, **262**, 744–747.

Voet, D. and Voet, J.G. (2004) *Biochemistry*, 3rd edition, Wiley, New York, Chichester, 1591 pp.

Williams, R.J.P. (2006) The evolution of calcium biochemistry, *Biochim. Biophys. Acta*, **1763**, 1139–1146.

– 12 –

Zinc: Lewis Acid and Gene Regulator

INTRODUCTION

After iron, zinc is the second most abundant trace element in the human body: an average adult has ~3 g of Zn, corresponding to a concentration of about 0.6 mM. Some 95% of zinc is intracellular. It is essential for growth and development in all forms of life, and has been proposed to have beneficial therapeutic and preventative effects on infectious diseases, including a shortening of the length of the common cold in man.

Zinc is found in more than 300 enzymes, where it plays both a catalytic and a structural role. It is the only metal to have representatives in each of the six fundamental classes of enzymes recognized by the International Union of Biochemistry: *oxidoreductases*, such as alcohol dehydrogenase and superoxide dismutase; *transferases*, such as RNA polymerase and aspartate transcarbamoylase; *hydrolases*, such as carboxypeptidase A and thermolysin; *lyases*, such as carbonic anhydrase and fructose-1,6-bisphosphate aldolase; *isomerases*, such as phosphomannose isomerase; and *ligases*, such as pyruvate carboxylase and amino acyl-tRNA synthetases. Not only is zinc involved in enzymes, where it plays both a catalytic and a structural role, but there are also growing numbers of nucleic acid-binding proteins with essential Zn atoms. This demonstrates that Zn is also widely involved in the regulation of the transcription and translation of the genetic message.

The bioinorganic chemistry of zinc is dominated by a number of factors, the most pertinent of which are summarized here. The divalent zinc ion is redox inactive, in contrast, for example, to manganese, iron and copper. Its d^{10} configuration means that not only does it have no *d-d* transitions, and therefore no absorption spectroscopy, but also its complexes are not subject to ligand field stabilization effects such that Zn^{2+} has no ligand field constraints on its coordination geometry. Coordination number and geometry are therefore dictated only by ligand size and charge. This means that zinc can, in principle, adopt highly flexible coordination geometry. However, in most zinc proteins there is a strong preference for tetrahedral coordination, frequently slightly distorted, which enhances both the Lewis acidity of the zinc centre and the acidity of a coordinated water molecule. Only Cu(II) is a better Lewis acid. A few cases of zinc in five-coordinate distorted trigonal bipyramidal geometry have been reported. Since zinc is of borderline hardness, it can bind oxygen (Asp, Gu, H_2O), nitrogen (His) and sulfur (Cys) ligands.

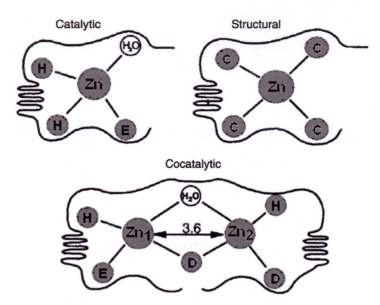

Figure 12.1 Zinc-binding sites in enzymes can be catalytic, structural or cocatalytic. The protein ligands are indicated by smaller filled circles. (From Auld, 2001. With kind permission of Springer Science and Business Media.)

Three types of zinc-binding sites have been recognized in zinc enzymes (Figure 12.1)—catalytic sites, structural sites and cocatalytic sites. Many of these zinc enzymes are peptidases and amidases, involved in the cleavage of amide bonds—they include peptidases, such as thermolysin and carboxy-peptidases; β-lactamases, which destroy the four-member β-lactam rings in penicillins; and matrix metalloproteinases, which degrade extracellular matrix components such as collagen. Zinc enzymes also participate in the cleavage of the phosphodiester bonds in both DNA and RNA, and their role extends beyond catalysis of hydrolytic reactions to include the important lyase, carbonic anhydrase and the oxidoreductase, alcohol dehydrogenase.

We consider, successively, the catalytic role of several classes of mononuclear Zn^{2+} enzymes and then discuss enzymes with di- and tri-nuclear cocatalytic zinc centres, some of which include a metal ion other than zinc. We conclude with a presentation of some of the zinc-based motifs found in proteins involved in the regulation of nucleic acid and protein synthesis.

MONONUCLEAR ZINC ENZYMES

The first zinc enzyme to be discovered was carbonic anhydrase in 1940, followed by carboxypeptidase A some 14 years later. They both represent the archetype of mono-zinc enzymes, with a central catalytically active Zn^{2+} atom bound to three protein ligands, and the fourth site occupied by a water molecule. Yet, despite the overall similarity of catalytic zinc sites with regard to their common tetrahedral $[(XYZ)Zn^{2+}-OH_2]$ structure, these mononuclear zinc enzymes catalyse a wide variety of reactions, as pointed out above. The mechanism of action of the majority of zinc enzymes centres around the zinc-bound water molecule,

Table 12.1

Coordination motifs in catalytic sites of some typical mononuclear zinc enzymes

Carbonic anhydrase	His-X	His-X_{22}	His
β-Lactamase	His-X	His-X_{121}	His
Thermolysin	His-X_3	His-X_{19}	Glu
Carboxypeptidase	His-X_2	Glu-X_{123}	His
Alcohol dehydrogenase	Cys-X_{20}	His-X_{106}	Cys
Alkaline phosphatase	Asp-X_3	His-X_{80}	His
Adenosine deaminase	His-X	His-X_{196}	His

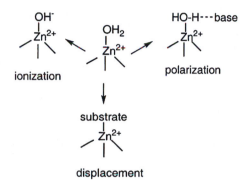

Figure 12.2 The zinc-bound water can be ionized to zinc-bound hydroxide, polarized by a general base to generate a nucleophile for catalysis or displaced by the substrate. (From McCall et al., 2000. Copyright (2000) the American Society for Nutritional Sciences.)

which is best represented as Zn^{2+}-OH_2. What determines the catalytic properties of each enzyme is not only the nature of the donor ligands, but also the distance that separates them in the amino acid sequence of the protein. Typically (Table 12.1), two of the ligands are separated by only 1–3 amino acids, whereas the third ligand is separated by a longer spacer of between 5 and 196 residues.

The mechanism of action of mononuclear zinc enzymes depends on the Zn^{2+}-OH_2 centre, which can participate in the catalytic cycle in three distinct ways (Figure 12.2) – either by ionization, to give zinc-bound hydroxyl ion (in carbonic anhydrase), polarization by a general base (in carboxypeptidase) or displacement of the –OH_2 ligand by the substrate (in alkaline phosphatase). In the first two examples of mononuclear zinc enzymes which we consider, the lyases (carbonic anhydrase) and the hydrolases (carboxypeptidase), the zinc ion functions as a powerful electrophilic catalyst by providing some or all of the following properties: (i) an activated water molecule for nucleophilic attack, (ii) polarization of the carbonyl of the bond to be cleaved and (iii) stabilization of the negative charge that develops in the transition state.

CARBONIC ANHYDRASE

The carbonic anhydrases of mammalian erythrocytes have been the object of extensive study for the last 68 years, and can be considered as the prototype of zinc enzymes that use

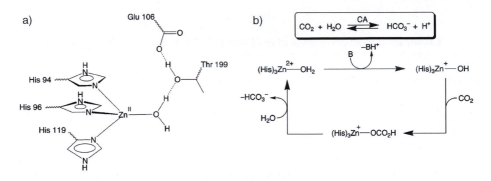

Figure 12.3 (a) The active site of human carbonic anhydrase and (b) a simplified mechanism of action for the enzyme: B = general base, probably His64. (Reprinted with permission from Parkin, 2004. Copyright (2004) American Chemical Society.)

the hydroxyl ion generated by ionization of the Zn^{2+}-OH_2 as the nucleophile for hydrolysis or hydration reactions. The Zn^{2+} ion lies at the bottom of a 15 Å conical cavity and is coordinated to the protein by three invariant His residues, with the remaining tetrahedral site occupied by a water molecule. The water molecule is involved in a hydrogen bond with a Thr residue, which in turn is hydrogen bonded to a Glu residue (Figure 12.3a).

The main features of the mechanism of carbonic anhydrase are illustrated in Figure 12.3b, and involve the following steps: (i) deprotonation of the coordinated water molecule with a pK_a ~7, in a process facilitated by general base catalysis involving His 64. This residue is too far away from the Zn^{2+}-bound water to directly remove its proton, but it is linked to it by two intervening water molecules, forming a hydrogen-bonded network which acts as a proton shuttle. (ii) The zinc-bound hydroxide then carries out a nucleophilic attack on the carbon dioxide substrate to generate a hydrogen carbonate intermediate [(His)$_3$Zn-OCO_2H]$^+$ that (iii) is displaced by H_2O to release bicarbonate and complete the catalytic cycle. The key to understanding the role of the Zn^{2+} ion is that its charge makes the bound water molecule more acidic than free H_2O. This generates a source of nucleophilic zinc-bound OH^-, even at neutral pH.

CARBOXYPEPTIDASES AND THERMOLYSINS

As mentioned earlier, by far the largest number of zinc enzymes are involved in hydrolytic reactions, frequently associated with peptide bond cleavage. Carboxypeptidases and thermolysins are, respectively, exopeptidases, which remove amino acids from the carboxyl terminus of proteins, and endopeptidases, which cleave peptide bonds in the interior of a polypeptide chain. However, they both have almost identical active sites (Figure 12.4) with two His and one Glu ligands to the Zn^{2+}. It appears that the Glu residue can be bound in a mono- or bi-dentate manner. The two classes of enzymes are expected to follow similar reaction mechanisms.

Following myoglobin and lysozyme, bovine carboxypeptidase A was the third protein to have its 3-D structure solved at high resolution. The active site zinc is bound to His69, Glu72 and His196 (Figure 12.4), and to a water molecule, which is displaced when a

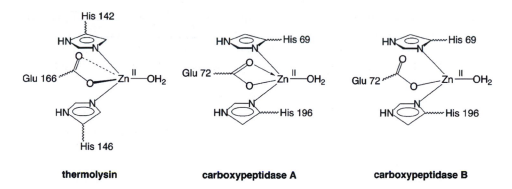

thermolysin   carboxypeptidase A   carboxypeptidase B

Figure 12.4 Active sites of thermolysin and carboxypeptidases A and B. (Reprinted with permission from Parkin, 2004. Copyright (2004) American Chemical Society.)

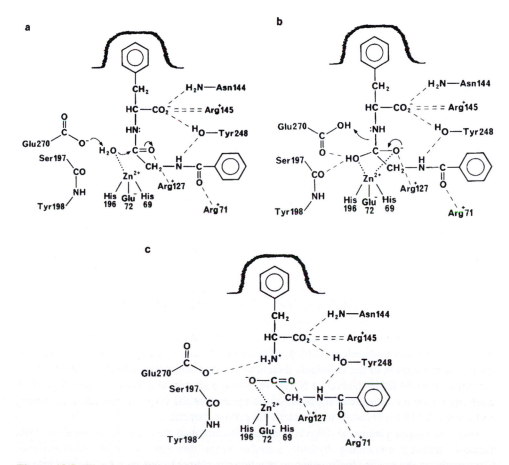

Figure 12.5 'Zinc-hydroxide' reaction mechanism for peptide hydrolysis by carboxypeptidase A. (Reprinted with permission from Lipscomb and Sträter, 1996. Copyright (1996) American Chemical Society.)

bidentate ligand such as glycyltyrosine binds to Zn^{2+} through its carbonyl oxygen and $-NH_2$ terminus. The zinc-bound water molecule is itself hydrogen bonded to Glu270. The mechanism of carboxypeptidase has been controversial for almost 50 years. Two proposals have been advanced, which differ as to whether hydrolysis of the peptide bond occurs by attack of a zinc-bound hydroxide anion or by replacement of the zinc-bound water molecule by the oxygen atom of the peptidyl carbonyl group. In the first of these (Figure 12.5), the so-called 'general base Zn-OH$^-$ pathway', peptide hydrolysis by carboxy-peptidase A can be described in the following steps: (a) in the Michaelis complex the substrate's carbonyl oxygen is hydrogen bonded to Arg127, which facilitates nucleophilic attack by a water molecule promoted by zinc and assisted by Glu270; (b) this generates a tetrahedral intermediate that is stabilized by both Zn^{2+} and Arg127. Glu270 then accepts a proton from a zinc-bound water molecule and transfers the proton to the leaving NH group of the scissile peptide bond to give (c) the final product complex.

ALCOHOL DEHYDROGENASES

Alcohol dehydrogenases are a class of zinc enzymes, which catalyse the oxidation of primary and secondary alcohols to the corresponding aldehyde or ketone by the transfer of a hydride anion to NAD^+ with release of a proton:

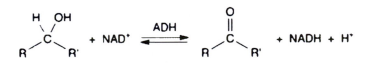

The most extensively studied alcohol dehydrogenases are those of mammalian liver. They are dimeric proteins, with each subunit binding two Zn^{2+} ions, only one of which is catalytically active. This catalytic Zn^{2+} ion has distorted tetrahedral geometry, coordinated to one histidine and two cysteine residues. The non-catalytic zinc plays a structural role and is coordinated tetrahedrally to four cysteine residues.

The essential features of the catalytic cycle are summarized in Figure 12.6. After binding of NAD^+ the water molecule is displaced from the zinc atom by the incoming alcohol substrate. Deprotonation of the coordinated alcohol yields a zinc alkoxide intermediate, which then undergoes hydride transfer to NAD^+ to give the zinc-bound aldehyde and NADH. A water molecule then displaces the aldehyde to regenerate the original catalytic zinc centre, and finally NADH is released to complete the catalytic cycle.

Thus, the role of zinc in the dehydrogenation reaction is to promote deprotonation of the alcohol, thereby enhancing hydride transfer from the zinc alkoxide intermediate. Conversely, in the reverse hydrogenation reaction, its role is to enhance the electrophilicity of the carbonyl carbon atom. Alcohol dehydrogenases are exquisitely stereo specific; and by binding their substrate via a three-point attachment site (Figure 12.7), they can distinguish between the two-methylene protons of the prochiral ethanol molecule.

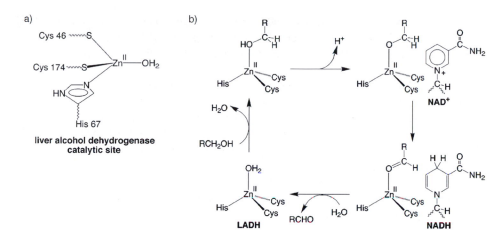

Figure 12.6 (a) The catalytic site of liver alcohol dehydrogenase and (b) the essential features of its catalytic cycle. (Reprinted with permission from Parkin, 2004. Copyright (2004) American Chemical Society.)

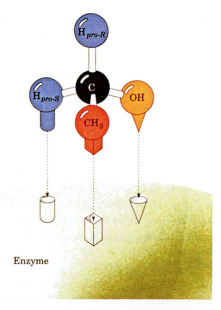

Figure 12.7 Specific attachment of a prochiral centre to an enzyme-binding site enables the enzyme to distinguish between prochiral methylene protons in ethanol. (From Voet and Voet, 2004. Reproduced with permission from John Wiley & Sons., Inc.)

OTHER MONONUCLEAR ZINC ENZYMES

We have already seen the diversity of function in the lyases, hydrolases and oxidoreductases. Several other types of zinc coordination are found in a number of other enzymes, illustrated in Figure 12.8. These include enzymes with the coordination motif $[(His)_2(Cys) Zn^{2+}\text{-}OH_2]$, illustrated by the lysozyme of bacteriophage T7; this group also includes a peptidyl deformylase.

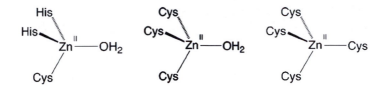

Figure 12.8 Some other active-site coordination motifs in mononuclear zinc enzymes: from left to right bacteriophage T7 lysozyme, 5-aminolaevulinate dehydratase, Ada DNA repair protein. (Reprinted with permission from Parkin, 2004. Copyright (2004) American Chemical Society.)

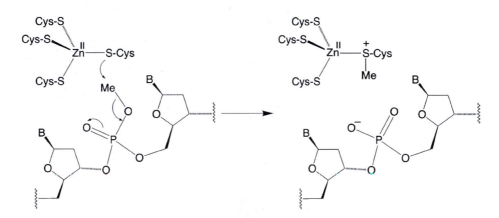

Figure 12.9 Repair of damaged DNA by sacrificial alkylation of one of the zinc cysteine thiolate ligands of the Ada DNA repair protein. (Reprinted with permission from Parkin, 2004. Copyright (2004) American Chemical Society.)

The 5-aminolaevulinate dehydratase (or porphobilinogen synthase), which catalyses the condensation of two molecules of 5-aminolaevulinate to form the pyrrole precursor of the porphyrins (haem, chlorophyll, cobalamines), has the motif [(Cys)$_3$ Zn^{2+}-OH$_2$]. As pointed out earlier (see Chapter 1), this enzyme is the target for saturnism, the Pb toxicity frequently observed among inner city children.

Tetrahedral structural sites typically only involve coordination by the protein, frequently by cysteine residues, as illustrated by the [Cys$_4$Zn^{II}] structural site in liver alcohol dehydrogenase. More recently, a class of zinc proteins and enzymes with tetrahedral 'non-aqua' functional zinc sites have started to emerge in which the activity centres upon the reactivity of a zinc thiolate linkage rather than that of a zinc-bound water molecule. The first to be discovered was the Ada DNA repair protein (Figure 12.8), which has a [(Cys)$_4$Zn] motif and whose function is to repair damage to DNA as a result of methylation. The Ada protein achieves the repair by undergoing sacrificial alkylation of one of the zinc cysteine thiolate ligands (Figure 12.9); this indicates that Ada is not acting as an enzyme, but rather as a reagent (hence its description as a DNA repair protein). Other examples that involve reactivity of zinc cysteine thiolate linkages include methionine synthase and the farnesyl- and geranylgeranyl-transferases, which participate, respectively, in the transfer of farnesyl and geranylgeranyl groups to target proteins.

MULTINUCLEAR AND COCATALYTIC ZINC ENZYMES

It was clear for some time that a number of zinc enzymes required two or more metal ions for full activity, but in the absence of X-ray structural data the location of these metal centres with regard to one another was often uncertain. When the first 3-D structures began to appear, it became clear that the metals were in close proximity. A particular feature of many of these enzymes was the presence of a bridging ligand between two of the metal sites, usually an Asp residue of the protein, which is occasionally replaced by a water molecule. While some of the sites contain only Zn ions, several contain Zn in combination with Cu (in cytosolic superoxide dismutases) Fe (in purple acid phosphatases) or Mg (in alkaline phosphatase and the aminopeptidase of lens).

We will discuss the Cu-Zn superoxide dismutases in greater detail in Chapter 14. Suffice to say here that this is the only cocatalytic site to have a bridging His ligand, and that the role of the Zn ion is thought to be structural, whereas the Cu undergoes redox cycling during catalysis. The importance of the Zn atom is underlined by the observation that the zinc-deficient enzyme is thought to participate in both the sporadic and familial forms of the neurodegenerative disease amyotrophic lateral sclerosis, which is discussed in Chapter 18.

Among the enzymes that contain dinuclear zinc sites we can include the metallo-β-lactamases, a number of aminopeptidases (Figure 12.10) and alkaline phosphatase. β-lactams are the most important class of antibiotics, but bacterial resistance to β-lactams is increasingly observed, severely compromising their efficacy. In many cases, the bacterial resistance is achieved by the production of β-lactamases, enzymes that cleave the four-membered β-lactam ring of many classes of β-lactam antibiotics, including penicillins, cephalosporins and carbapenems. There are four classes of β-lactamases, the most recent to be discovered are the metallo-β-lactamases which have a dinuclear zinc centre. One of the zinc atoms is tetra-coordinate with a $[(His)_3(Zn(\mu\text{-}OH)]$ motif in which the hydroxide ion serves as a bridge to the second Zn site, which has trigonal bipyramidal geometry with a $[(His)(Asp)(X)Zn(OH_2)(\mu\text{-}OH)]$ motif. Whereas most β-lactamases have maximum activity with both Zn sites occupied, the enzyme from *Aeromonas hydrophila* is fully active with only one Zn site occupied. This zinc site has a dissociation constant lower than 20 nM. Binding of a second zinc atom inhibits the enzyme, with a K_i of 46 µM. The mechanism

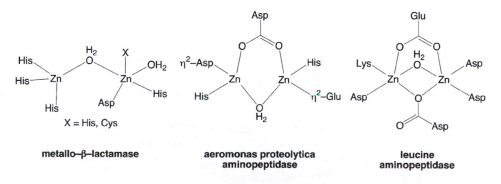

metallo–β–lactamase   **aeromonas proteolytica aminopeptidase**   **leucine aminopeptidase**

Figure 12.10 Metal coordination sites in dinuclear zinc enzymes. (Reprinted with permission from Parkin, 2004. Copyright (2004) American Chemical Society.)

of β-lactamases remains unknown, but it is thought to be analogous to that described above for carboxypeptidases, with the presence of one Zn^{2+} ion that has the characteristics of a catalytic zinc site, while the role, and even the essentiality of the second zinc atom, is not clear.

Aminopeptidases are counterparts to carboxypeptidases, removing N-terminal amino acids. However, unlike the carboxypeptidases, they seem to require two Zn^{2+} ions, which are typically linked by a bridging carboxylate ligand (Figure 12.10).

Several zinc enzymes that catalyse the hydrolysis of phosphoesters have catalytic sites, which contain three metal ions in close proximity (3–7 Å from each other). These include (Figure 12.11) alkaline phosphatase, phospholipase C and nuclease P1. In phospholipase C and nuclease P1, which hydrolyse phosphatidylcholine and single-stranded RNA (or DNA), respectively, all three metal ions are Zn^{2+}. However, the third Zn^{2+} ion is not directly associated with the dizinc unit. In phospholipase C, the Zn–Zn distance in the dizinc centre is 3.3 Å, whereas the third Zn is 4.7 and 6.0 Å from the other two Zn^{2+} ions. All three Zn^{2+} ions are penta-coordinate. Alkaline phosphatase, which is a non-specific phosphomonoesterase, shows structural similarity to phospholipase C and P1 nuclease; however,

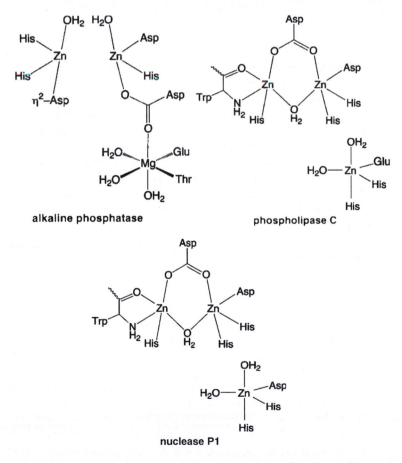

Figure 12.11 Metal coordination sites in trinuclear zinc enzymes. (Reprinted with permission from Parkin, 2004. Copyright (2004) American Chemical Society.)

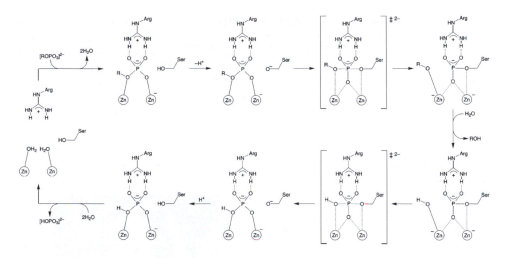

Figure 12.12 Principal steps in the mechanism of alkaline phosphatase. (Reprinted with permission from Parkin, 2004. Copyright (2004) American Chemical Society.)

the third metal ion is Mg^{2+}. One of the Zn^{2+} sites shares a common Asp ligand with the Mg^{2+} site, which is typically hexa-coordinate. The main features of the mechanism of alkaline phosphatase are illustrated in Figure 12.12.

The first step involves binding of the substrate monophosphate $[ROPO_3]^{2-}$, which is coordinated by both Zn^{2+} ions, accompanied by dissociation of the two bound water molecules. The two remaining oxygen atoms of $[ROPO_3]^{2-}$ that are not coordinated to the Zn^{2+} ions interact with an Arg residue. The phosphorus of the substrate undergoes nucleophilic attack by a Ser-O^- in an S_N2 manner, cleaving the P–OR bond with the formation of a phosphoserine intermediate. It is thought that the two Zn^{2+} ions play several roles—by coordinating the RO group one of the zinc atoms activates the RO–P bond, while by coordinating the Ser–OH group, the other zinc centre facilitates deprotonation to generate the highly nucleophilic Ser–O^-. The zinc alkoxide group is hydrolysed, releasing ROH and generating a zinc hydroxide species. This carries out a nucleophilic attack on the phosphoenzyme intermediate, cleaving the phosphorus–enzyme bond with the formation of a bridging phosphate $[HOPO_3]^{2-}$ complex. This catalytic cycle is completed by the displacement of $[HOPO_3]^{2-}$ by water. The importance of the Ser residue is underlined by the observation that its mutation to Gly or Ala decreases activity by 10^4 or 10^5 fold. The two-step reaction sequence explains the retention of configuration, in agreement with experimental observations.

Finally we should briefly mention the purple acid phosphatases, which, unlike the alkaline phosphatases, are able to hydrolyse phosphate esters at acid pH values. Their purple colour is associated with a Tyr to Fe(III) charge transfer band. The mammalian purple acid phosphatase is a dinuclear Fe(II)-Fe(III) enzyme, whereas the dinuclear site in kidney bean purple acid phosphatase (Figure 12.13) has a Zn(II), Fe(III) centre with bridging hydroxide and Asp ligands. It is postulated that the iron centre has a terminal hydroxide ligand, whereas the zinc has an aqua ligand. We do not discuss the mechanism here, but it must be different from the alkaline phosphatase because the reaction proceeds with inversion of configuration at phosphorus.

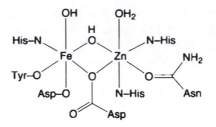

Figure 12.13 Coordination of the dinuclear site in kidney bean purple acid phosphatase. (Reprinted with permission from Parkin, 2004. Copyright (2004) American Chemical Society.)

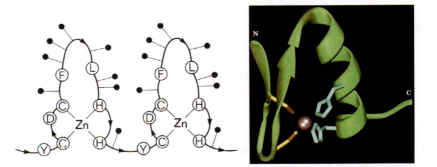

Figure 12.14 (Left) Schematic representation of tandemly repeated zinc finger motif with their tetrahedrally coordinated Zn^{2+} ions. Conserved amino acids are labelled, and the most probable DNA-binding side chains are indicated by balls (from Klug and Rhodes, 1988). (Right) A ribbon diagram of a single zinc finger motif in a ribbon diagram representation. (From Voet and Voet, 2004. Reproduced with permission from John Wiley & Sons., Inc.)

ZINC FINGERS – DNA- AND RNA-BINDING MOTIFS

Aaron Klug discovered the first of the eukaryotic DNA-binding motifs in *Xenopus* transcription factor IIIA (TFIIIA), a protein that binds to the 5S rRNA gene. The resulting complex subsequently binds two other transcription factors and RNA polymerase III, which leads to the initiation of transcription of the 5S rRNA gene. The TFIIIA molecule contains 9 similar ~30 residue long, tandemly repeated modules. Each of these modules contains two invariant Cys residues, two invariant His residues and several conserved hydrophobic residues (Figure 12.14), and a Zn^{2+} ion, which is tetrahedrally coordinated by the invariant Cys and His residues. These so-called *Cys$_2$–His$_2$ Zinc Fingers* occur from 2 to at least 37 times each in a family of eukaryotic transcription factors. In some zinc fingers, the invariant His residues are replaced by Cys residues (*Cys$_2$–Cys$_2$ Zinc Fingers*), while in others six Cys residues bind two Zn^{2+} ions (*Dinuclear Cys$_6$ Zinc Fingers*). Structural diversity is a hallmark of zinc finger proteins, and it appears that the Zn^{2+} ion(s) allow(s) formation of a relatively compact globular DNA-binding domain, precluding the requirement for a much larger hydrophobic core. The *zinc finger proteins* constitute a super family, and ~1% of all mammalian proteins contain this motif.

The global structural details of how zinc fingers bind to double-stranded DNA are well understood. One isolated Cys$_2$-His$_2$ zinc finger, consisting of two β-strands joined by a β-bend followed by an α-helix (Figure 12.14) held together by a tetrahedrally coordinated Zn^{2+} ion, can span three or four consecutive base pairs of the DNA sequence. The multiple zinc fingers follow a right-handed helical path as they wrap around the double helix, with multiple contacts being made with particular nucleotide bases in the major groove. The contacts are frequently made by the side chains of amino acid residues at positions -1, $+2$, $+3$ and $+6$ of the α-helix.

Among the strong preferences that have been observed, it seems that Arg prefers binding to guanosine, Asp to adenosine and cytosine and Leu to thymidine. However, we are not yet sufficiently advanced to define a set of coding rules (i.e. to define the amino acid sequence of one or more zinc fingers, which would bind to a specific DNA sequence).

The *Xenopus* transcription factor IIIA not only acts as an essential RNA polymerase transcription factor for the expression of the 5S rRNA gene, it also binds to the 5S rRNA to form a 7S ribonucleoprotein particle that stabilizes the RNA until it is required for ribosome assembly and facilitates nuclear export of the 5S rRNA. Indeed, it was originally shown to be the protein component associated with 5S rRNA in the 7S particle in *Xenopus* oocytes before it was recognized as a transcription factor. How, we may ask, can this protein not only recognize specific DNA sequences in the 5S rRNA gene upstream region, but also recognize different, but equally specific, sequences in 5S rRNA?

Both biochemical and X-ray crystallographic data show that binding of TFIIIA to the 5S rRNA gene internal control region utilizes all but the fourth and sixth of the nine zinc fingers of the transcription factor (Figure 12.15).

Fingers 1–3 bind to a 10-base-pair 'box C' sequence, wrapping around the major groove of the DNA. Finger 5 binds to a 3-base-pair 'intermediate element' (IE) sequence. Fingers 4 and 6 act as non-DNA-binding spacers to allow recognition of the separated elements. This allows fingers 7–9 to interact with an 11-base-pair 'box A' sequence.

By using a very clever strategy, Klug and his colleagues were able to design a truncated 5S rRNA that binds to zinc fingers 4–6. The resulting X-ray structure shows that finger 4 binds to sequences in loop E, finger 5 binds to backbone atoms in helix V, while finger 6

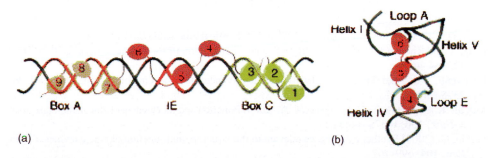

(a) (b)

Figure 12.15 (a) TFIIIA binds to 5S rRNA promoter sequences using zinc fingers 1–3, 5 and 7–9 which recognize box C (green), the IE sequences (red) and box A (orange), respectively. (b) TFIIIA binds to 5S rRNA using primarily zinc fingers 4–6. Finger 4 binds to sequences in loop E, finger 5 to backbone atoms in helix V and finger 6 to sequences in loop A. (From Tanaka Hall, 2005. Copyright 2005, with permission from Elsevier.)

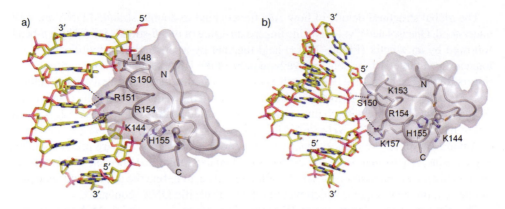

Figure 12.16 DNA and RNA recognition by the fifth zinc finger of TFIIIA. (a) The zinc finger recognizes bases in the major groove of 5S rRNA promoter DNA. (b) The finger recognizes the phosphate groups of 5S rRNA. (From Tanaka Hall, 2005. Copyright 2005, with permission from Elsevier.)

binds to sequences in loop A. The three fingers are associated with identical sequences in their DNA and RNA complexes, but in quite different ways. This is illustrated in Figure 12.16, which compares the binding of the unique member of the three zinc fingers, finger 5, which binds to both DNA and RNA, in its complexes, respectively, with the 5S rRNA promoter DNA and with the 5S rRNA. Whereas in the DNA structure, the interactions are mostly in the major groove and include both specific interactions with bases of the DNA (e.g. Leu148 with a thymine residue of the DNA), in the RNA interactions there are no direct interactions with the bases, and the interactions are essentially with the phosphate groups of the 5S rRNA. Some residues in finger 5, such as Ser150, Lys144, Arg154 and His155, bind to both RNA and DNA, but to different sites in each case. This is in marked contrast to the zinc finger 4 in which a His binds directly to a guanosine and possible also to a second guanosine, and finger 6 in which a Trp stacks onto an adenosine.

REFERENCES

Auld, D.S. (2001) Zinc coordination sphere in biochemical zinc sites, *BioMetals*, **14**, 271–313.

Klug, A. and Rhodes, D. (1988) 'Zinc fingers': a novel protein motif for nucleic acid recognition, *TIBS*, **12**, 464–469.

Lipscomb, W.N. and Sträter, N. (1996) Recent advances in zinc enzymology, *Chem. Rev.*, **96**, 2375–2433.

McCall, K.A., Huang, C.-C. and Fierke, C.A. (2000) Function and mechanism of zinc metalloenzymes, *J. Nutr.*, **130**, 1437S–1446S.

Parkin, G. (2004) Synthetic analogues relevant to the structure and function of zinc enzymes, *Chem. Rev.*, **104**, 699–767.

Tanaka Hall, T.M. (2005) Multiple modes of RNA recognition by zinc finger proteins, *Curr. Opin. Struct. Biol.*, **15**, 367–373.

Voet, D. and Voet, J.G. (2004) *Biochemistry*, 3rd edition, Wiley, New York, Chichester, 1360 pp.

– 13 –

Iron: Essential for Almost All Life

GOLD is for the mistress—silver for the maid—
Copper for the craftsman cunning at his trade."
"Good!" said the Baron, sitting in his hall,
"But Iron—Cold Iron—is master of them all."

Rudyard Kipling "Cold Iron"

What characterizes the Three Ages of Man (Stone, Bronze and Iron) is that they are defined by the materials out of which weapons were manufactured. The Stone Age began ~ 2 million years ago, and in the absence of metals, represented the use of stone as the sole means of making martial instruments. It was superseded by the Bronze Age during which metals, initially copper, began to be used for the manufacture of weapons. The use of copper spread from eastern Anatolia (modern-day Turkey) through Mesopotamia and the Middle East from 4000 to 3000 BC. True bronze, an alloy[1] of copper and tin, was used only rarely initially, but during the second millennium the use of true bronze increased greatly; the tin deposits of Cornwall were extensively exploited and were responsible for a large part of the bronze objects produced during this period. The Bronze Age was also marked by important inventions—notably the wheel and the ox-drawn plough. However, by around 1200 BC the ability to heat and forge another metal, iron, brought the Bronze Age to an end. The final technological and cultural stage in the three ages of man, the Iron Age, took off between 1200 and 1000 BC, and the export of knowledge of iron metallurgy and with it the production of iron objects expanded rapidly and widely. However, while we have innumerable artefacts from both the Stone and the Bronze Ages, little remains of the Iron Age on account of the poor stability of iron in the face of oxygen and water (rust is not a very practical way of preserving historical relics!).

INTRODUCTION

Iron, element 26 in the periodic table, is the fourth most abundant element of the earth's crust and, after aluminium, the second most abundant metal. In the middle of the first transition

---

[1] Alloy—an amalgam typically involving the mixture of a less valuable (base) metal with a more valuable (noble) metal, as in the decreasing amounts of silver and gold in coinage.

series, iron has the possibility of existing in various oxidation states (from $-II$ to $+VI$), the principal being II (d^6) and III (d^5), although a number of iron-dependent monooxygenases generate high-valent Fe(IV) or Fe(V) intermediates during their catalytic cycle. The suitability of iron for catalysis in living organisms comes from the extreme variability of the Fe^{2+}/Fe^{3+} redox potential, which can be fine-tuned by an appropriate choice of ligands, to encompass almost the entire biologically significant range of redox potentials, from approximately -0.5 V to approximately $+0.6$ V. Fe^{3+} is quite insoluble in water ($K_{sp} = 10^{-39}$ M and at pH 7.0, $[Fe^{3+}] = 10^{-18}$ M) and significant concentrations of water-soluble Fe^{3+} species can be attained only by strong complex formation, whereas in contrast, Fe^{2+} is extremely water soluble. Fe(III) with an ionic radius of 0.067 nm and a charge of 3^+ is a "hard" acid and will prefer "hard" oxygen ligands such as phenolate and carboxylate rather than imidazole or thiolate. Fe(II) with an ionic radius of 0.083 nm and a charge of only 2^+ is on the borderline between "hard" and "soft" favouring nitrogen (imidazole and pyrrole) and sulfur ligands (thiolate and methionine) over oxygen ligands.

Coordination number 6, associated with octahedral geometry, is most frequently found for both Fe(II) and Fe(III), although four- (tetrahedral) and particularly five-coordinate complexes (trigonal bipyramidal or square pyrimidal) are also found. For octahedral complexes, two different spin states can be observed. Whereas strong-field ligands, such as F^- and OH^-, where the crystal field splitting is high and hence electrons are paired, give low-spin complexes, weak-field ligands, such as CO and CN^-, favour a maximum number of unpaired electrons and give high-spin complexes. Changes of spin state affect ion size of both Fe(II) and Fe(III), the high-spin ion being significantly larger than the low-spin ion. For both oxidation states only high-spin tetrahedral complexes are formed. Both oxidation states are Lewis acids, particularly the ferric state. For more details of iron biochemistry see Crichton, 2001.

IRON AND OXYGEN

When life began on earth, with its essentially reducing atmosphere, the natural abundance and bioavailability of iron (mostly in the ferrous state) and its redox properties predisposed it to play a crucial role in the first stages of evolution. However, with the appearance of photosynthetic cyanobacteria $\sim 10^9$ years ago, dioxygen was evolved into the earth's atmosphere. It probably required 200–300 million years for oxygen to attain a significant concentration in the atmosphere, since at the outset the oxygen produced by photosynthesis would have been consumed by the oxidation of ferrous ions in the oceans. Once dioxygen had become a dominant chemical entity, iron became poorly bioavailable due to the precipitation of ferric hydroxides as is clearly indicated by the Precambrian deposits of red ferric oxides laid down in the geological strata at that time. In parallel with the loss of iron bioavailability, copper became more available as the oxidation of insoluble Cu(I) led to formation of soluble Cu(II). The aerobic world now required a new redox-active metal with E_oM^{n+1}/M^n from 0 to 0.8 V. Copper was ideally suited for this role, and, as we will see in the next chapter, began to be used in enzymes with higher redox potentials (such as the dicopper centre in laccase and the mixed iron–copper centre in cytochrome oxidase) to take advantage of the oxidizing power of dioxygen. The interaction of iron (and copper) centres and oxygen is of paramount importance in biological inorganic chemistry, and we have summarized some of the main features in Figure 13.1.

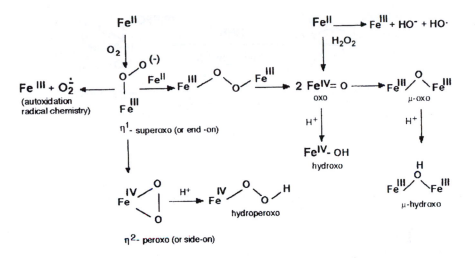

Figure 13.1 Iron–oxygen chemistry. (From Crichton and Pierre, 2001. With kind permission of Springer Science and Business Media.)

When a single electron is accepted by the ground-state O_2 molecule, it will form the superoxide radical, O_2^-. Addition of a second electron to O_2^- gives the peroxide ion O_2^{2-} with no unpaired electrons. At physiological pH O_2^{2-} will immediately protonate to give hydrogen peroxide, H_2O_2. The third reactive oxygen species found in biological system is the hydroxyl free radical. As Fenton first observed in 1894 (Fenton, 1894), a simple mixture of H_2O_2 and an Fe(II) salt produces the ·OH radical (13.1):

$$Fe^{2+} + H_2O_2 \rightarrow Fe^{3+} + \cdot OH + OH^- \tag{13.1}$$

In the presence of trace amounts of iron, superoxide can then reduce Fe^{3+} to molecular oxygen and Fe^{2+}. The sum of this reaction (13.2) plus the Fenton reaction (13.1) produces molecular oxygen, hydroxyl radical and hydroxyl anion from superoxide and hydrogen peroxide, in the presence of catalytic amounts of iron. This is the Haber–Weiss reaction (13.3), originally described by Haber and Weiss (1934), but manifestly impossible from thermodynamical considerations in the absence of catalytic amounts of redox metals such as iron and copper:

$$Fe^{3+} + O_2^- \rightarrow Fe^{2+} + O_2 \tag{13.2}$$

$$O_2^- + H_2O_2 \rightarrow O_2 + \cdot OH + OH^- \tag{13.3}$$

This capacity of iron (and copper) to transform oxygen into highly toxic products is the origin of the so-called *oxygen paradox*. As we saw in Chapter 5, the advent of respiratory pathways instead of fermentation represented an almost 20-fold increase in the energetic yield of intermediary metabolism, which enabled organisms with access to such pathways to flourish. The down side (inexorably, for everything which brings advantages, there are inevitable disadvantages) was that the utilization of the burning capacity of oxygen came along with a serious fire warning, namely the potential havoc that reactive oxygen species

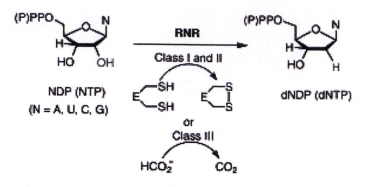

Figure 13.2 The reaction catalysed by ribonucleotide reductase (RNR). (From Stubbe et al., 2001. Copyright 2001, with permission from Elsevier.)

would wreak. We will return to the toxicity of reactive oxygen species particularly in neurodegenerative diseases in Chapter 18 (metals in brain function).

THE BIOLOGICAL IMPORTANCE OF IRON

It would be difficult to underestimate the biological importance of iron for almost all living organisms, but it can be illustrated by three simple examples. *E. coli* has almost 50 genes for proteins involved in iron uptake—six distinct siderophore-mediated Fe^{3+} transport systems, one for iron uptake from ferric citrate, and one Fe^{2+} transport system, yet it synthesizes only one siderophore, enterobactin (as we saw in Chapter 7). When blue-green algal blooms occur in lakes the determining factor in which algal species takes over is the efficacy of its capacity to chelate iron. And when a clinician wishes to determine the potential for growth of a mammalian tumour, he measures the density of transferrin receptors, which are required for iron uptake and hence cellular growth and division. While iron readily undergoes electron transfer and acid–base reactions, it also has the capacity to participate in one electron transfer (i.e. free radical) reactions. One such free radical reaction, essential for DNA synthesis, is the reduction of ribonucleotides to the corresponding deoxyribonucleotides, catalysed by ribonucleotide reductases (RNRs), all of which depend on different metal cofactors for their catalytic activity. Since all known cellular life forms store their genetic information in DNA, RNRs must be present in all growing cells of all living organisms. They all catalyse the conversion of adenine, uracil, cytosine and guanine nucleotides to deoxynucleotides, cleaving a 2′ carbon–hydroxyl bond with formation of a 2′ carbon–hydrogen bond (Figure 13.2). The hydrogen is derived from water and replaces the hydroxyl with retention of configuration. There are three classes of RNRs: class I, which has a stable tyrosyl radical adjacent to a diferric cluster (present in a great number of organisms including *E. coli*, yeast and man); class II, which uses adenosylcobalamine, a Co(III) metallo-cofactor (for example, in Lactobacillus species[2]); and class III, which is

_____

[2] This may explain why this family of bacteria is found in dairy products where the presence of lactoferrin makes iron availability problematic. Class II RNRs is also found in some archaebacteria.

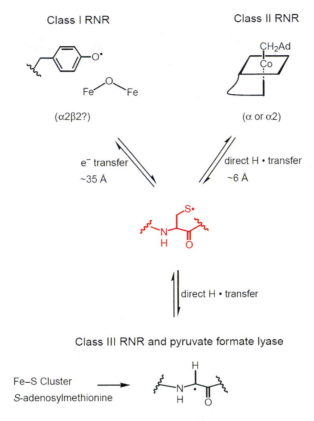

Figure 13.3 The three classes of RNRs utilize their metallo-cofactors to generate an active site thiyl radical (S). The diferric-tyrosyl radical in Class I enzymes is on the R2 subunit, at best 35 Å from the thiyl radical (on subunit R1). (From Stubbe et al., 2001. Copyright 2001, with permission from Elsevier.)

found in *E. coli* grown under strictly anaerobic conditions and has a glycyl radical, which is generated by an FeS cluster together with S-adenosyl methionine (SAM) (Figure 13.3).

In spite of the diversity of the metallo-cofactors, it has become apparent that all three classes of RNR have a common active site, and that the role of the metallo-cofactors is to generate a transient thiyl radical which then initiates the radical-dependent ribonucleotide reduction process (Figure 13.3). One of the fascinating particularities of the Class I enzymes is that, even in the most favourable models, the tyrosyl radical (in the R2 subunit) is at least 35 Å from the putative thiyl radical (in the adjacent R1 subunit) at the substrate binding site. This is well beyond the reach of pure electron tunnelling processes found in most biological electron transfer reactions. It has been proposed that the radical is transferred by proton-coupled electron transfer via a chain of hydrogen-bonded amino acid side chains. Pyruvate formate lyase, described later, generates a radical from S-adenosyl methionine. For a review of ribonucleotide reductases see Stubbe et al., 2001; Kohlberg et al., 2004.

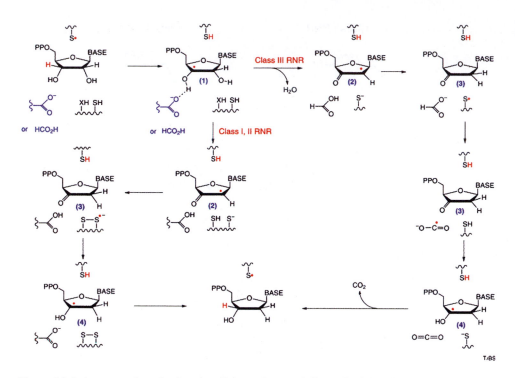

Figure 13.4 A proposed mechanism for all three classes of ribonucleotide reductases. Classes I and II RNRs require an active site Glu residue and a pair of redox-active Cys. Class III RNRs lack the Glu and one of the Cys, and use formate as the reductant. (From Stubbe et al., 2001. Copyright 2001, with permission from Elsevier.)

A working model for the mechanism of ribonucleotide reduction by all three classes of RNRs is presented in Figure 13.4. In all cases, the metallo-cofactor generates a transient $S^{\cdot}$ in the active site, which initiates the reaction by abstracting a hydrogen atom from the 3′-position of the substrate, producing a 3′-nucleotide radical (1). Loss of water generates the 3′-keto-2′-deoxynucleotide radical (2). In Classes I and II RNRs, this intermediate is reduced in a two-electron-coupled proton transfer process by a pair of Cys residues. This first gives a 3′-keto-2′-deoxynucleotide intermediate (3), which is thought to generate a 3′-deoxynucleotide radical (4) in a step which is assisted by a Glu residue, which protonates the 3′-ketone group concomitant with its reduction. In Class III RNRs, intermediate (2) is reduced by formate in two-electron-coupled proton transfers to generate (4) with production of CO_2. The intermediate (4) in all classes of RNRs is then reduced to give the ultimate deoxynucleotide product, regenerating the thiyl radical at the active site.

BIOLOGICAL FUNCTIONS OF IRON-CONTAINING PROTEINS

We have chosen here a classification of iron metalloproteins based on the coordination chemistry of the metal. This enables us to more easily appreciate the diversity of biochemical functions that iron can play viewed through the ligands, which bind it to the protein. We consider successively:

(i)   haemoproteins in which an iron porphyrin is incorporated into different apo-proteins to give O_2 carriers, O_2 activators or alternatively electron transfer proteins.
(ii) iron–sulfur proteins (ISP), many of which are involved in electron transfer.
(iii) non-haem, non-iron sulfur, iron-containing proteins, which include proteins of iron storage and transport, already described in Chapter 8.

Since the roles played by iron are so diverse, we clearly cannot cover all of them, so we have chosen only to give a small sample of selected illustrations.

HAEMOPROTEINS

Oxygen transport

Oxygen transport and storage in multi-cellular oganisms, whether they be mammals, insects or worms, is assured by haemoglobins and myoglobins. These were the first proteins to have their X-ray crystal structures determined by John Kendrew and Max Perutz, for which they received the Nobel Prize for Chemistry in 1962; shortly after when the structures of insect and lamprey haemoglobins were determined, it became clear that all these oxygen-binding proteins share a common tertiary structure, known as the globin fold. This is illustrated in Figure 13.5 by sperm whale myoglobin. However, whereas the monomeric myoglobin with a single haem has a hyperbolic oxygen-binding curve, the tetrameric haemoglobin with four haem groups has a sigmoidal oxygen-binding curve (Figure 13.5). This reflects the cooperativity of oxygen binding—the fourth O_2 molecule binds with 100-fold greater affinity than the first. We know that, like other allosteric proteins, haemoglobin

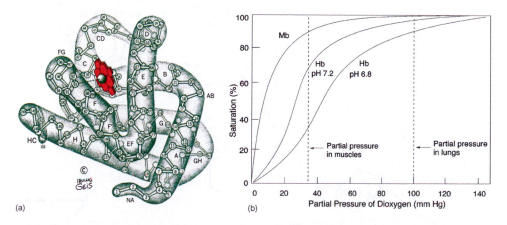

(a)   (b)

Figure 13.5 (a) The structure of sperm whale myoglobin. (From Voet and Voet, 2004. Reproduced with permission from John Wiley & Sons., Inc.) (b) The oxygen-binding curves of myoglobin and haemoglobin. (Reprinted with permission from Collman et al., 2004. Copyright (2004) American Chemical Society.)

exists in two distinct and different conformations, corresponding to the T (deoxy) and R (oxy) states. Indeed the conformations of oxy- and deoxyhaemoglobins are so great that crystals of deoxyhaemoglobin break when oxygen is introduced. But since the haem groups are so far apart in the haemoglobin structure, the positive cooperativity must be transmitted by the protein itself. What might be the trigger that would signal to a neighbouring subunit that oxygenation had taken place?

The haem is tightly bound to the protein in a hydrophobic pocket formed principally by helices E and F and by a single coordinate bond between the imidazole of His F8, termed the *proximal* histidine (Figure 13.6), and the ferrous iron, which is some 0.6 Å out of the plane of the domed porphyrin ring. A second His residue, His E7 (the *distal* histidine), is too far away from the iron atom to coordinate with it in the deoxy state.

A comparison of the deoxy- and oxyhaemoglobin structures reveals a number of important differences. Whereas in the T (deoxy) state the Fe atom is out of the haem plane, on oxygenation it moves into the plane of the now undomed porphyrin, pulling the proximal His F8 and the F helix, to which it is attached (Figure 13.7), as we will see shortly, thereby triggering the T to R transition. The major differences between R and T conformations are

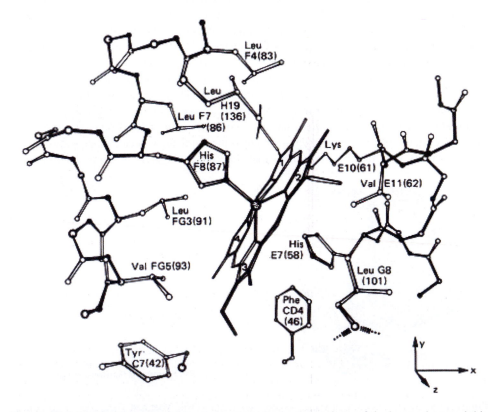

Figure 13.6 The haem group and its environment in the deoxy form of the human haemoglobin α-chain. Only selected side chains are shown, and the haem 4 propionate is omitted for clarity. (From Gelin and Karplus, 1977. Copyright (1993) National Academy of Sciences, USA.)

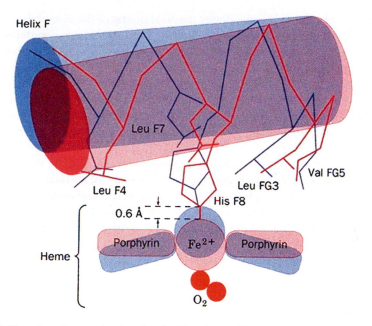

Figure 13.7 The triggering mechanism for the T to R transition in haemoglobin. (From Voet and Voet, 2004. Reproduced with permission from John Wiley & Sons., Inc.)

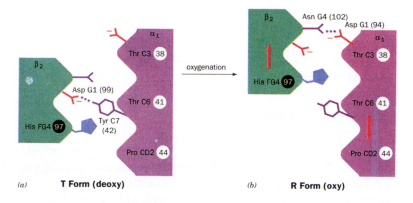

Figure 13.8 The α_1–β_2 interface in (a) human deoxyhaemoglobin and (b) oxyhaemoglobin. (From Voet and Voet, 2004. Reproduced with permission from John Wiley & Sons., Inc.)

at the α_1–β_2 (and the corresponding α_2–β_1) subunit interfaces, which consist of the C helix of α-subunits and the FG interface of the β-subunits. These fit to one another in two distinct conformations, which correspond to a 6 Å relative shift at the interface. In the T state His FG4 is in contact with Thr C6, whereas in the R state the same His is in contact with Thr C3, one turn further back along the C helix (Figure 13.8). Another series of very important differences concern a network of salt bridges at subunit–subunit interfaces that stabilize the T state, but are broken in the more relaxed R state.

In oxymyoglobin and oxyhaemoglobin, the N–H proton of the distal histidine E7 in the O_2 binding pocket (Figure 13.6) forms a hydrogen bond with the iron-coordinated dioxygen molecule and imposes an angular bend on the dioxygen molecule. In carbon monoxide adducts of myoglobin and haemoglobin, the steric hindrance caused by the distal histidine results in a less favourable binding geometry (CO prefers a linear coordination). Thus, CO, a poison present both in tobacco smoke and in automobile exhausts, but also produced in the normal biological degradation of haem, binds only ~ 250 times more tightly than O_2 to both myoglobin and haemoglobin, whereas the affinity of free haem for carbon monoxide is much greater.

What is responsible then for the change in coordination geometry at the iron atom upon oxygenation? Resonance Raman spectroscopy of oxyhaemoglobin shows an O–O stretching band at ~1105 cm^{-1} which is characteristic of coordinated superoxide ion. This implies electron transfer from iron (II) to dioxygen, such that we could consider oxyhaemoglobin and oxymyoglobin as ferric-superoxide complexes, in which the superoxide is stabilized by hydrogen bonding to the distal histidine proton. In deoxyhaemoglobin Fe(II) is high spin, and its covalent radius is too large to allow it to fit into the plane of the tetrapyrrole, which explains both the doming of the porphyrin and the out-of-plane location of the iron. The hexa-coordinate Fe(III)-superoxo is low spin, with a decreased covalent radius, and the iron atom can now move into the plane of the porphyrin. Through the movement of the proximal His F8 and the F helix the T to R transition is set in motion, perhaps first involving rupture of intersubunit salt bridges and progressively the high-affinity R state predominates. In this mechanism the salt bridges play three roles—(i) they stabilize the T quaternary structure relative to R; (ii) they lower the oxygen affinity in the T state because of the energy required to break them on oxygen binding; (iii) they release protons when they are broken, which explains the almost century old effect discovered by the physiologist father of the atomic physicist Niels Bohr, Christian, namely that the affinity of haemoglobin for oxygen is lowered when the pH decreases (Figure 13.5b).

Activators of molecular oxygen

A large number of haem enzymes which include cytochrome oxidases, peroxidases, catalases and cytochrome P-450s are characterized by a penta-coordinate geometry in which the sixth site of the metal centre can either bind molecular oxygen, hydrogen peroxide or, in the case of cytochrome P-450s, even form iron–carbon bonds with the substrate. The high-spin iron–porphyrin system can go to the radical cation state at a redox potential close enough to that of the couple Fe(IV)/Fe(III) to allow a ferryl type of iron to participate in chemical reactions such as the activation of oxygen or oxidation of molecules at the expense of hydrogen peroxide. We have chosen to illustrate this type of enzyme by the haem-copper oxidases, the terminal components of the respiratory chains in aerobic organism. Cytochrome c oxidases (CcOs) are membrane-bound enzymes which couple electron transfer reactions, associated with the reduction of dioxygen to water (13.4), to proton pumping across the membrane.

$$O_2 + 4H^+ + 4e^- \rightarrow 2H_2O \qquad (13.4)$$

In mammals, CcO spans the mitochondrial inner membrane and catalyses the reduction of molecular dioxygen to water at the rate of up to 250 molecules of O_2 per second. The energy released in this process is coupled to the translocation of protons, which in turn contributes to the chemiosmotic gradient required for ATP synthesis. Since the electrons and protons are taken up from opposite sides of the membrane, the reaction results in a net charge separation across the membrane which, together with the coupled proton pumping corresponds to the overall translocation of two positive charges across the membrane per electron transferred to O_2 from the negative (N) side to the positive (P) side of the membrane. In CcOs the electron donor is cytochrome c, which donates an electron to a dinuclear copper centre (Cu_A). Electrons from Cu_A are then transferred consecutively to the catalytic site of CcO, the haem group of haem a, and to the catalytic site of CcO, the dinuclear haem-copper centre (haem a_3-Cu_A). A tyrosine residue, Y (I-288), which is covalently cross-linked to His240, one of the Cu_B ligands, is thought to be redox active, and is also included as part of the active site. The structure of the four-subunit CcO from *Rhodobacter sphaeroides* is presented in Figure 13.9a, while a more detailed view of the redox-active cofactors and amino acid residues on the proton transfer pathways is given in Figure 13.9b.

Part of the energy released in the redox reaction is conserved by vectorial transfer of protons across the membrane from the N-side to the P-side, thereby maintaining an electrochemical proton gradient that is used for synthesis of ATP. Two proton transfer pathways leading from the N-side surface toward the binuclear centre have been identified. In cytochrome c oxidase from *R. sphaeroides*, one of the pathways (D-pathway) starts with Asp(I-132) and leads to Glu(I-286). Since the D-pathway is used for both the substrate protons, which are transferred to the catalytic site and pumped protons, which are transferred to a proton-accepting group in the exit pathway, there must be a branching point within the pathway from where protons can be transferred either toward the dinuclear centre or toward the output side of the enzyme. This is thought to be at Glu(I-288). The other pathway (K-pathway) starts at the N-side surface at Glu(II-101) and leads via a highly conserved Lys(I-362) and Thr(I-359) to the dinuclear centre.

The individual steps of oxygen binding and its subsequent reduction have been followed spectroscopically in a time-resolved manner using flash-flow techniques. The initial binding of dioxygen to ferrous haem a_3 takes place within 10 μs (Figure 13.10), forming the intermediate state **A**. In the next step, the O–O bond is broken, forming an oxoferryl state on haem a_3 (denoted $\mathbf{P_R}$) and a hydroxide ion at Cu_B with a time constant of 50 μs. In the course of this step, an electron is transferred from haem a to the catalytic site. The other three electrons required for dioxygen cleavage are already in the catalytic site, as is the proton required for transfer from Y (I-288) to generate Cu_B^{2+}-OH. In the $\mathbf{P_R}$ state there is an uncompensated negative charge at the catalytic site, which is neutralized by a proton in the next step, the formation of the intermediate **F** with a time constant of 100 μs. Concomitantly with internal proton transfer from the bulk solution there is electron transfer from Cu_A to haem a. This results, with a time constant of ~1.2 ms, in the formation of the fully oxidized enzyme, **O**, which requires proton transfer to the oxoferryl group at haem a_3, to form a hydroxyl group, proton transfer from the bulk solution and electron transfer from Cu_A and haem a to the catalytic dinuclear site.

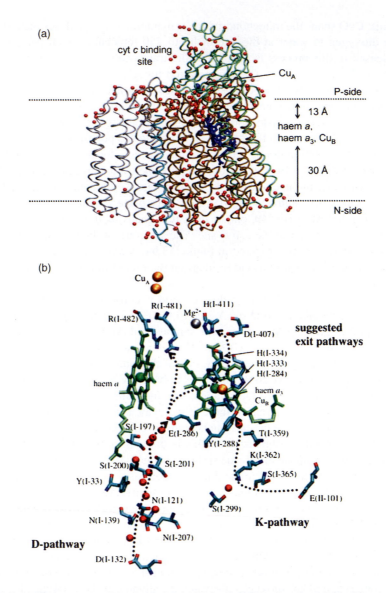

Figure 13.9 (a) The structure of the four subunits of the CcO from *R. sphaeroides*; (b) a more detailed view of the redox-active cofactors and amino acid residues in the proton transfer pathways (dotted arrows). (From Namslauer and Brzezinski, 2004. Copyright 2004, with permission from Elsevier.)

Electron transport proteins

The third class of haemoproteins, with hexa-coordinate low-spin iron, are the cytochromes. First discovered by McMunn in 1884, they were rediscovered in 1925 by David Keilin. Using a hand spectroscope he observed the characteristic absorption (Soret) bands of the three cytochromes *a*, *b* and *c* in respiring yeast cells, which disappeared upon oxygenation.

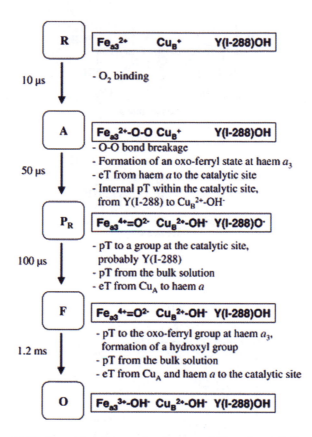

Figure 13.10 A scheme illustrating the reaction between fully reduced CcO and oxygen. (From Namslauer and Brzezinski, 2004. Copyright 2004, with permission from Elsevier.)

He correctly concluded that they transferred electrons from substrate oxidation to the terminal oxidase, cytochrome c oxidase as we know it today, and which we have discussed in the section above. The cytochromes vary in the nature of their haem group—b-type haems have protoporphyrin IX, as in haemoglobin, c-type cytochromes have protoporphyrin IX in which the vinyl groups form covalent thioether bonds with cysteine residues of the protein, while a-type haems contain a long hydrophobic tail of isoprene units attached to the porphyrin, as well as a formyl group in place of a methyl substituent. The axial ligands of haem iron vary with cytochrome type (Figure 13.11). In cytochromes *a* and *b*, both ligands are usually His residues, whereas in cytochrome *c*, one is His and the other is often Met. They have a wide distribution, functioning as electron transporters in mitochondria, chloroplasts, endoplasmic reticulum and bacterial redox chains. The iron in all cytochromes can alternate between an oxidized Fe(III) low-spin state with a single unpaired electron and a formal charge of $+1$, and a reduced Fe(II) low-spin form with no unpaired electrons and a net charge of zero. Since the iron remains low spin, electron transfer is greatly facilitated.

Cytochromes, as components of electron transfer chains, must interact with the other components, accepting electrons from reduced donor molecules and transferring them to

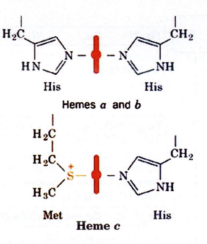

Hemes *a* **and** *b*

Heme *c*

Figure 13.11 Axial ligands to the haem groups of cytochromes a, b and c. (From Voet and Voet, 2004. Reproduced with permission from John Wiley & Sons., Inc.)

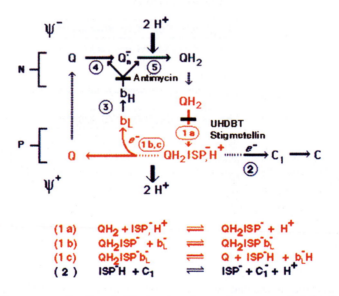

(1 a)	$QH_2 + ISP,H^+$	$\rightleftharpoons$	$QH_2ISP^- + H^+$
(1 b)	$QH_2ISP^- + b_L^-$	$\rightleftharpoons$	$QH_2ISP^-b_L^-$
(1 c)	$QH_2ISP^-b_L^-$	$\rightleftharpoons$	$Q + ISP^-H + b_L^-H$
(2)	$ISP^-H + C_1$	$\rightleftharpoons$	$ISP^- + C_1^- + H^+$

Figure 13.12 The protonmotive Q cycle. Electron transfer reactions are numbered and circled. Dashed arrows designate movement of ubiquinol or ubiquinone between centres N and P and of the ISP between cytochrome b and cytochrome c_1. Solid black bars indicate sites of inhibition by antimycin, UHDTB and stigmatellin. (From Hunte et al., 2003. Copyright 2003, with permission from Elsevier.)

appropriate acceptors. In the respiratory chain of the mitochondria, the cytochrome bc_1 (ubiquinol:cytochrome c oxidoreductase) complex transfers electrons coming from Complex I (and II) to cytochrome c. The bc_1 complex oxidizes a membrane-localized ubiquinol: the redox process is coupled to the tranlocation of protons across the membrane, in the so-called protonmotive Q cycle, which is presented in a simplified form in Figure 13.12. This cycle was first proposed by Mitchell (1975) some years ago and substantially confirmed experimentally since then. The Q cycle in fact consists of two turnovers of QH_2 (Figure 13.12). In both turnovers, the lipid-soluble quinol transfers one electron to the ISP, one electron to one of the two cytochrome b haems (b_L), and two protons to the intermembrane space. In both of the Q cycles, the cytochrome b_L reduces cytochrome b_H while the Reiske iron–sulfur protein (ISP) reduces cytochrome c_1. The cytochrome c_1 in turn reduces the water-soluble cytochrome c, which transfers its electrons to the terminal oxidase, cytochrome c oxidase, described above. In one of the two Q cycles, reduced cytochrome b_H reduces Q to the semiquinone, which is then reduced to QH_2 by the second reduced cytochrome b_H. The protons required for this step are derived from the matrix side of the membrane. The overall outcome of the two CoQ cycles (13.5) (m—matrix; ims—intermembrane space) is:

$$CoQH_2 + 2cytc_1(Fe^{3+}) + 2H + (m) \rightarrow CoQ + 2cytc_1(Fe^{2+}) + 4H^+(ims) \qquad (13.5)$$

The cytochrome bc_1 complex in eukaryotes is a homodimeric, multi-subunit entity (Figure 13.13a). Each monomer has three catalytic subunits: a cytochrome b, with two b-type haems, one Rieske ISP containing a [2Fe-2S] cluster and one cytochrome c_1, with

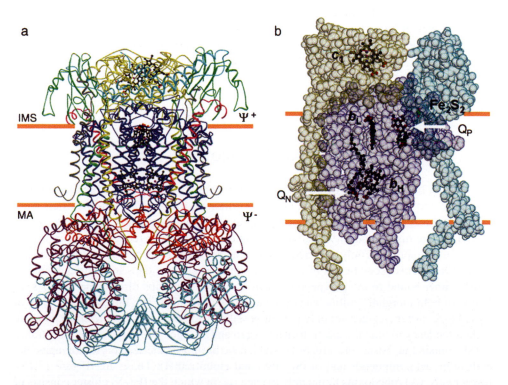

Figure 13.13 The structure of the yeast cytochrome bc_1 complex: (a) the homodimeric complex of the catalytic subunits cytochrome b (blue), Rieske protein (green) and cytochrome c_1 (yellow) with their cofactors and the six additional subunits. Part b shows the catalytic subunits of one functional unit in the same orientation. (From Hunte et al., 2000. Copyright 2003, with permission from Elsevier.)

a c-type haem. In addition, there are up to eight additional protein subunits. The central domain of the complex is formed by eight transmembrane helices of cytochrome *b* per monomer. Both cytochrome c_1 and the Rieske protein have their catalytic domains located in the intermembrane space as shown in Figure 13.13b. It has been proposed that the catalytic domain of the Rieske protein acts as a mobile electron shuttle between the cytochromes *b* and c_1. In the first step of the Q cycle (Figure 13.12) ubiquinol is oxidized at the centre P in a concerted reaction that transfers the two electrons from ubiquinol to the ISP and the cytochrome b_L haem. The ionizable proton of His181 of the ISP is replaced by a hydroxyl from ubiquinol (Figure 13.12, reaction 1a), which forms a second hydrogen bond with Glu272 of cytochrome *b* (Figure 13.12, reaction 1b), as illustrated in Figure 13.14a. This allows simultaneous electron transfer to the Rieske cluster and the b_L haem (reaction 1c) and release of ubiquinone. In reaction 2 (Figure 13.12) the reduced ISP cluster oscillates to within electron transfer distance of cytochrome c_1 accompanied by electron transfer. Two protons are released from centre P concomitant with ubiquinol oxidation. In reaction 3, electron transfer from haem b_L to haem b_H is followed by reduction of ubiquinone to the semiquinone (reaction 4). After the oxidation of the second ubiquinol at centre P and reduction of the b haems, haem b_H reduces the semiquinone to ubiquinol (reaction 5), accompanied by the uptake of two protons at the centre N. Glu272, which is conserved in all mitochondrial cytochrome b, is thought to move to a position proximal to the propionate of haem b_L, with a water molecule hydrogen bonded between Glu272 and the haem propionate (Figure 13.14b). This allows direct proton transfer from the primary proton acceptor, Glu272, to the propionate. A functionally similar but structurally much simpler version of the bc_1 complex is found in the plasma membrane of many bacteria, where it participates among other processes in respiration, denitrification, nitrogen fixation and cyclic photosynthetic electron transfer.

IRON–SULFUR PROTEINS

For the first billion years of evolution the environment was anaerobic, which meant that, since iron and sulfur were abundant, proteins containing iron–sulfur (Fe–S) clusters were probably abundant, and therefore were among the first catalysts that Nature had available to it. They are distributed in virtually all living organisms; their recognition as a distinct class of metalloproteins only occurred after characteristic EPR spectra in the oxidized state in the 1960s. This second class of iron-containing proteins contains iron atoms bound to sulfur, either bound to the polypeptide chain by the thiol groups of cysteine residues, or else with both inorganic sulfide and cysteine thiols as ligands. The biochemical utility of these Fe–S clusters resides not only in their possibility to easily transfer electrons, but also in their tendency to bind the electron-rich oxygen and nitrogen atoms of organic substrates.

ISP contain four basic core structures which have been characterized crystallographically both in model compounds and in ISP (Rao and Holm, 2004). These are (Figure 13.15), respectively, (A) rubedoxins found only in bacteria, in which the [Fe–S] cluster consists of a single Fe atom bound to four Cys residues—the iron atom can be in the $+2$ or $+3$ valence; (B) rhombic two-iron–two-sulfide [$Fe_2–S_2$] clusters—typical stable cluster oxidation states are $+1$ and $+2$ (the charges of the coordinating cysteinate residues are not considered);

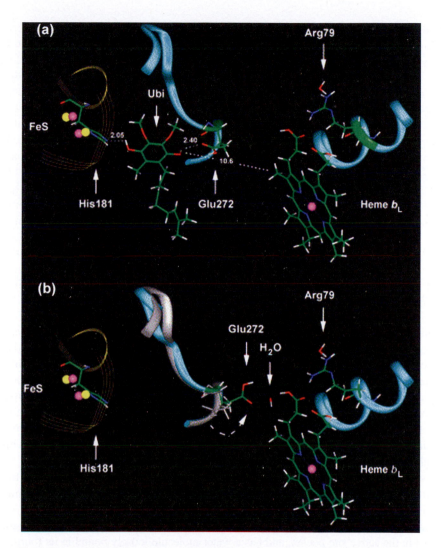

Figure 13.14 Structural basis for electron and proton transfer at centre P: part a shows ubiquinol hydrogen bonded to His181 of ISP and Glu272 of cytochrome b. Part b shows the ubiquinol-binding pocket after movement of Glu272. (From Hunte et al., 2000. Copyright 2003, with permission from Elsevier.)

(C) cuboidal three-iron–four-sulfide [Fe_3–S_4] clusters—stable oxidation states are 0 and +1; and (D) cubane four-iron–four-sulfide [Fe_4–S_4] clusters—stable oxidation states are +1 and +2 for ferredoxin-type clusters and +2 and +3 for "HIPIP" clusters. Electrons can be delocalized, such that the valences of individual iron atoms lie between ferrous and ferric forms. Low-molecular-weight proteins containing the first and the last three types are referred to as rubredoxins (Rd) and ferredoxins (Fd), respectively. The protein ligands are frequently Cys residues, but a number of others are found, notably His, which replaces two of the thiol ligands in the [Fe_2–S_2] Rieske proteins. In addition to these, discrete Rd

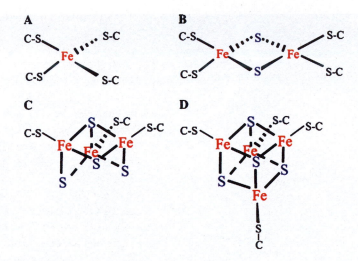

Figure 13.15 Structures of common iron–sulfur centres. (From Imlay, 2006. Reproduced with permission of Blackwell Publishing Ltd.)

and Fd electron transfer proteins which are found in electron transfer chains and as electron donors to enzymes, such centres are often found within redox enzymes where they act as wires (Figure 13.16) delivering electrons one at a time between redox couples which are physically separated. In the *E. coli* quinol-fumarate reductase (Figure 13.16) which transfers electrons from a membrane-bound quinone to cytosolic fumarate, the electrons are transferred to a covalently bound flavin adenine nucleotide at the active site through three distinct iron–sulfur clusters and ultimately are used to reduce fumarate. More complex structures are found in specialized redox enzymes, including hybrid or mixed metal clusters, through metal substitution and/or bridges between simpler clusters (Rees, 2002).

However, the biological activity of Fe–S proteins is not restricted to one-electron transfer reactions. A completely different role for Fe–S clusters is found in a family of dehydratases, of which the best known is aconitase. They all have a [4Fe–4S] cluster (Figure 13.17). However, only three of the four iron atoms have thiolate ligands; the fourth is solvent exposed in the active site pocket, and has a water molecule loosely bound in its fourth coordination site. Binding of the substrate (citrate in the case of aconitase) occurs by smooth switching of the iron from tetrahedral to octahedral (six-coordinate) geometry, binding the substrate by both a carboxylate residue and the hydroxyl group to be abstracted. An adjacent base then deprotonates a methylene group simultaneously with removal of the hydroxyl group by the cationic iron atom, acting as a Lewis acid, accomplishing the net dehydration of the substrate. The iron–sulfur cluster does not function to transfer electrons, but rather to assist in substrate binding and to provide a local positive charge to effect the catalysis.

Studies on three different iron–sulfur enzyme systems, which all require S-adenosyl methionine—lysine 2,3-aminomutase, pyruvate formate lyase and anaerobic ribonucleotide reductase—have led to the identification of SAM as a major source of free radicals in living cells. As in the dehydratases, these systems have a [4Fe–4S] centre chelated by only three cysteines with one accessible coordination site. The cluster is active only in the reduced

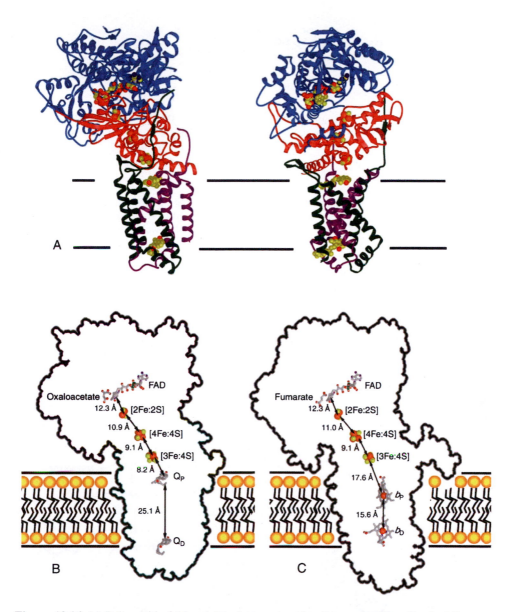

Figure 13.16 (a) Polypeptide fold and (b) electron transfer distances in *E. coli* quinol-fumarate reductase, (c) intercofactor distances in the *Wolinella succinogenes* enzyme. (From Iverson et al., 2002. Reproduced by permission of the Journal of Biological Chemistry.)

state $[4Fe–4S]^{1+}$ and appears to combine the two roles described previously, serving both as a ligand for substrate binding and as a redox catalyst (Figure 13.18). Their mechanism again requires that the exposed iron atom of the cluster shifts toward octahedral geometry as it binds both the amino and carboxylate group of SAM. An electron is then transferred from the low-potential reduced cluster onto SAM generating the adenosyl radical, and possibly

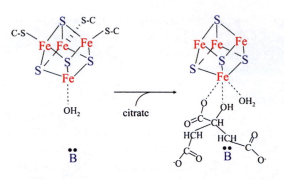

Figure 13.17 Role of clusters in substrate binding—in aconitase the cluster geometry shifts from 4- to 6-coordination on substrate binding. The coordinating iron atom abstracts the hydroxide anion during dehydration. (From Imlay, 2006. Reproduced with permission of Blackwell Publishing Ltd.)

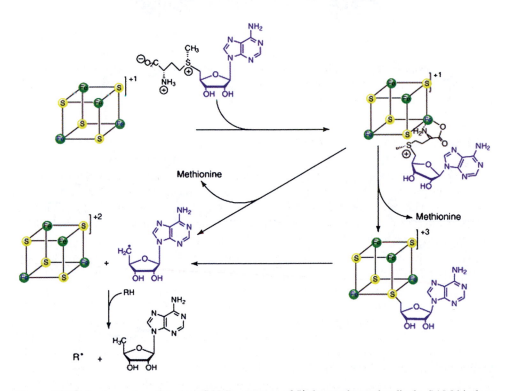

Figure 13.18 S-adenosyl methionine (SAM), a source of 5′-deoxyadenosyl radicals. SAM binds to the subsite iron (in blue) of the reduced [4Fe–4S] cluster via its α-aminocarboxylate group. The 5′-deoxyadenosine radical is formed by electron transfer which occurs either (a) by outer-sphere mechanism or (b) by μ-sulfide alkylation followed by homolytic cleavage of the 5′-S-CH$_2$Ado bond. In both cases, methionine is released. (From Fontecave et al., 2004. Copyright 2004, with permission from Elsevier.)

also coordinating the liberated thiolate. The binding of the liberated sulfur atom to the remaining coordination site of the iron may help to drive the energetically unfavourable electron transfer step. The highly reactive 5′-deoxyadenosyl radical then initiates the enzyme reaction by abstracting a hydrogen atom from the substrate RH to generate the free radical R and 5′-deoxyadenosine. These radical SAM enzymes, as they are now called, constitute a large family involved in many metabolic pathways, including biotin synthase and lipoate synthase. These enzymes catalyse the insertion of sulfur atoms into aliphatic substrates. After activation of the organic substrate by the adenosyl radical, a second Fe–S cluster is the source of the sulfur atoms for insertion into the dethioprecursor.

Finally, we remind our reader that in its apoform, cytoplasmic aconitase is active as an ion regulatory protein (see Chapter 8), binding to iron regulatory elements in the mRNAs of ferritin and transferrin receptor and regulating their translation.

OTHER IRON-CONTAINING PROTEINS

There are many other proteins that contain iron in a form that is neither in haem nor in iron–sulfur clusters. We have already encountered the iron storage and transport proteins, ferritin and transferrin (see Chapter 8). We propose to discuss here two other classes of iron-containing proteins, those with mononuclear non-haem iron centres and those with dinuclear non-haem iron centres.

Mononuclear non-haem iron enzymes

The mononuclear non-haem iron enzymes include a large number of enzymes involved in oxygen activation and insertion into organic substrates. They catalyse a wide variety of reactions and have been classified into five families (Figure 13.19). The extradiol-cleaving catechol dioxygenases catalyse the oxidative aromatic ring cleavage of catechols at the C–C bond adjacent to the enediol group through a four-electron oxidation which incorporates both atoms of the dioxygen molecule into the product. Rieske dioxygenases (so-called because they contain a [2Fe–2S] in addition to the mononuclear iron centre) catalyse the *cis*-dihydroxylation of arene double bonds in which NADH is the source of two electrons and both dioxygen atoms are incorporated into the *cis*-diol product. An extremely versatile group of enzymes, including the proline hydroxylase involved in collagen biosynthesis, constitute the third group. They typically require α-ketoglutarate as electron source, which subsequently undergoes oxidative decarboxylation. These enzymes use the oxidative equivalents produced by this reaction to carry out hydroxylation of C–H bonds, oxygen atom transfers, heterocyclic ring formation or desaturation reactions. The fourth class are the aromatic amino acid hydroxylases, which use tetrahydrobiopterin as cofactor for the hydroxylation of the aromatic amino acids Phe, Tyr and Trp. The latter two hydroxylases are the rate-limiting steps in the biosynthesis of serotonin and the catecholamines, both important for neuronal function. Finally, there is a mixed bag of oxidases, illustrated in Figure 13.19 by isopenicillin synthase (which catalyses cyclization of the heterocyclic β-lactam ring), ethylene formation in plants and many other diverse reactions. However, a

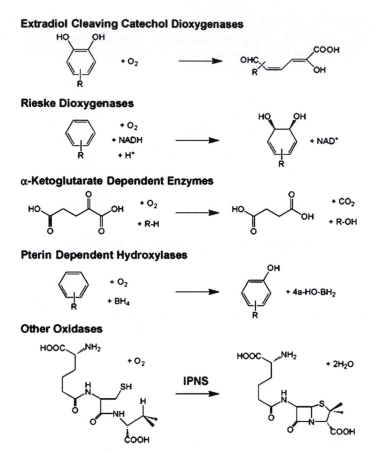

Figure 13.19 Reactions catalysed by each of the five families of mononuclear non-haem iron enzymes with a 2-His-1-carboxylate facial triad. Dioxygen is labelled to indicate the fate of each oxygen atom. (From Koehntop et al., 2005. With kind permission of Springer Science and Business Media.)

huge advance has been made in the last few years by the recognition that behind all of this apparent diversity, the five classes of enzymes which catalyse these reactions all turn out to have at their active site a common metal-binding motif, in the form of a 2-His-1-carboxylate facial triad (Figure 13.20). This structural motif consists of three protein ligands, two His residues and one carboxylate, either from Glu or Asp, arranged at the vertices of one triangular face of an octahedron, locking the iron to the enzyme. In contrast to haem enzymes, which have a fourth equatorial ligand, this leaves three coordination sites on the opposite side of the 2-His-1-carboxylate facial triad available for binding exogenous ligands, such as O_2, substrate, and/or cofactor, enabling the protein the flexibility with which to tune the reactivity of its Fe(II) centre.

The active site structure of Figure 13.20, with three solvent molecules occupying the opposite face of the triad, is unreactive towards dioxygen. However, structural studies of members of each enzyme family show that formation of an enzyme–substrate complex

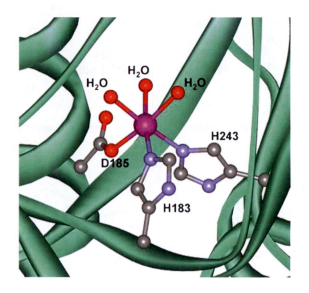

Figure 13.20 The 2-His-1-carboxylate facial triad illustrated for the resting state of deacetoxy-cephalosporin C synthase, an α-ketoglutarate-dependent mononuclear non-haem iron enzyme. (From Koehntop et al., 2005. With kind permission of Springer Science and Business Media.)

affects the coordination chemistry of the active site iron atom. For the extradiol-cleaving catechol dioxygenase, 2,3-dihydroxybiphenyl 1,2-dioxygenase and isopenicillin N-synthase (Figure 13.21a, c), the substrates, DHBP (2,3-dihydroxybiphenyl) and ACV (δ-(L-α-aminoadipoyl)-L-cysteinyl-D-valine), bind directly to the iron (II) centre. In contrast, for the α-ketoglutarate-dependent enzymes, such as clavaminate synthase, the cofactor, α-ketoglutarate, (α-KG) coordinates the metal centre while the substrate, proclavaminate (PCV), is bound close by (Figure 13.21d). In the Rieske dioxygenases, such as naphthalene dioxygenases (Figure 13.21b), and in the pterin-dependent hydroxylases, such as pheny-lalanine hydroxylase, neither substrate nor cofactor binds directly to the metal centre. However, in both of these enzyme classes, the carboxylate of the facial triad functions as a bidentate ligand with a water molecule occupying the fifth coordination position. In Figure 13.21e, both the substrate analogue 3-(2-thienyl)-L-alanine (THA) and the cofactor, tetrahydrobiopterin (BH_4), are shown. In all five structures represented in Figure 13.21, the consequence of substrate and/or cofactor binding is the formation of a five-coordinate iron (II) centre, which is poised for dioxygen binding. The only O_2 complex to have been char-acterized crystallographically is that of the naphthalene dioxygenase with both oxygen and the substrate analogue, indole, bound (Figure 13.22), in which the oxygen molecule is bound side-on, ready to attack the substrate. This is in contrast to the results observed using the O_2 analogue NO, which binds end-on to the five-coordinate iron centres of catechol dioxygenase, isopenicillin N-synthase and clavaminate synthase. However, the structure of naphthalene dioxygenase with NO and the substrate analogue, indole, show the NO bound end-on, suggesting that NO may not be as appropriate a surrogate for oxygen as was orig-inally thought.

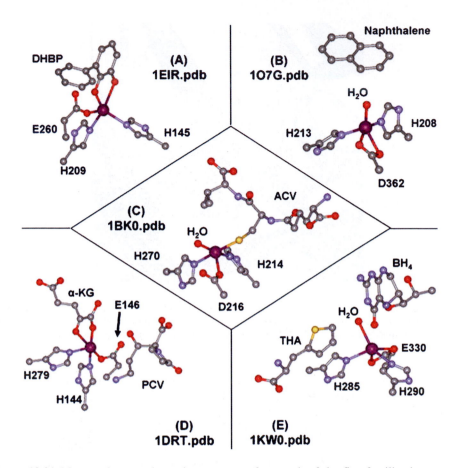

Figure 13.21 Mononuclear non-haem iron enzymes from each of the five families in structures which are poised for attack by O_2. (a) The extradiol-cleaving catechol dioxygenase, 2,3-dihydroxy-biphenyl 1,2-dioxygenase; (b) the Rieske dioxygenase, naphthalene 1,2-dioxygenase; (c) isopenicillin N-synthase; (d) the α-ketoglutarate dependent enzyme clavaminate synthase and (e) the pterin-dependent phenylalanine hydroxylase. (From Koehntop et al., 2005. With kind permission of Springer Science and Business Media.)

A general mechanism can be formulated for this superfamily of enzymes which involves (Figure 13.23) substrate and/or cofactor binding to the resting form of the enzyme with concomitant formation of a five-coordinate iron (II) centre (b) which is now primed for dioygen binding (c). In all enzyme classes O_2 is reduced to the peroxide level of oxidation in the next step of the mechanism (d), as illustrated for the Rieske dioxygenases in Figure 13.22, with the electron coming from the nearby Rieske [2Fe–2S] cluster. Cleavage of the O–O bond occurs in the next step, generating a high-valent oxidizing species, which is responsible for transforming the primary substrate into product, which is then released from the enzyme with regeneration of the resting state of the enzyme.

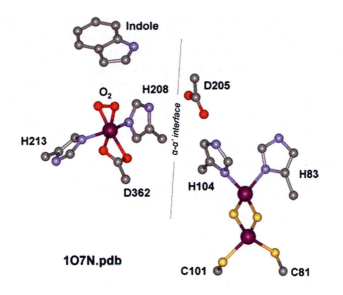

Figure 13.22 The O_2 complex of naphthalene dioxygenase with the substrate analogue indole bound. Asp205 connects the Rieske [2Fe–2S] centre (located in a neighbouring subunit) to the mononuclear iron active site. (From Koehntop et al., 2005. With kind permission of Springer Science and Business Media.)

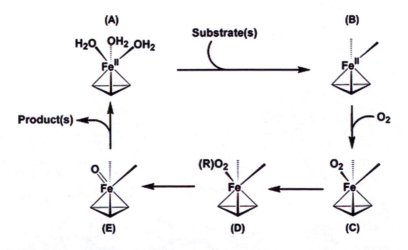

Figure 13.23 The general mechanistic pathway catalysed by mononuclear non-haem iron enzymes with a 2-His-1-carboxylato facial triad, shown in its resting state in (a). (From Koehntop et al., 2005. With kind permission of Springer Science and Business Media.)

Dinuclear non-haem iron enzymes

The final class of iron proteins that we consider here are a family that contain a carboxylate-bridged diiron centre. They carry out a variety of functions, which have the common link that they react with dioxygen as part of their functional processes. The dimetallic centre is incorporated into a four-helix bundle domain (see Figure 3.9a), which seems to represent a preferred biological scaffold for the binding and activation of dioxygen. In many members of the family, four of the iron-binding ligands are provided by two E(D/H)XXH motifs. Among their diverse functions we find ferritins, which store iron—the H-chains have the diiron ferroxidase centre (see Chapter 8), while the L-chains function to nucleate the mineral iron core. Haemerythrins transport O_2 in a number of marine invertebrates, although the protein has been recently identified in an archaebacteria. The RNR-R2 protein of Class I RNRs use their diiron centre to generate a tyrosyl radical, which in turn is used to generate the active centre thiyl radical (see earlier in this chapter). Rubrerythrins (the name is a contraction of rubredoxin and hemerythrin, reflecting its two types of iron sites, namely a rubredoxin-type $[Fe(Cys)_4]$ and a diiron, respectively) found in air-sensitive bacteria and archaebacteria function as a peroxide scavenger. The stearoyl-acyl carrier protein Δ^9 desaturases introduce double bonds into saturated fatty acids. Finally, the bacterial multi-component monoxygenases (BMMs) catalyse hydroxylation of a variety of hydrocarbon substrates, including alkanes, alkenes and aromatics. Members of the family include methane/butane monooxygenases (sMMOs) and four-component alkene/aromatic monooxygenases (TMOs), which transform small alkanes to their corresponding alcohols. Other family members specialize in the regiospecific hydroxylation of aromatics.

The best characterized of the BMMs are the sMMOs (Figure 13.24), which are the only members of the family capable of activating the inert C–H bond of methane, one of the most difficult reactions in nature to achieve. Like most members of the BMM superfamily, sMMO requires three protein components, the hydroxylase MMOH, which contains the carboxylate-bridged diiron centre, a regulatory protein MMOB and a [2Fe–2S]- and FAD-containing reductase (MMOR) which shuttles electrons from NADH to the diiron centre.

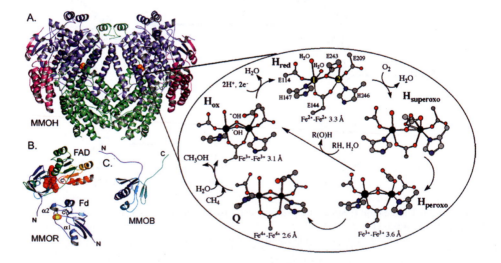

Figure 13.24 Structures of sMMO components and proposed reaction cycle. (a) MMOH; (b) the MMOR FAD and ferredoxin (Fd) domains; (c) MMOB. In MMOH the α, β and γ subunits are coloured blue, green and purple, respectively. Iron, sulfur and FAD are coloured orange, yellow and red, respectively and are depicted as spheres. The MMO reaction cycle is shown on the right, with atoms coloured [Fe (black), C (grey), O (red) and N (blue)]. (Reprinted with permission from Sazinsky and Lippard, 2006. Copyright (2006) American Chemical Society.)

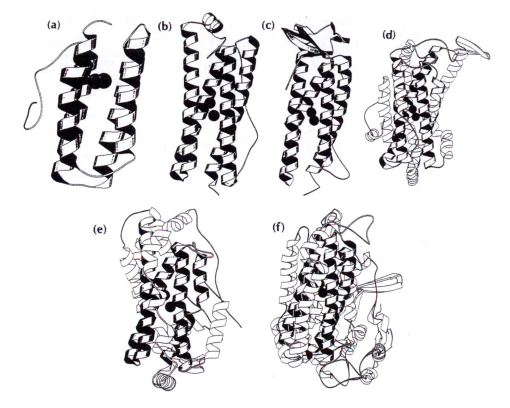

Figure 13.25 Three-dimensional structures of diiron proteins. The iron-binding subunits of (a) haemery-thrin, (b) bacterioferritin, (c) rubryerythrin (the FeS centre is on the top), (d) ribonucleotide reductase R2 subunit, (e) stearoyl-acyl carrier protein Δ^9 desaturase, (f) methane monooxygenase hydroxylase α-subunit. (From Nordlund and Eklund, 1995. Copyright 1995, with permission from Elsevier.)

The hydroxylase component (MMOH) is composed of two αβγ protomers which use pro-tein contacts between each of the α and β subunits along a two-fold symmetry axis to form an $\alpha_2\beta_2\gamma_2$ heterodimer. The interface between the two protomers forms a large cleft within which the MMOR is thought to dock. The diiron centre is located within a four-helix bundle made up of helices B, C, E and F of the α subunit. Helices E and F are on the surface of the hydroxylase, forming part of the rim of the cleft with the diiron centre some 12 Å beneath this rim. Figure 13.25 compares the four-helix bundles that contain the diiron cen-tres in six representatives of the superfamily.

In the proposed reaction cycle for MMOH, the resting enzyme, with both iron atoms in the ferric state, is reduced by the MMOR to the diiron(II) form. The bridging hydroxyls are expelled and Glu243 shifts to become a bridging ligand while remaining bound to Fe2, and a water molecule coordinates weakly to Fe1. The Fe–Fe distance lengthens, and the open coordination position that forms on Fe2 facing the active site pocket can now bind dioxygen, which proceeds to a peroxo-intermediate designated $\mathbf{H}_{\text{peroxo}}$. $\mathbf{H}_{\text{peroxo}}$ and can itself carry out oxygen insertion reactions with some substrates. However, the key intermediate in

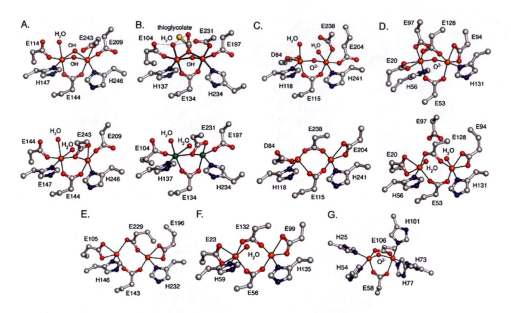

Figure 13.26 Dioxygen-utilizing carboxylate-bridged diiron centres: (a) Oxidized (top) and reduced (bottom) MMOH; (b) oxidized (top) and Mn^{II}-reconstituted ToMOH (bottom); (c) oxidized (top) and reduced (bottom) RNR-R2; (d) oxidized (top) and reduced (bottom) rubrycrythrin; (e) reduced stearoyl-acyl carrier protein Δ^9 desaturase; (f) reduced bacterioferritin; (g) methaemerythrin. Fe1 is on the left and Fe2 on the right. (Reprinted with permission from Sazinsky and Lippard, 2006. Copyright (2006) American Chemical Society.)

MMOs is **Q**, which has been characterized spectroscopically and proposed to have an Fe$_2^{IV}$ (μ-O)$_2$ core, and an Fe–Fe distance of only 2.6 Å, in marked contrast to the distance of 3.6 Å in **H$_{peroxo}$**. There is good evidence that **Q** reacts directly with methane at a rate that depends on methane concentration.

The diiron centres of MMOH, ToMOH (toluene monooxygenase hydroxylase), RNR-R2, rubreythrin, stearoyl-acyl carrier protein Δ^9 desaturases, bacterioferritin and methaemery-thrin (Figure 13.26) all have the same three-amino-acid structural motif on one side of the diiron site. This is made up by a bridging Glu and two His residues, coordinated in positions distal to the active site pocket. The remaining ligands in the two hydroxylases MMOH and ToMOH are quite different from those in the other five. In the resting state Fe1 is coordinated by a monodentate Glu and a water molecule and Fe2 by two monodentate Glu, with bridging hydroxide ions completing the octahedral geometry around the iron atoms. The structural observation of product alcohols and monoanions, occupying the bridging position facing the hydrophobic active site pocket, strongly implies this site in C–H and O$_2$ activation. As we mentioned above, concomitant with two-electron reduction of MMOH, Glu243 shifts to a bridging position, the Fe–Fe distance increases to 3.3 Å, and an open coordination site is formed on Fe2. Theoretical calculations and synthetic model studies suggest that this is the primary site of O$_2$ activation by sMMO and perhaps by all BMMs.

In the diiron active sites of RNR-R2, Δ^9 desaturase, bacterioferritin and rubrerythrin the flanking carboxyl ligands on the opposite side of the diiron centre are all quite different, and some of them change their coordination geometry upon reduction (Figure 13.26), though not in the way MMOH and ToMOH do. The very varied chemistry carried out by these proteins no doubt is reflected in the active site geometry, but we are as yet unable to predict what changes in ligands might have what consequences for biological activity (not forgetting of course the rest of the protein with its binding sites for other substrates, etc.). Two members of the family do not appear to have highly flexible iron ligands that change position and orientation with metal oxidation state, and variable Fe–Fe distances that fluctuate during the catalytic cycle, namely bacterioferritin (and ferritin) and haemerythrin. This may reflect the fact that unlike the others, they do not have a 10-residue stretch of π-helix within the four-helix bundle which contributes a Glu ligand to the diiron centre and several residues to the active site pocket. This may reflect greater rigidity due to the entirely α-helical nature of the four-helix bundle which preludes conformational changes necessary for function.

REFERENCES

Collman, J.P., Boulatov, R., Sunderland, C.J. and Fu, L. (2004) Functional analogues of cytochrome c oxidase, myoglobin and hemoglobin, *Chem. Rev.*, **104**, 561–588.

Crichton, R.R. (2001) *Inorganic Biochemistry of Iron Metabolism: From Molecular Mechanisms to Clinical Consequences*, Wiley, Chichester, 326 pp.

Crichton, R.R. and Pierre, J.-L. (2001) Old iron, young copper: from Mars to Venus, *Biometals*, **14**, 99–112.

Fenton, H.J.H. (1894) *Trans. Chem. Soc.,* **65**, 899–910.

Fontecave, M., Atta, M. and Mulliez, E. (2004) S-adenosylmethionine: nothing goes to waste, *TIBS*, **29**, 243–249.

Gelin, B.R. and Karplus, M. (1977) Mechanism of tertiary structural change in hemoglobin, *Proc. Natl. Acad. Sci. U.S.A.*, **74**, 801–805.

Haber, F. and Weiss, J. (1934) The catalytic decomposition of hydrogen peroxide by iron salts, *Proc. Roy. Soc. Ser. A.*, **147**, 332–351.

Hunte, C., Koepke, J., Lange, C., Rossmanith, T. and Michel, H. (2000) Structure at 2.3 Å resolution of the cytochrome bc_1 complex from the yeast *Saccharomyces cerevisiae* with an antibody Fv fragment, *Structure*, **8**, 669–684.

Imlay, J.A. (2006) Iron-sulphur clusters and the problems with oxygen, *Mol. Microbiol.*, **59**, 1073–1082.

Iverson, T.M., Luna-Chavez, C., Croal, L.R., Cecchini, G. and Rees, D.C. (2002) Crystallographic studies of the *Eschericia coli* quinol-fumarate reductase with inhibitors bound to the quinol-binding site, *J. Biol. Chem.*, **277**, 16124–16130.

Koehntop, K.D., Emerson, J.P. and Que, L. Jr. (2005) The 2-His-1-carboxylate facial triad: a versatile platform for dioxygen activation by mononuclear non-heme iron(II) enzymes, *J. Biol. Inorg. Chem.*, **10**, 87–93.

Kolberg, M., Strand, K.R., Graff, P. and Andersson, K.K. (2004) Structure, function and mechanism of ribonucleotide reductases, *Biochim. Biophys. Acta*, **1699**, 1–34.

Mitchell, P. (1975) Protonmotive redox mechanism of the cytochrome b-c_1 complex in the respiratory chain: protonmotive ubiquinone cycle, *FEBS Lett.*, **56**, 1–6.

Namslauer, A. and Brzezinski, P. (2004) Structural elements involved in electron-coupled proton transfer in cytocrome c oxidase, *FEBS Lett.*, **567**, 103–110.

Nordlund, P. and Eklund, H. (1995) Di-iron-carboxylate proteins, *Curr. Opin. Struct. Biol.*, **5**, 758–766.

Rao, P.V. and Holm, R.H. (2004) Synthetic analogues of the active sites of iron–sulfur proteins, *Chem. Rev.*, **104**, 527–559.

Rees, D.C. (2002) Great metalloclusters in enzymology, *Annu. Rev. Biochem.*, **71**, 221–246.

Sazinsky, M.H. and Lippard, S.J. (2006) Correlating structure with function in bacterial multicomponent monooxygenases and related diiron proteins, *Acc. Chem. Res.*, 39, 558–566.

Stubbe, J., Ge, J. and Yee, C.S. (2001) The evolution of ribonucleotide reduction revisited, *TIBS*, **26**, 93–99.

Voet, D. and Voet J.G. (2004) *Biochemistry*, 3rd edition, Wiley, Hoboken, NJ, 1591pp.

– 14 –

Copper: Coping with Dioxygen

INTRODUCTION

While iron, on account of the solubility of its ferrous form, was widely available in the reducing environment of the early Earth, copper, which was present as highly insoluble cuprous sulfides, must have been poorly bioavailable. In contrast, once photosynthetic *Cyanobacteria* set off the first major irreversible pollution of our environment with the production of dioxygen, copper became much more bioavailable in its oxidized cupric form. Whereas the enzymes involved in anaerobic metabolism were designed to act in the lower portion of the range of redox potentials, the presence of dioxygen created the need for new redox systems with standard redox potentials in the range from 0 to 0.8 V; copper proved eminently suitable for this role. For aerobic metabolism, enzymes and proteins with higher redox potentials came to be utilized to take advantage of the oxidizing power of dioxygen. However, whereas the early evolution of life was the 'iron age', it is clear that the subsequent was the 'copper age', or rather the 'iron–copper age', where both metals are involved together. This is well illustrated by ceruloplasmin, the principal copper-binding protein in serum that plays an important role in iron metabolism, and by the terminal oxidase of the mitochondrial respiratory chain, cytochrome c oxidase (CcOx), which requires both haem iron and copper for its activity.

Copper is present in a large number of enzymes, many involved in electron transfer, activation of oxygen and other small molecules such as oxides of nitrogen, methane and carbon monoxide, superoxide dismutation, and even, in some invertebrates, oxygen transport. The routinely encountered oxidation states are Cu(I) and Cu(II), and as with iron, the reduced form can catalyse Fenton chemistry with hydrogen peroxide. Cu(I) can form complexes with coordination numbers 2, 3 or 4, while Cu(II) prefers coordination numbers 4, 5 or 6. Whereas four-coordinate complexes of Cu(II) are square-planar, the corresponding Cu(I) complexes are tetrahedral. Among the divalent elements of the transition series, Cu(II) forms the most stable complexes. In terms of the HSAB classification Cu(II) is 'hard', while Cu(I) is 'soft' underlined by its preference for sulfur ligands. Both forms have fast ligand exchange rates. It appears that throughout the living world intracellular concentrations of 'free' copper are maintained at extremely low levels, most likely because intracellular copper metabolism is characterized by the use of copper chaperone proteins to transport copper towards their target proteins (cytochrome oxidase, superoxide dismutase (SOD) and the multi-copper oxidases, whose copper is inserted in the Golgi apparatus).

In copper-containing proteins, three types of copper, classified on the basis of their visible, UV and EPR spectra, as originally proposed by one of the pioneers of copper biochemistry, Bo Malmström, are found. Types 1 and 2 have one copper atom, which has an intense blue colour in Type 1 and is almost colourless in Type 2; whereas Type 3 has an EPR-silent dicopper centre. More details of the three types are as follows:

> Type 1 Cu(II): intense ($\varepsilon > 3000$ M^{-1} cm^{-1}) blue ($\lambda_{max}\sim 600$ nm) optical absorption band; EPR spectrum with an uncommonly small hyperfine splitting in g_{II} region.
> Type 2 Cu(II): weak absorption spectrum; EPR spectrum characteristic of square-planar Cu(II) complexes.
> Type 3 Cu(II): dicopper centre; strong absorption in the near UV ($\lambda_{max}\sim 330$ nm); no EPR spectra, the two coppers are anti-ferromagnetically coupled.

The Type 1 copper ions are normally coordinated by three strong ligands, a cysteine and two histidines, and often have one or two weaker ligands such as methionine sulfur or oxygen. Type 3 coppers are usually each coordinated by three histidines, with a bridging ligand such as oxygen or hydroxyl anion.

We will first consider the electron-transport role of copper in the Type 1 blue copper proteins. Then will discuss the involvement of copper with dioxygen in enzymes containing, respectively, mononuclear Type 2 copper ions, dinuclear Type 3 copper and the multi-copper oxidases, such as ascorbate oxidase, which have all three types of copper centres. Next we describe the role of copper in cytochrome oxidase, the terminal oxidase of the mitochondrial electron-transport chain, in SOD and in enzymes that involve other low-molecular weight, mostly inorganic substrates. The important human diseases of copper metabolism first described by Wilson in 1912 and by Menkes half a century later, which bear their names are discussed in Chapter 18. We conclude this chapter devoted to the metal, which the alchemists designated as Venus, the Roman goddess of love and beauty, by considering its interaction with iron, designated as Mars, the Roman god of war[1] (Crichton and Pierre, 2001).

BLUE COPPER PROTEINS INVOLVED IN ELECTRON TRANSPORT

One of the intriguing questions in the biochemistry of copper is how it can function in rapid electron-transfer reactions[2] when Cu(I) and Cu(II) have such drastically different preferences in coordination geometry. As we have already pointed above, four-coordinate Cu(II) complexes are square-planar, while the corresponding Cu(I) complexes tend to be

---

[1] As has been pointed out elsewhere, Mars and Venus were no strangers to one another according to classical mythology, as a number of wall paintings from Pompei clearly illustrate. Despite being married to Vulcan, the god of fire, the encounters of Venus with Mars were not just getting together for a congenial chat.

[2] An important concept in enzymology is that while catalysis involving bond cleavage and formation requires conformational change, and is relatively slow (maximum ~10^8 s^{-1}), electron transfer is much more rapid (10^{12} s^{-1}), which does not allow much time for conformational change.

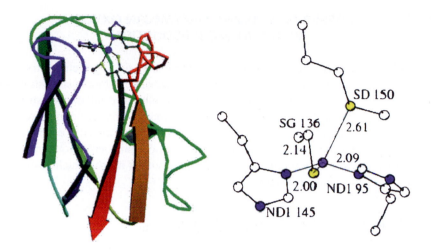

Figure 14.1 (Left) X-ray structure of plastocyanin from poplar leaves as a ribbon diagram with the metal ion and its ligands highlighted, PDB code 1PLC (right) Type 1 Cu site in Cu(II)-nitrite reductase from *Alcaligenes faecalis*, PDB code 1AS6. (From Messerschmidt et al., 2001. Reproduced with permission from John Wiley & Sons., Inc.)

more tetrahedral. When the Type 1 copper centre in plastocyanin was first characterized by X-ray crystallography (Figure 14.1), it revealed a copper-binding site that was virtually the same in the apoprotein and in the copper-containing protein, whether the copper was Cu(I) or Cu(II). In other words, the protein imposes a binding site geometry on the metal, which is in reality closer to that of Cu(I) than of Cu(II), such that Cu(II) has no possibility to rearrange towards its preferred geometry. The copper-coordination site (Figure 14.1) is highly distorted with two His nitrogen and one Cys sulfur donors lying almost in a plane with the metal ion, together with a long, out of the plane axial bond, between the sulfur of a Met residue and the Cu atom. This structure in many ways is the convincing proof of the idea that proteins can fine-tune the properties of bound metal centres, imposing what Vallee and Williams (1968) called the '*entatic state*', 'closer to a transition state that to a conventional, stable molecule'. The entatic state, or strain induced by metal binding to proteins, both on the metal and the protein itself, is a useful concept for explaining the generation of metal sites in electron-transfer proteins, such as the blue copper proteins, which are designed for rapid electron transfer. The trigonal pyramidal structure with three strong equatorial ligands (one Cys and two His), which the protein imposes, provides a favourable geometry for both cuprous and cupric oxidation states, and facilitates rapid electron transfer. A number of other copper electron-transfer proteins that contain Type 1 copper centres (azurin, ceruloplasmin, laccase, nitrite reductase, rusticyanin and stellacyanin) are known. They all have three coordination positions contributed by two His and one Cys, similar to the copper-coordination chemistry in plastocyanin—yet they span a range of redox potential from less than 200 mV to more than 1000 mV. One of the challenges for the future will be to determine what programmes this fine-tuning of redox properties of Type 1 copper centres.

COPPER-CONTAINING ENZYMES IN OXYGEN ACTIVATION AND REDUCTION

There has been enormous activity in the field of copper(I)–dioxygen chemistry in the last 25 years, with our information coming from both biochemical–biophysical studies and to a very important extent from coordination chemistry. This has resulted in the structural and spectroscopic characterization of a large number of copper dioxygen complexes, some of which are represented in Figure 14.2. The complex F, first characterized in a synthetic system was subsequently established to be present in oxy-haemocyanin, and is found in derivatives of tyrosinase and catechol oxidase, implying its involvement in aromatic hydroxylations in both enzymes and chemical systems.

Type 2 Copper Oxidases and Oxygenases

A number of X-ray structures are available for oxidases and oxygenases containing Type 2 copper sites including amine oxidases, galactose oxidase, lysyl oxidase and peptidyl-glycine α-hydroxylating monooxygenase. The latter, a ubiquitous enzyme in the nervous system of higher organisms, catalyses the oxidative cleavage of the terminal acetyl group from prohormone peptides with a C-terminal glycine residue (Figure 14.3) to give the corresponding amide, glyoxylate and water, and it has been established that dioxygen binds to one of the two Type 2 copper atoms in an 'end-on mode'. A copper–dioxygen complex has been trapped by freezing crystals of the enzyme, which had been soaked with a slowly reacting substrate, N-acetyl-diiodo-tyrosyl-D-threonine (IYT), in the presence of oxygen and ascorbate (to keep the copper reduced). The structure of a pre-catalytic complex in Figure 14.4 suggests a likely mechanism in which electron transfer from the second Cu atom, some 11 Å away, and hydrogen abstraction from the substrate leads to the formation of an α-hydroxyglycine intermediate, which is subsequently converted to the amide product, glyoxylic acid and water.

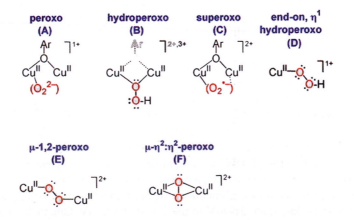

Figure 14.2 Cu_nO_2 structural types. (From Hatcher and Karlin, 2004. With kind permission of Springer Science and Business Media.)

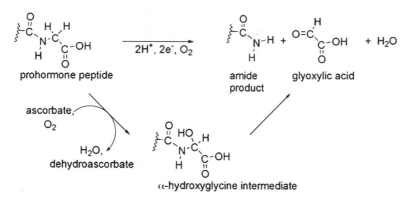

Figure 14.3 The reaction catalysed by peptidylglycine α-hydroxylating monooxygenase.

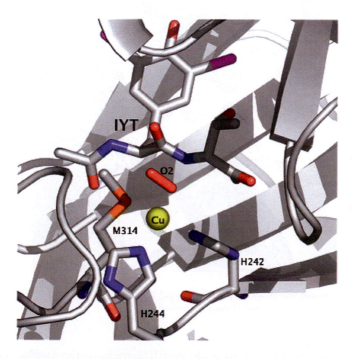

Figure 14.4 The dioxygen site in peptidylglycine α-hydroxylating monooxygenase incubated in the presence of IYT. (From Bento et al., 2006. With kind permission of Springer Science and Business Media.)

Dinuclear Type 3 Copper Proteins

Catechol oxidase and tyrosinase are among the enzymes that have Type 3 dinuclear centres. However, the prototype of this class of proteins is the invertebrate oxygen transport protein, haemocyanin (Figure 14.5), for which structures of the oxy and deoxy forms have been determined at high resolution and confirm, as predicted from model compounds, that the

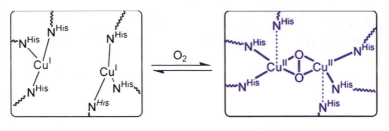

deoxy-Hemocyanin   **oxy-Hemocyanin**

Figure 14.5 Copper-coordination sites in deoxy- and oxy-haemocyanin. (From Hatcher and Karlin, 2004. With kind permission of Springer Science and Business Media.)

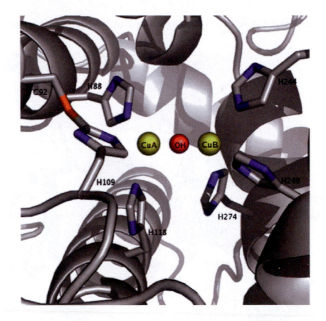

Figure 14.6 The active site of sweet potato catechol oxidase. (From Bento et al., 2006. With kind permission of Springer Science and Business Media.)

dioxygen molecule is bound in a peroxo–dicopper(II) complex, corresponding to Figure 14.2F. The Cu–Cu distance in the oxy-form is 3.6 Å, 1 Å less than in the deoxy form. In catechol oxidase from sweet potato, in the 'resting state', the copper ions are separated by 2.9 Å and bridged by what is thought to be a hydroxide ion (Figure 14.6), and on reduction of the enzyme this bridging ligand is lost and the Cu–Cu separation increases to 4.4 Å. Interestingly, one of the three His ligands to Cu_A forms a direct covalent thioether link to a Cys residue. Despite this, the catalytic core has an overall structure very similar to the copper-binding centre in haemocyanin. Catechol oxidase catalyses the oxidation of two molecules of catechol to two molecules of the corresponding diketone, through a

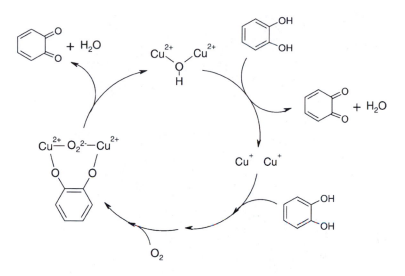

Figure 14.7 Simplified version of the reaction mechanism for catechol oxidase. (From Granata et al., 2004. With kind permission of Springer Science and Business Media.)

four-electron reduction of a molecule of dioxygen to two molecules of water. A simplified version of the reaction mechanism is shown in Figure 14.7. The first stage involves interaction of the catechol substrate with a hydroxyl-bridged $Cu^{2+}- OH^-- Cu^{2+}$ species to give the diketone product, a molecule of water and a dinuclear species in which both coppers are reduced. The second stage involves reaction of a second molecule of catechol, with this species in the presence of dioxygen to form a $Cu^{2+}- O_2^{2-}- Cu^{2+}$ species in which the catechol bridges the two copper ions by two hydroxyl groups. A second molecule of diketone and of water is released, and the enzyme returns to its bridged hydroxyl state.

Multi-Copper Oxidases

An important family of multi-copper enzymes, comprising laccase, ascorbate oxidase and ceruloplasmin, contain at least four copper ions, representing Types 1, 2 and 3 within the same molecule. The blue Type 1 site is usually located at some 12–13 Å distance from a trinuclear site, which has the two Type 3 coppers, linked by a bridging oxygen and one Type 2 copper. We illustrate this class of oxidases with laccase, which is involved in the oxidation of phenols, including uroshiol in the production of Japanese lacquer[3]. Figure 14.8 is an overall view of the CotA laccase showing the locations of the copper centres (yellow) and the entrance and exit channels for dioxygen and water, respectively (stippled red).

---

[3] The biotechnological potential of laccases—working with air and producing water as sole by-product—has led to applications from the textile to the pulp and paper industries, from food applications to bioremediation, as well as their use in organic synthesis.

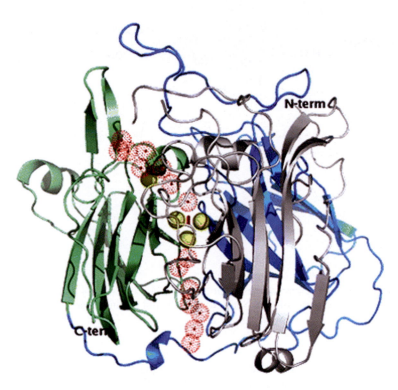

Figure 14.8 The overall structure of the CotA laccase from *Bacillus subtilis* showing the entrance and exit channels for dioxygen and water, above and below the trinuclear cluster, respectively. (From Bento et al., 2006. With kind permission of Springer Science and Business Media.)

It is suggested that dioxygen binds to the trinuclear centre, diffusing through a well-defined channel, normally occupied by solvent molecules. In this 'resting state' all the coppers will be in the oxidized $+2$ state (the blue colour of this intermediate confirms that the Type I copper is certainly oxidized). Two electrons are then transferred from the substrate molecules through the Type 1 copper to the Type 3 copper of the trinuclear site, where dioxygen is reduced to give a peroxo intermediate. Two further electrons are then assimilated in a similar manner and result in splitting of the peroxo intermediate into two hydroxyl groups. Subsequent addition of protons, provided by acidic residues in the exit channel, will result in the successive release of the two molecules of water. While the Type 3 coppers are involved in oxygen binding and electron transfer to the dioxygen and peroxo intermediate, the role of the Type 2 copper seems to be to help anchor the dioxygen molecule to the trinuclear cluster prior to reduction, and to temporarily bind the hydroxyl groups arising from the reduction of the peroxo intermediate prior to their release as water molecules in the exit channel.

Cytochrome c Oxidases

We have already discussed the terminal oxidase of the respiratory chain, CcOx, in the previous chapter. Here we focus on the role of copper in this key metabolic enzyme.

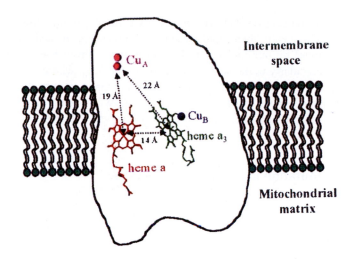

Figure 14.9 Schematic representation of the redox metals of beef heart CcOX with their relative distances. (From Brunori et al., 2005. Copyright 2005, with permission from Elsevier.)

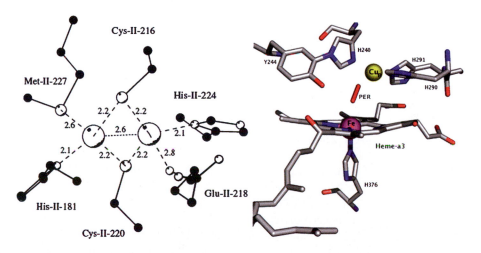

Figure 14.10 (Left) The Cu$_A$ site in bacterial cytochrome oxidase. (From Messerschmidt et al., 2001. Reproduced with permission from John Wiley & Sons., Inc.) (Right) The haem-a_3/Cu$_B$ site in the 'resting' form of oxidized bovine heart cytochrome c oxidase showing peroxide bound between the haem Fe and Cu$_B$. (From Bento et al., 2006. With kind permission of Springer Science and Business Media.)

The disposition of the different metal centres of bovine heart CcOx is represented in Figure 14.9. The dimetallic Cu$_A$ site receives electrons directly from cytochrome c, and is located in a globular domain of subunit II, which protrudes into the intermembrane space (the periplasmic space in bacteria). This centre, which was widely believed to be mononuclear is a dicopper site (Figure 14.10) in which the coppers are bridged by two cysteine sulfurs: each copper in addition has two other protein ligands. In the one electron-reduced form, the electron is fully delocalized between the two Cu atoms, giving rise to a $[Cu^{+1.5} \cdots Cu^{+1.5}]$ state. The Cu_A centre then rapidly reduces the haem a, located some 19 Å away (metal–metal distance) by intramolecular electron transfer. From haem a, electrons are transferred intramolecularly to the active site haem a_3 and Cu_B, where oxygen binds. The Cu_B centre (Figure 14.10) involves coordination of the copper atom to three His ligands, and the Fe–Cu_B distance in the oxidized enzyme is 4.5 Å, with one of the His ligands covalently linked to a nearby Tyr residue. The mechanism of oxygen reduction has been discussed in Chapter 13.

Superoxide Dismutation in Health and Diseases

Superoxide is generated by a number of enzymes in the course of their reaction cycles, but by far the greatest production of superoxide anion and the reactive oxygen species that can be derived from it is the respiratory chain within the mitochondria. SODs lower the levels of superoxide by catalysing the transformation of two superoxide ions into dioxygen and hydrogen peroxide. The CuZnSOD is widely distributed, located in the periplasmic space in bacterial cells and in both the cytosol and the mitochondrial intermembrane space in eukaryotic cells. The reaction is a two-step process in which a molecule of superoxide reduces the oxidized (Cu^{2+}) form of the enzyme to give dioxygen and the reduced (Cu^+) enzyme, which subsequently reduces a second molecule of superoxide, giving hydrogen peroxide and restoring the oxidized form of the enzyme. An active site channel

$$2O_2^- + 2H^+ \rightarrow O_2 + H_2O_2$$

$$Cu^{2+}ZnSOD + O_2^- \rightarrow Cu^+ZnSOD + O_2$$

$$Cu^+ZnSOD + O_2^- + 2H^+ \rightarrow Cu^{2+}ZnSOD + H_2O_2$$

leading to the copper atom is constructed ideally for small anionic species such as superoxide, allowing nearly diffusion-controlled rates of enzyme catalysis (rate constants $\sim 2 \times 10^9 \, M^{-1} \, s^{-1}$). The human CuZnSOD, SOD1 is a 32 kDa homodimer, each subunit made up of an eight-stranded β-barrel (Figure 14.11), as we saw in Chapter 4, with one copper and one zinc site, and contains an intra-subunit disulfide.

Amyotrophic lateral sclerosis (ALS, also referred to as motor neurone disease or Lou Gehrig's disease[4]) is a fatal disease that targets essentially the motor neurones (which bring messages from the brain to the muscles). As they are destroyed, denervation and muscular atrophy causes weakness and finally paralysis. Despite the degradation of physical ability, mental activity is not usually affected due to the selectivity of the disease for motor neurones. Most cases of ALS, which affects worldwide 0.5–3 in 100,000 people per year, have no genetic factor implicated (known as sporadic). Of the 5–10% remaining familial cases, around 20–25% map to the SOD1 gene. More than 100 mutations have been identified and

---

[4] Henry Louis Gehrig a New York Yankees first baseman, inducted into the American Baseball Hall of Fame in 1939, died 2 years later. The disease is so rare that it became known due to him and is widely known as 'Lou Gehrig's disease'. For more informtion on metals and neurodegenerative diseases see Crichton and Ward, 2006.

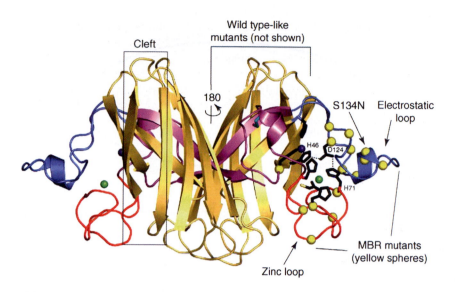

Figure 14.11 Human SOD1 showing the disulfide bond in blue in the β-strand which forms the bulk of the dimer interface (in blue) with the Cu and Zn ions as blue and green spheres. In the right monomer, MBr mutations, are shown as yellow spheres. (From Hart, 2006. Copyright 2006, with permission from Elsevier.)

can be divided into two groups (i) wild-type like, in which the metal content is similar to that found in wild-type protein and (ii) 'metal-binding region' (MBR) because they are in metal-binding ligands, or in the electrostatic and zinc loop elements intimately involved in metal binding. Positions of MBR mutants are shown by yellow spheres in Figure 14.11. Purified MBR mutants are deficient in copper and zinc, and with few exceptions, the loss of metal ions as well as reduction of the intrasubunit disulfide bond destabilize the protein, ultimately leading to the formation of aggregates of SOD1 protein which may play a role in the pathogenesis of ALS.

COPPER ENZYMES INVOLVED WITH OTHER LOW-MOLECULAR WEIGHT SUBSTRATES

Copper enzymes are involved in reactions with a large number of other, mostly inorganic substrates. In addition to its role in oxygen and superoxide activation described above, copper is also involved in enzymes that activate methane, nitrite and nitrous oxide. The structure of particulate methane mono-oxygenase from the methanotrophic bacteria *Methylococcus capsulatus* has been determined at a resolution of 2.8 Å. It is a trimer with an $\alpha_3\beta_{33}$ polypeptide arrangement. Two metal centres, modelled as mononuclear and dinuclear copper, are located in the soluble part of each β-subunit, which resembles CcOx subunit II. A third metal centre, occupied by Zn in the crystal, is located within the membrane.

While nitrite reductases in many bacteria are haem proteins, some are copper containing homotrimers, which bind three Type I and three Type II copper centres. The Type I copper

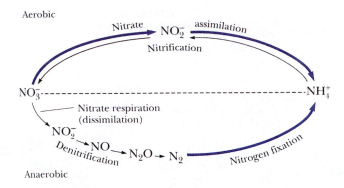

Figure 14.12 The biological nitrogen cycle. Living organisms acquire nitrogen from incorporation of NH_4 into organic molecules.

centre serves to transfer electrons from donor proteins (cupredoxins and cytochromes) to the Type II centre that has been proposed to be the site of substrate binding.

Nitrous oxide reductases, which catalyse the final step in the denitrification[5] process by reducing N_2O to N_2, are particularly interesting. Organisms that carry out denitrification use oxidized forms of nitrogen instead of oxygen as the terminal electron acceptors for anaerobic respiration (Figure 14.12), which is coupled, via proton-pumping, to ATP synthesis. N_2O reductase is also of environmental interest, since not only is N_2O the third most important greenhouse gas (after CO_2 and CH_4) but also a potentially attractive oxo-transfer reagent to oxidize organic substrates in a green reaction, where the only by-product is N_2! Nitrous oxide reductase contains two copper sites designated Cu_A and Cu_Z. The Cu_A site is the well-characterized mixed-valence dinuclear electron-transfer site with two coppers bridged by two Cys ligands, which we already encountered in cytochrome oxidase (and which is also found in NO reductase). The structure of the Cu_Z proved to be quite unusual, namely a μ_4-sulfide bridged tetranuclear copper cluster (Figure 14.13). The Cu_Z centre is located in the N-terminal domain of the dimeric enzyme, whereas the Cu_A centre is located in the C-terminal domain of each subunit. Thus in the dimeric protein structure the neighbouring Cu_A and Cu_Z sites are contributed by different subunits. While the $[Cu_4S]$ cluster has approximate two-fold symmetry, with very similar Cu–S bond lengths, the Cu–Cu distances are very different with three copper centres, designated as Cu_{II}, Cu_{III} and Cu_{IV}, closer to one another, with Cu_I further away. The entire $[Cu_4S]$ cluster is coordinated by seven His ligands to the protein, with an additional as yet unidentified oxygen ligand at the Cu_I/Cu_{IV} edge. This Cu_I/Cu_{IV} edge is thought to be the substrate-binding site. The catalytically relevant form of the Cu_Z is the fully reduced state with Cu^I at each of the four coppers. In the proposed mechanism (Figure 14.14), the reduction of N_2O to N_2 is assumed to involve binding of the N_2O substrate at the Cu_I/Cu_{IV} edge where it could interact with Cu_I and Cu_{IV} in a bridged-binding mode. Simultaneous donation of electrons from Cu_I and

---

[5] In denitrification, part of the biological nitrogen cycle, nitrate in the soil is converted via four enzymatic reactions stepwise to nitrite, nitric oxide and nitrous oxide to finally yield gaseous nitrogen.

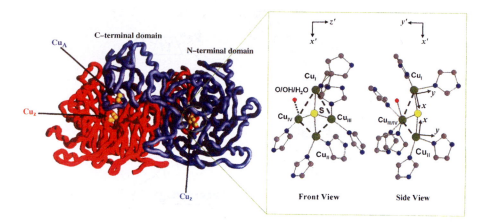

Figure 14.13 Structure of the subunits of the homodimer nitrous oxide reductase (red and blue) and of the Cu_Z site. (From Chen et al., 2004. Reproduced with permission from John Wiley & Sons., Inc.)

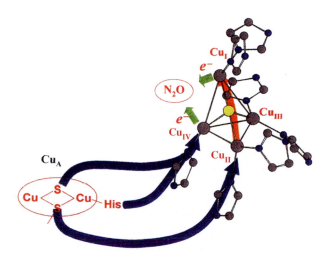

Figure 14.14 Reduction of N_2O at the Cu_Z site. (From Chen et al., 2004. Reproduced with permission from John Wiley & Sons., Inc.)

Cu_{IV} would allow the two-electron reduction of N_2O. Good electron-transfer pathways exist from the neighbouring Cu_A centre in the second subunit of the dimeric protein to Cu_{II} and Cu_{IV}, to allow rapid re-reduction of the Cu_Z centre.

MARS AND VENUS: THE ROLE OF COPPER IN IRON METABOLISM

Very early studies established that copper deficiency is associated with anaemia in a number of animals. However, the key to understanding the interaction between copper and iron came

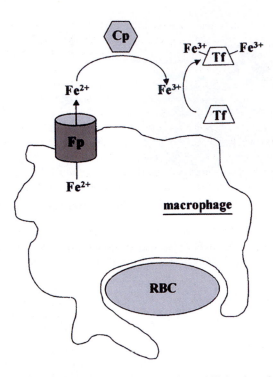

Figure 14.15 Representation of the role of ceruloplasmin in mobilizing iron from reticuloendothelial cells. (From Hellman and Gitlin, 2002. Reprinted with permission from Annual Reviews.)

from the observations that in yeast mutations affecting copper metabolism blocked the high affinity iron uptake system. Whether the mutations were in the plasma membrane copper transporters or in the copper chaperone P-type ATPase Atx1, which inserts iron into the Fet3 oxidase, the outcome was the same, and for the same reason a multi-copper oxidase is required for high affinity iron uptake into yeast. It then came as no surprise to find that in the rare human neurological disease, aceruloplasminaemia, iron accumulated in brain and liver indicated that a key role of ceruloplasmin was in tissue iron mobilization (Figure 14.15). This was convincingly shown by studies in which the yeast Fet3 oxidase was shown to restore iron homeostasis in aceruloplasminaemic mice. The likely mechanism is shown in Figure 14.15, in which the export of iron via the Fe^{2+} transporter ferroportin is thought to require the ferroxidase activity of ceruloplasmin to ensure its incorporation into apotransferrin. For more information on iron-copper interactions see Crichton, 2001.

REFERENCES

Bento, I., Carrondo, M.A. and Lindley, P.F. (2006) Reduction of dioxygen by enzymes containing copper, *J. Biol. Inorg. Chem.*, **11**, 539–547.
Brunori, M., Giuffrè, A. and Sarti, P. (2005) Cytochrome c oxidase, ligands and electrons, *J. Inorg. Biochem.*, **99**, 324–336.

Chen, P., Gorelsky, S.I., Ghosh, S. and Solomon, E.I. (2004) N_2O reduction by the μ_4-sulfide-bridged tetranuclear Cu_Z cluster active site, *Angew. Chem. Int. Ed.*, **43**, 4132–4140.

Crichton, R.R. (2001) *Inorganic Biochemistry of Iron Metabolism: From Molecular Mechanisms to Clinical Consequences*, Wiley, Chichester, 326 pp.

Crichton, R.R. and Pierre, J.-L. (2001) Old iron, young copper: from Mars to Venus, *BioMetals*, **14**, 99–112.

Crichton, R.R. and Ward, R.J. (2006) *Metal-Based Neurodegeneration from Molecular Mechanisms to Therapeutic Strategies*, Wiley, Chichester, 227 pp.

Granata, A., Monzani, E. and Casella, L. (2004), Mechanistic insight into the catechol oxidase activity by a biomimetic dinuclear copper complex, *J. Biol. Inorg. Chem.*, **9**, 903–913.

Harris, Z.L., Davis-Kaplan, S.R., Gitlin, J.D. and Kaplan, J. (2004) A fungal multicopper oxidase restores iron homeostasis in aceruloplasminemia, *Blood*, **103**, 4672–4673.

Hart, P.J. (2006) Pathogenic superoxide dismutase structure, folding, aggregation and turnover, *Curr. Opin. Chem. Biol.*, **10**, 131–138.

Hatcher, L.Q. and Karlin, K.D. (2004) Oxidant types in copper-dioxygen chemistry: the ligand coordination defines the Cu_n-O_2 structure and subsequent reactivity, *J. Biol. Inorg. Chem.*, **9**, 669–683.

Hellman, N.E. and Gitlin, J.D. (2002) Ceruloplasmin metabolism and function, *Annu. Rev. Nutr.*, **22**, 439–458.

Lieberman, R.L. and Rosenzweig, A.C. (2005) Crystal structure of a membrane-bound metalloenzyme that catalyses the biological oxidation of methane, *Nature*, **434**, 177–182.

Messerschmidt, A., Huber, R., Poulos, T. and Weighardt, K. (eds.) (2001) *Handbook of Metalloproteins*, Wiley, Chichester, 227 pp.

Potter, S.Z. and Valentine, J.S. (2003) The perplexing role of copper–zinc superoxide dismutase in amyotrophic lateral sclerosis (Lou Gehrig's disease), *J. Biol. Inorg. Chem.*, **8**, 373–380.

Riva, S. (2006) Laccases: blue enzymes for green chemistry, *TIBS*, **24**, 219–226.

Vallee, B.L. and Williams R.J. (1968) Metalloenzymes: the entatic nature of their active sites. Proc. Natl. Acad. Sci. USA, **59**, 498–505.

– 15 –

Nickel and Cobalt: Evolutionary Relics

INTRODUCTION: COMPARISON OF NICKEL AND COBALT

When one examines the kinds of reactions catalysed by nickel and cobalt enzymes and their evolutionary distribution, one arrives at the conclusion that these two elements were particularly important in the metabolism of chemicals particularly abundant in the pre-oxygen evolutionary era, such as methane, carbon monoxide and hydrogen. This is reflected in the high levels of both elements in a number of anaerobic bacteria. In contrast the level of both metals in mammalian serum is less than 100-fold that of zinc, iron or copper. Nonetheless, cobalt, through its involvement in a number of important vitamin B_{12}-dependent enzymes continued to be used in higher organisms, including mammals. In contrast, with the exception of the plant enzyme urease, nickel proteins are virtually unknown in higher eukaryotes.

Both nickel and cobalt, together with iron, have the characteristic that they are electron-rich. Furthermore, in lower oxidation states some of their 3d electrons are forced into exposed σ-(or π-) orbitals: the outcome is that tetragonal Co(II) or Ni(III) are reactive free radicals, able to give or take a single electron, in the same way as σ-organic free radicals. So cobalt functions in free-radical reactions, such as the transformation of ribonucleotides into their corresponding deoxy derivatives, just like iron. The participation of cobalt or nickel in acid–base chemistry could easily be replaced by zinc, while any redox functions in a post-oxygen world could readily be substituted by iron, copper or manganese, all of which were much more bioavailable. So nickel, in particular, but also cobalt became the lost leaders of the post-photosynthetic supermarket shelf of biometals.

We begin by considering nickel enzymes, and then move on to cobalt, concentrating on enzymes with cobalamine cofactors, including also some consideration of a number of recently discovered non-corrin cobalt enzymes. For reviews see Bannerjee and Ragsdale, 2003; Hegg, 2004; Mulrooney and Hausinger, 2003; Ragsdale, 1998, 2004, 2006.

NICKEL ENZYMES

Urease

Historically the earliest Ni-containing enzyme to be described was urease from jack bean meal, which was crystallized by James Sumner in 1926[1]. However, analytical techniques did not allow urease to be recognized as a Ni-containing enzyme until 50 years later. Urease catalyses the hydrolysis of urea to ammonia and carbamate, which spontaneously hydrolyses to give carbonic acid and a second molecule of ammonia.

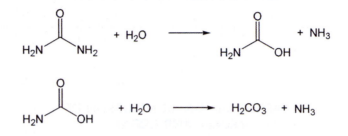

It plays a key role in nitrogen metabolism in plants and microbes whereas land dwelling animals excrete urea as the end product of their nitrogen metabolism; clearly, they do not produce urease. The active site (Figure 15.1) contains two Ni ions, ~3.5 Å apart which are bridged by a carbamylated lysine residue. Both Ni ions are coordinated by two His nitrogen atoms, oxygen from the bridging carbamyl group and an oxygen from bound water.

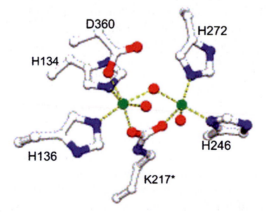

Figure 15.1 Dinuclear Ni active site of urease. Ni atoms are shown in green, metal-bound waters as red spheres; the carbamylated Lys is K217*. (From Mulrooney and Hausinger 2003. Copyright 2003, with permission from Elsevier.)

---

[1] James Sumner received the Nobel Prize for Chemistry in 1946 for the crystallization of proteins. Richard Willstätter, the 1915 Chemistry prizewinner, had proposed that proteins were not enzymes, and that the protein in urease was simply a scaffold for the veritable catalyst. Since urease is inactive without Ni, he was not so far wrong!

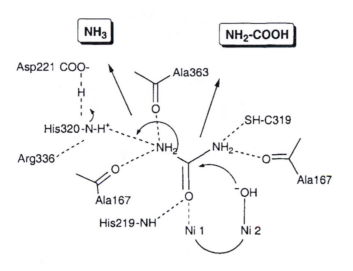

Figure 15.2 Reaction mechanism of urease. Ni 1 binds urea and acts as a Lewis acid to polarise the carbonyl group, making its carbon more electrophilic, while Ni 2 facilitates deprotonation of a bound water molecule to generate a nucleophilic hydroxyl species. (From Ragsdale, 1998. Copyright 1998, with permission from Elsevier.)

One of the Ni atoms in addition has an oxygen ligand from an Asp residue. CO_2 is required for formation of the carbamylated Lys bridge between the two Ni atoms, and mutation of this Lys results in loss of activity. A proposed reaction mechanism is presented in Figure 15.2. The large kinetic barrier to urea hydrolysis is presumed to be lowered by (i) coordination of the carbonyl group of urea to Ni 1, making the carbonyl more electrophilic, (ii) binding of water to Ni 2 to generate an activated hydroxyl species and (iii) hydrogen bonding interactions of all four of the protons of urea with electrophilic groups of the protein. Attack of the metal-activated hydroxyl would generate a tetrahedral intermediate. Protonation of this intermediate would eliminate ammonia, leaving carbamate bridged between the two Ni atoms. Dissociation of carbamate from the dimetallic site would be followed by spontaneous hydrolysis to carbonate and a second molecule of ammonia, with a protonated His residue acting as a general acid to promote ammonia release.

Ni–Fe–S Proteins

Three of the eight Ni proteins that are known as hydrogenase, CO dehydrogenase (CODH), and acetylCoA synthase (ACS), are Ni–Fe–S proteins. Hydrogenases play an important role in microbial energy metabolism by catalysing the reversible oxidation of hydrogen:

$$H_2 \rightleftharpoons 2H^+ + 2e^-$$

In some anaerobic microorganisms, production of hydrogen serves as a mechanism to get rid of excess reducing potential, while in many others hydrogen consumption is coupled

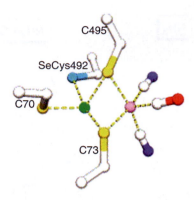

Figure 15.3 The dinuclear Ni–Fe site of the hydrogenase from *D. norvegium* in the reduced form. The diatomic ligands bound to the Fe are CO and two CN⁻molecules; Ni, *green*; Fe, *pink*; S, *yellow*; and Se, *light blue*. (From Mulrooney and Hausinger 2003. Copyright 2003, with permission from Elsevier.)

to the reduction of carbon dioxide, oxygen, sulphate or other electron acceptors and at the same time used to generate a proton gradient for use in ATP production. Three kinds of hydrogenases have been described: namely, the [Ni–Fe] hydrogenases (including the sub-family of [Ni–Fe–Se] hydrogenases), the [Fe] hydrogenases, and the metal-free hydrogenases[2]. The [Ni–Fe] hydrogenases are the most abundant, and are characterized by a quite unusual Ni–Fe active site (Figure 15.3). It required a combination of spectroscopic and crystallo-graphic studies to identify the three non-protein diatomic ligands to the Fe as one CO and two CN⁻ molecules. Whereas in *Desulfomicrobium norvegium* one of the Ni ligands is a selenocysteine, in most other Ni–Fe hydrogenases, the four protein ligands to the Ni atom are all Cys residues. Two of them are bridging ligands to the iron atom. While the mecha-nism of the Ni–Fe hydrogenases remains uncertain, it seems clear that different redox states of Ni are involved, while other components of the catalytic site may also play a sig-nificant role in catalysis. The Ni–Fe hydrogenases also contain multiple Fe–S clusters which channel electrons to the catalytic site.

Carbon monoxide dehydrogenase (CODH)/acetylCoA synthases (ACS) are thought to be ancient enzymes, which possibly allowed primitive organisms to live in the anaerobic, CO_2-rich atmosphere. Microorganisms that contain CODH/ACS enzymes are found in all locations where anaerobic metabolism is the only means of survival, from peat bogs to cow rumens to human intestine. The so-called C-cluster of CODHs allow organisms to use CO as a source of energy and carbon, while other acetogenic and methanogenic bacteria use bifunctional CODH/ACS enzymes to convert the greenhouse gas CO_2 to acetylCoA. Collectively CODH/ACS enzymes play a key role in the C1 metabolism of anaerobic organ-isms and represent a major component of the global carbon cycle.

CODHs catalyse the oxidation of carbon monoxide in a reversible, two-electron process. They are homodimeric enzymes with five metal clusters, two C-clusters that catalyse the oxidation of CO to CO_2 and three typical [Fe_4S_4] cubane clusters (Figure 15.4).

---

[2] Recently, it has been found that 'metal-free' hydrogenase do contain functionally active Fe.

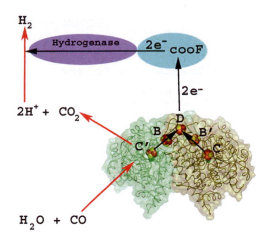

Figure 15.4 The coupling of CODH activity in *R. rubrum* with hydrogenase activity. (From Drennan et al., 2004. With kind permission of Springer Science and Business Media.)

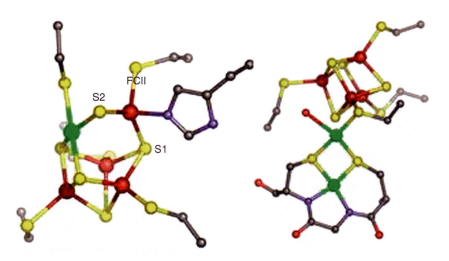

Figure 15.5 (Left) The C-cluster of CODH from *C. hydrogenoformans* and (right) the A-cluster from *C. hydrogenoformans*. Fe, *rust*; S, *yellow*; Ni, *green*; N, *blue*; C, *grey*; O, *red*. (From Drennan et al., 2004. With kind permission of Springer Science and Business Media.)

In *Rhodospirillum rubrum*, electrons are transferred from the D-cluster to a membrane-associated Fe–S protein designated CooF, which transfers electrons to a hydrogenase, coupling CO oxidation with H_2 production. The C-cluster is an unusual Fe–Ni–Fe_3S_{4-5}, which can be best viewed as a $[Fe_3S_4]$ cluster bridged to a dinuclear Ni–Fe centre (Figure 15.5). The catalytic efficacy of the C-cluster from *Carboxydothermus hydrogenoformans* is remarkable—for the oxidation of CO the turnover number is 39,000 s^{-1} and the k_{cat}/K_m greater than $10^9 M^{-1}s^{-1}$.

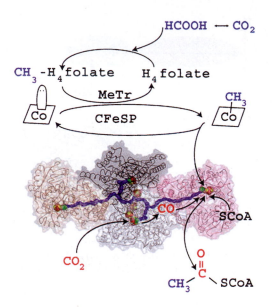

Figure 15.6 The Wood–Ljungdahl pathway. One molecule of CO_2 (blue) is converted to formate and then reduced to a methyl group, which is then transferred to the corrinoid-iron-sulphur protein CFeSP. CFeSP transfers the methyl group to the A-cluster of the bifunctional CODH/ACS. The other molecule of CO_2 (red) is reduced to CO by the C-cluster of the CODH subunit. The CO is then transferred to the A-cluster through a long channel, some 70 Å long, where with the methyl group and CoA it forms acetylCoA. (From Drennan et al., 2004. With kind permission of Springer Science and Business Media.)

Many anaerobic microorganisms which produce acetate or methane do so by reducing CO_2 to CO and then using CO as a metabolic intermediate. This fixation of CO_2 is known as the Wood–Ljungdahl pathway, and involves the overall conversion of two molecules of CO_2 to the methyl and carbonyl groups of acetylCoA, which is then used for energy production (Figure 15.6). CODH/ACS is responsible for the reduction of CO_2 to CO and the subsequent formation of acetylCoA. The C-cluster of CODH reduces CO_2 to CO, which is then converted by the A-cluster of the ACS to acetylCoA together with a methyl group and coenzyme A. The methyl group is derived from CO_2 in a series of reductive steps and incorporated into methyltetrahydrofolate (CH_3H_4folate). It is then transferred from methyltetrahydrofolate by a methyltransferase (MeTr) to a corrinoid iron–sulphur protein (CFeSP), which in turn transfers the methyl group to the A-cluster. The active site A-cluster of the ACS is unusual in that it consists of a [4Fe–4S] cubane unit linked via a bridging cysteine residue to a proximal metal ion which, in turn, is connected to a square-planar distal Ni via two cysteine bridges as shown in Figure 15.5. The identity of the catalytically active metal ion in the proximal site has been the subject of some debate, but it now seems to be agreed that the active enzyme has Ni in this site (Svetlitchnyi et al., 2004). Cluster-A binds both CO, the methyl group from CoFeSP and CoA. Two mechanisms of acetylCoA synthesis at cluster-A have been proposed. In the mononuclear mechanism, both CO and the methyl group bind at the proximal Ni, generating an acetyl group; subsequent attack of the

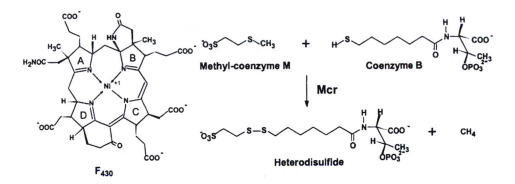

Figure 15.7 The Ni-containing tetrapyrrole F_{430} of methylCoM reductase and the reaction catalysed by methylCoM reductase (Mcr). (From Shima et al., 2002. Copyright 2002, with permission from Elsevier.)

carbonyl carbon by deprotonated CoA–S⁻ then gives acetylCoA. In the binuclear mechanism CO binds at the proximal Ni and the methyl group at the distal Ni.

METHYL-COENZYME M REDUCTASE

It is estimated that more than 10^9 tons of methane per year are generated by methanogenic archaebacteria functioning in anaerobic environments. The final step in methane formation is catalysed by methylCoM reductase (Mcr), and the structure of the enzyme from *Methanobacterium thermoautotrophicum* has been determined at high resolution in two states with substrate bound. Ni is present in the enzyme in a corrinoid cofactor designated F_{430} on account of its absorption maximum at 430 nm (Figure 15.7). The enzyme is a heterotrimer with two active sites, each with a Ni-containing tetrapyrrole. The cofactor F_{430} is non-covalently, but tightly bound, deeply buried in the protein, but connected to the surface by a 30 Å long channel through which the substrates enter. The reaction involves the substrates, methyl-S-coenzyme M (CH_3–S–CoM) and N-7-mercaptoheptanoylthreonine phosphate, or coenzyme B (CoB–SH), which are converted into methane and the heterodisulphide CoB–S–S–CoM.

COBALAMINE AND COBALT PROTEINS

Vitamin B_{12}, identified as the antipernicious anaemia factor in 1925, is a tetrapyrrole cofactor in which the central hexacoordinate cobalt atom is coordinated by four equatorial nitrogen ligands donated by the pyrroles of the corrin ring (Figure 15.8). The fifth Co ligand is a nitrogen atom from a 5,6-dimethylbenzimidazole nucleotide (Dmb) covalently linked to the corrin D ring. The sixth ligand in vitamin B_{12} is –CN. In the coenzyme B_{12} (AdoCbl) this ligand is 5′-deoxyadenosine, while in the other biologically active alkylcobalamine (MeCbl), it is a methyl group. This sixth ligand is unusual in that it forms a C–Co bond–carbon-metal bonds are rare in biology. The free cofactor can exist in the

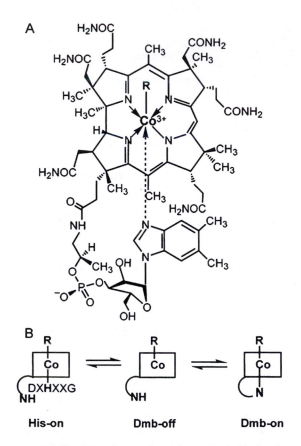

Figure 15.8 (a) Structure and (b) alternative conformations of cobalamine found in B_{12}-dependent enzymes. The functional group R is deoxyadenosine in AdoCbl, methyl in MeCbl and –CN in vitamin B_{12}. (From Bannerjee and Ragsdale, 2003. Reprinted with permission from Annual Reviews.)

base-on (Dmb-on) or base-off (Dmb-off) conformations, with the Dmb-on form predominant at physiological pH. In some B_{12}-dependent enzymes an active site His residue replaces the dimethylbenzimidazole (the so-called His-on form). In the corrinoid iron–sulphur protein (CFeSP) involved in the CODH/ACS system described earlier, the cofactor is in the Dmb-off conformation and a protein ligand does not appear to occupy the lower axial position. The reactive C–Co bond participates in all three classes of enzymes that use cobalamine cofactors, namely the adenosylcobalamine-dependent isomerases, the methylcobalamine-dependent methyltransferases and the reductive dehalogenases. We will discuss the first two classes in greater detail here in addition to a number of non-corrin cobalt-containing enzymes.

B_{12}-DEPENDENT ISOMERASES

Isomerases are the largest subfamily of B_{12}-dependent enzymes found in bacteria, which play important roles in fermentation pathways. The only exception is methylmalonylCoA

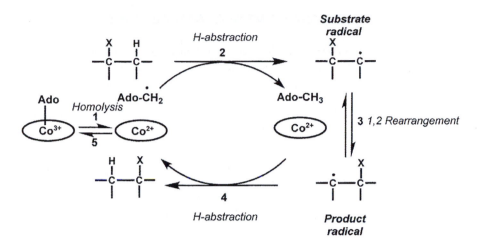

Figure 15.9 General reaction mechanism for AdoCbl-dependent isomerases. (From Bannerjee and Ragsdale, 2003. Reprinted with permission from Annual Reviews.)

mutase, an enzyme required for the metabolism of propionylCoA in man as well as in bacteria. The general reaction mechanism (Figure 15.9) for AdoCbl-dependent isomerases involves homolytic formation of a 5′-deoxyadenosyl radical (step 1) followed by H abstraction to generate a substrate radical (step 2). Once the substrate radical has been formed it can undergo 1,2 rearrangement (step 3) to generate the product radical. Hydrogen abstraction will then result in the product and the 5′-deoxyadenosyl radical (step 4), which can revert to the initial B_{12} coenzyme (step 5). As pointed out in Chapter 13, some microorganisms, such as *Lactobacillii* have B_{12}-dependent Class II ribonucleotide reductases, where a thiyl radical is generated rather than a substrate radical by the deoxyadenosyl radical. The thiyl radical is common to both the diiron-tyrosyl radical-dependent and the B_{12}-dependent ribonucleotide reductases, and is responsible for in turn generating the substrate radical. It is interesting to point out that unlike the other B_{12}-dependent isomerases, the B_{12}-dependent Class II ribonucleotide reductase has a different fold for binding B_{12}, which is similar to the corresponding structural elements used in the Class I diiron-tyrosyl ribonucleotide reductases.

The Co–carbon bond in AdoCbl is stable in water, but is inherently labile, with bond dissociation energy of around 30–35 kcal mol^{-1}. This instability is exploited by the AdoCbl-dependent isomerases to effect radical-based rearrangements that, as pointed out above, are initiated by homolytic cleavage of the Co–carbon bond. In the absence of substrate, the homolysis products are not observed, yet in their presence the homolytic cleavage rate is considerably accelerated. It is proposed that the homolysis equilibrium favours recombination, whereas in the presence of substrate, the high-energy **dAdo** abstracts a hydrogen atom from the substrate to generate a more stable substrate-centred radical intermediate. This has the net effect of shifting the overall homolysis equilibrium from recombination to radical propagation.

B$_{12}$-DEPENDENT METHYLTRANSFERASES

B$_{12}$-dependent methyltransferases are involved in C1 metabolism and, as we saw earlier, in CO$_2$ fixation in anaerobic microorganisms. They also play an important role in amino acid metabolism in many organisms, including humans. They catalyse the transfer of methyl groups from a methyl donor to a methyl acceptor (Figure 15.10a), with a B$_{12}$-containing protein acting as the intermediate carrier of the methyl group. The methyltransferases involve three protein components, each of which is localized on a different polypeptide or domain. The first, an MT1 component binds the methyl donor (CH$_3$–X) and transfers it to the B$_{12}$-containing protein, leading to the formation of an organometallic methylcobalt intermediate. The third component (MT2) catalyses the transfer of the Co-bound methyl group to the acceptor Y$^-$. The methyl donor can be any one of a number of molecules such as methyltetrahydrofolate, while the methyl acceptor can be, for example, homocysteine to give methionine, or the CODH/ACS bifunctional complex to form acetylCoA. The MT2 enzymes all appear to contain Zn, which both coordinates and activates the thiolate methyl acceptor (Figure 15.10b). However, in the transfer of a methyl group within the CODH/ACS system a different type of reaction is involved in which the methyl group is transferred from Co to Ni.

The best characterized B$_{12}$-dependent methyltransferases is methionine synthase (Figure 15.11) from *E. coli*, which catalyses the transfer of a methyl group from methyltetrahydrofolate to homocysteine to form methionine and tetrahydrofolate. During the catalytic cycle, B$_{12}$ cycles between CH$_3$–Co(III) and Co(I). However, from time to time, Co(I) undergoes oxidative inactivation to Co(II), which requires reductive activation. During this process, the methyl donor is S-adenosylmethionine (AdoMet) and the electron donor is flavodoxin (Fld) in *E. coli*, or methionine synthase reductase (MSR) in humans. Methionine synthase

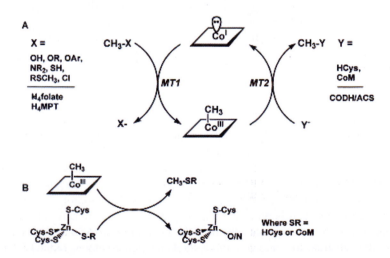

Figure 15.10 (a) The three components involved in the B$_{12}$-dependent methyltransferases. (b) The MT2 enzymes have a thiol group which activates the thiol acceptor. (From Bannerjee and Ragsdale, 2003. Reprinted with permission from Annual Reviews.)

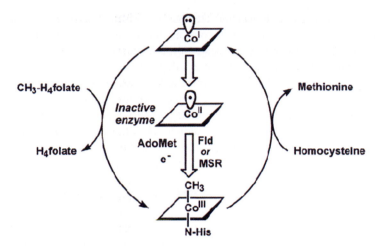

Figure 15.11 Reactions catalysed by cobalamin-dependent methionine synthase. (From Bannerjee and Ragsdale, 2003. Reprinted with permission from Annual Reviews.)

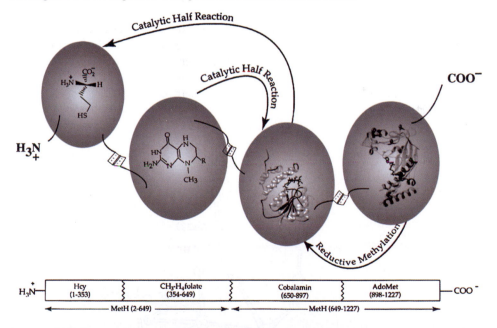

Figure 15.12 The modular structure of methionine synthase. The four domains are connected by flexible hinges, which allow the CH$_3$tetrahydrofolate-, AdoMet- or homocystein-binding domains to alternatively access the B$_{12}$-binding domain. (From Bannerjee and Ragsdale, 2003. Reprinted with permission from Annual Reviews.)

is a modular enzyme, with separate domains for binding of homocysteine, methyltetrahydrofolate, B$_{12}$ and AdoMet (Figure 15.12). The B$_{12}$ domain in its different oxidation states must interact with each of the other three domains: the Co(I) form with methyltetrahydrofolate, the inactive Co(II) form with the AdoMet binding domain, and the CH$_3$–Co(III) form with the homocysteine binding domain. When the cobalamin binds, the lower axial

Dmb ligand is replaced by His to generate the His-on conformation (Figure 15.8B). This His residue is part of a catalytic triad which controls the coordination state of cobalt (His-on/His-off) by modulating the protonation state of the histidine.

As pointed out earlier a third class of B_{12}-dependent enzymes, present in anaerobic microbes, carry out reductive dehalogenation reactions, which play an important role in the detoxification of chlorinated aliphatic and aromatic compounds, among which are many important man-made pollutants. The role of B_{12} in this class of enzymes is not clear— possibly by formation of an organocobalt adduct, as in the case of methyltransferases or alternatively by the corrinoid serving as an electron donor.

NON-CORRIN Co-CONTAINING ENZYMES

Non-corrin cobalt has a number of interesting applications in the chemical industry, for example in the hydroformylation (OXO) reaction between CO, H_2 and olefins. A number of non-corrin Co-containing enzymes have been described, including methionine aminopeptidase, prolidase, nitrile hydratase and glucose isomerase. We describe the best characterized of these, namely the *E. coli* methionine aminopeptidase, a ubiquitous enzyme, which cleaves N-terminal methionine from newly translated polypeptide chains. The active site of the enzyme (Figure 15.13) contains two Co(II) ions that are coordinated by the side-chain atoms of five amino acid residues. The distance between the two Co^{2+} is similar to that between the two Zn^{2+} atoms in leucine aminopeptidase, and indeed the catalytic mechanism of methionine aminopeptidase shares many features with other metalloproteases, in particular leucine aminopeptidases.

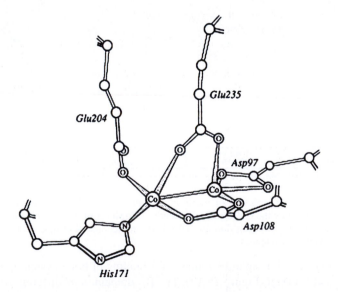

Figure 15.13 Structure of the bimetallic core of *E. coli* methionine aminopeptidase. (From Kobayashi and Shimizu, 1999. Reproduced with permission of Blackwell Publishing Ltd.)

Prolidases are widespread in bacteria, and specifically cleave Pro-containing dipeptides, where the Pro residue is C-terminal. The enzyme from the hyperthermophilic archaebacteria *Pyrococcus furiosus* has one tightly bound Co^{2+}, but requires a second Co^{2+} for catalytic activity, suggesting that it may have a similar mechanism of action as methionine aminopeptidase. Nitrile hydratase catalyses the hydration of nitriles to amides and is used industrially on the kiloton scale for the production of acrylamide and nicotinamide from the corresponding nitriles. It has an obligate requirement for Co, which appears to be bound to three Cys residues in a similar motif to that found in iron-containing nitrile hydratases from other microorganisms. Finally, one of the most widely used industrial enzymes, glucose isomerase (which in fact is a xylose isomerase), a Co^{2+}-requiring enzyme, is used in the production of fructose-high corn syrup from glucose[3]. However, it is important to reduce the amount of Co^{2+} utilized in high-fructose corn syrup production because of health hazards, and substitution of other metals for cobalt is considered preferable.

REFERENCES

Banerjee, R. and Ragsdale, S.W. (2003) The many faces of vitamin B_{12}: catalysis by cobalamin-dependent enzymes, *Annu. Rev. Biochem.*, **72**, 209–247.

Drennan, C.L., Doukov, T.I. and Ragsdale, S.W. (2004) The metalloclusters of carbon monoxide dehydrogenase/acetyl-CoA synthase: a story in pictures, *J. Biol. Inorg. Chem.*, **9**, 511–515.

Hegg, E.L. (2004) Unravelling the structure and mechanism of acetyl-coenzyme A synthase, *Acc. Chem. Res.*, **37**, 775–783.

Kobayashi, M. and Shimizu, S. (1999) Cobalt proteins, *Eur. J. Biochem.*, **261**, 1–9.

Mulrooney, S.B. and Hausinger, R.P. (2003) Nickel uptake and utilisation by microorganisms, *FEMS Microbiol. Rev.*, **27**, 239–269.

Ragsdale, S.W. (1998) Nickel biochemistry, *Curr. Opin. Chem. Biol.*, 2208–2215.

Ragsdale, S.W. (2004) Life with carbon monoxide, *Crit. Rev. Biochem. Mol. Biol.*, **39**, 165–195.

Ragsdale, S.W. (2006) Metals and their scaffolds to promote difficult enzymatic reactions, *Chem. Rev.*, **106**, 3317–3337.

Shima, S., Warkentin, E., Thauer, R.K. and Ermler, U. (2002) Structure and function of enzymes involved in the methanogenic pathway utilising carbon dioxide and molecular hydrogen, *J. Biosci. Bioeng.*, **93**, 519–530.

Svetlitchnyi, V., Dobbek, H., Meyer-Klauke, W., Meins, T., Thiele, B., Römer, P., Huber, R. and Meyer, O. (2004) A functional Ni-Ni-[4Fe-4S] cluster in the monomeric acetyl-CoA synthase from *Carboxydothermus hydrogenoformans*, *Proc. Natl. Acad. Sci. U.S.A.*, **101**, 446–451.

---

[3] Fructose has a much sweeter taste than glucose, hence the transformation of glucose derived from enzymatic hydrolysis of starch from corn, provides an alternative sweetener to sucrose (a disaccharide of glucose and fructose). This replaced the use of sugar cane by the US soft drinks and candy industry (and effectively destroyed the economy of Cuba in the process).

– 16 –

Manganese: Water Splitting, Oxygen Atom Donor

INTRODUCTION: MANGANESE CHEMISTRY

The importance of manganese for bacteria, such as that of Ni and to a lesser extent Co, as we saw in the last chapter, is considerable. Of course, as we will see shortly, it is also important in the tetranuclear Mn cluster that is involved in oxygen production in photosynthetic plants, algae and cyanobacteria, as well as in a number of mammalian enzymes such as arginase and mitochondrial superoxide dismutase. Most of manganese biochemistry can be explained on the one hand by its redox activity, and on the other by its analogy to Mg^{2+} (reviewed in Yocum and Pecoraro, 1999).

Manganese has access to three oxidation states of relevance to biology, Mn(II), Mn(III) and Mn(IV). A major difference with other redox active metals, such as iron, is that manganese has less reducing potential than iron under most biological conditions. While Fe^{3+} is stabilized with respect to Fe^{2+}, Mn^{2+} is stabilized relative to Mn^{3+}—this is because in both cases the half-filled d^5 shell of both Fe^{3+} and Mn^{2+} confers thermodynamic stability. Two important consequences of this redox chemistry are that, not surprisingly Mn^{2+} can participate in useful redox catalysis on many similar substrates to Fe^{3+}, whereas the higher redox potential of Mn^{2+} makes free Mn^{2+} innocuous under conditions where free Fe^{2+} would wreak havoc through the generation of hydroxyl radicals. This means that cells (notably bacterial cells) can tolerate very high cytoplasmic concentrations of Mn^{2+} with no negative consequences, which is certainly not the case with other biologically important redox metal ions such as iron and copper.

The other property of Mn^{2+}, which has important biochemical consequences, is that it is a close but not exact surrogate of Mg^{2+}. As we saw in Chapter 10, Mg^{2+} is confined to a strict octahedral coordination geometry, with ligand bond angles close to 90°, making it an ideal 'structural' cation, particularly for phosphorylated biological molecules. Mn^{2+} with its relatively similar ionic radius readily exchanges with Mg^{2+} in most structural environments, and exhibits much of the same labile, octahedral coordination chemistry. However, since Mn^{2+}-ligand bonds are generally much more flexible than Mg^{2+}-ligand bonds, when Mn^{2+} replaces Mg^{2+} in a catalytic environment, its flexibility is better at lowering the activation energy. It can more easily accommodate the distortions in coordination geometry in progressing from the substrate-bound to the transition state and from there to the bound product. Thus, substituting Mn^{2+} in the active site of a Mg^{2+}-enzyme often results in improved enzyme efficacy.

We begin this overview of manganese biochemistry with a brief account of its role in the detoxification of free radicals, before considering the function of a dinuclear Mn(II) active site in the important eukaryotic urea cycle enzyme arginase. We then pass in review a few microbial Mn-containing enzymes involved in intermediary metabolism, and conclude with the very exciting recent results on the structure and function of the catalytic manganese cluster involved in the photosynthetic oxidation of water.

Mn^{2+} AND DETOXIFICATION OF OXYGEN FREE RADICALS

Manganese is the cofactor for catalases, peroxidases and superoxide dismutases, which are all involved in the detoxification of reactive oxygen species (SOD). We consider here the widely distributed Mn SOD, and then briefly describe the dinuclear Mn catalases.

Mn superoxide dismutases are found in both eubacteria and archaebacteria as well as in eukaryotes, where they are frequently found in mitochondria. They (Figure 16.1) have considerable structural homology to Fe SODs: both are monomers of ~200 amino acid and occur as dimers or tetramers, and their catalytic sites are also very similar. They both catalyse the two-step dismutation of superoxide anion and, like the Cu–Zn SODs, avoid the difficulty of overcoming electrostatic repulsion between two negatively charged superoxide anions by reacting with only one molecule at a time. As in the case of Cu–Zn SOD, a first molecule of superoxide reduces the oxidized (Mn^{3+}) form of the enzyme, releasing

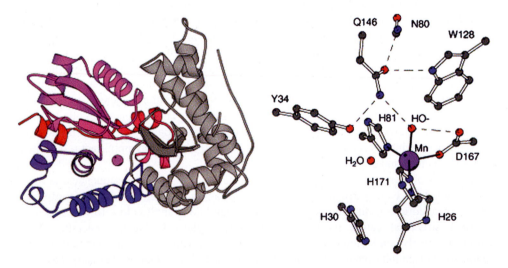

Figure 16.1 The protein fold of Mn- or Fe-SODs (left) and the active site (right). (From Miller, 2004. Copyright 2004, with permission from Elsevier.)

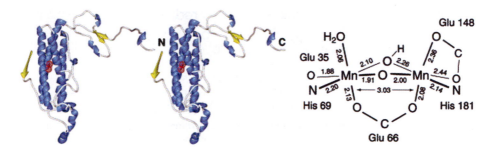

Figure 16.2 *Lactobacillus plantarum* Mn catalase: (left) stereo view of the secondary structure—the di-Mn unit as red spheres and (right) the detailed geometry of the di-Mn centre. (From Barynin et al., 2001. Copyright 2001, with permission from Elsevier.)

dioxygen, and the reduced (Mn^{2+}) form of the enzyme then reacts with a second superoxide anion and two protons, to give hydrogen peroxide, regenerating the oxidized form of the enzyme.

Catalases play an important protective role, catalysing the disproportionation of toxic hydrogen peroxide into O_2 and H_2O. In contrast to the haem-containing catalases, which are ubiquitous in aerobic organisms, a broad range of microorganisms, living in microaerophilic (almost oxygen-free) environments, including the lactic acid bacteria[1], have catalases that have a dinuclear manganese centre in their active site (Figure 16.2). These 'alternative' catalases are 4-helix bundle proteins, with the di-Mn centre located in the middle of the α-helical bundle. Like their haem-containing counterparts, the dinuclear Mn centre carries out a two-electron catalytic cycle, interconverting between reduced (Mn[II]Mn[II]) and oxidized (Mn[III]Mn[III]) states during turnover. A catalytic mechanism based on the structure of the active site has been proposed (Figure 16.3), which involves distinct pathways of reactivity in the oxidized and reduced half-reactions. H_2O_2 binds terminally to the oxidized cluster of the oxidative half-reaction, replacing the water molecule bound to one of the Mn ions (Figure 16.2); this is also the site at which azide binds in the catalase crystals. Two electron oxidation of H_2O_2 by the dinuclear Mn(III) results in release of the dioxygen product, facilitated by the oxygen bridges that electronically couple the Mn ions, allowing them to function as a unit. Glu178 is proposed to transfer the peroxide protons to active site bases, most likely the solvent bridges between the metal ions. In the reductive half-reaction, H_2O_2 is proposed to bind in a bridging position, to give a symmetric μ-bridging peroxide complex that will be activated to O–O bond cleavage and reoxidation of the di-Mn core. Glu178 could serve to protonate the non-bridging oxygen of the bound substrate. This mechanism results in replacement of one of the oxygen bridges of the cluster during each turnover cycle, with retention of a substrate oxygen atom between successive reactions.

---

[1] We mentioned earlier that this family of bacteria has adapted to its environment to function without iron, using Co and Mn instead.

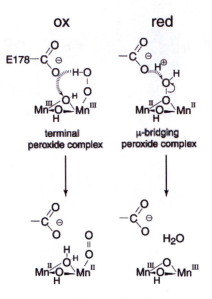

Figure 16.3 Proton transfers within the active site complex catalysed by Glu178. (From Barynin et al., 2001. Copyright 2005, with permission from Elsevier.)

NON-REDOX di-Mn ENZYMES: ARGINASE

In addition to the redox di-Mn catalases, there are a number of other enzymes with di-Mn centres, of which the best characterized are arginases that catalyse the divalent cation-dependent hydrolysis of L-arginine to form L-ornithine and urea. In mammals, hepatic arginase is the terminal enzyme of the urea cycle, which represents the major end product of nitrogen metabolism—the average adult human excretes some 10 kg of urea per year. The enzyme is not restricted to the liver, since ornithine is a precursor of the non-essential amino acid proline and a biosynthetic precursor of polyamines, required for rapidly dividing tissues. A common feature of arginases, whether eukaryotic or prokaryotic, is the requirement of divalent cations for activity, and in almost all arginases, they have two spin-coupled Mn(II)/subunit, which are some 3.3 Å apart (Figure 16.4). The dinuclear Mn(II) centre is located at the bottom of a 15 Å-deep active site cleft. An interesting feature is the hydrogen bond donated by a metal bridging hydroxide to the non-coordinating oxygen of Asp128: residues analogous to Asp128 are found in the active sites of a large number of other dimetallic hydrolases.

The mechanism that is consistent with biochemical, enzymological and structural data involves binding of arginine, in which the side-chain of Glu277 plays an important role, attack of the nucleophilic metal-bridged hydroxyl ion, formation of a neutral, tetrahedral intermediate which is stabilized by the dinuclear Mn(II) centre, and finally proton transfer from His141, followed by release of the two products (Figure 16.5).

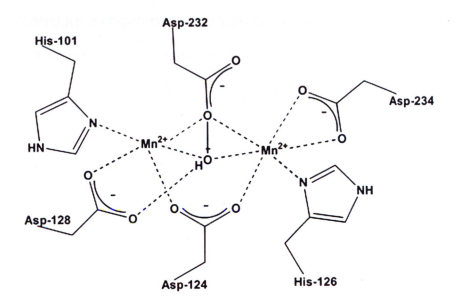

Figure 16.4 Dinuclear Mn centre of rat liver arginase. (From Ash, 2004. Copyright 2004, the American Society for Nutritional Science.)

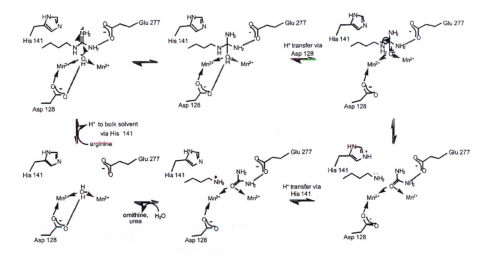

Figure 16.5 Proposed mechanism for the hydrolysis of arginine based on the crystal structure of rat liver arginase. The α-amino and α-carboxyl groups of the substrate have been omitted for clarity. (From Ash, 2004. Copyright 2004, the American Society for Nutritional Science.)

PHOTOSYNTHETIC OXIDATION OF WATER: OXYGEN EVOLUTION

The water-plastoquinone photo-oxidoreductase, also known as photosystem II (PSII), embedded in the thylakoid membrane of plants, algae and cyanobacteria, uses solar energy to power the oxidation of water to dioxygen by a special centre containing four Mn ions. The overall reaction catalysed by PSII is outlined below:

$$2Q + 2H_2O \xrightarrow{\text{light}} O_2 + 2QH_2$$

The special pair of chlorophyll molecules in PSII, often called *P680*, absorbs light at 680 nm and transfers an electron to a nearby pheophytin (chlorophyll with two H^+ in place of the central Mg^{2+}), from where it is transferred through other electron carriers to an exchangeable plastoquinone pool (Figure 16.6). The positive charge, which is formed on the special pair, *P680$^+$*, is a powerful oxidant. Each time a photon of light kicks an electron out of *P680*, *P680$^+$* extracts an electron from water molecules bound at the Mn centre, which are transferred through the redox-active TyrZ to reduce *P680$^+$* back to *P680* for yet another photosynthetic cycle. In classic experiments using an oxygen electrode and short flashes of light, it was established that four photochemical turnovers were required for every molecule of oxygen that was released, and the features of this were rationalized into a kinetic

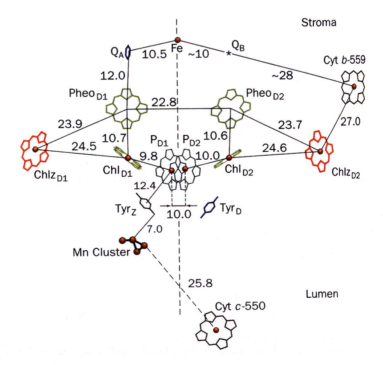

Figure 16.6 Arrangement of cofactors in the dimeric (D1 and D2) subunits of PSII. The dashed line running through the non-haem Fe is the two-fold axis of pseudosymmetry. Thin lines show the centre-to-centre distances in Å between cofactors. (From Voet and Voet, 2004. Reproduced with permission from John Wiley & Sons., Inc.)

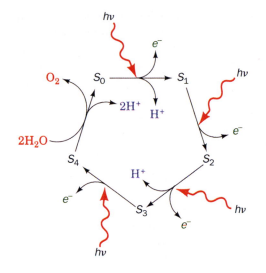

Figure 16.7 The S-state cycle model of O_2 generation. (From Voet and Voet, 2004. Reproduced with permission from John Wiley & Sons., Inc.)

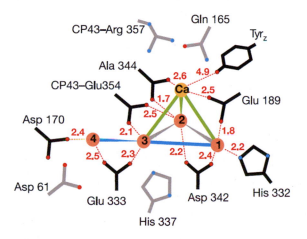

Figure 16.8 Schematic view of the Mn_4Ca cluster: amino acids of the first coordination sphere are in black, those of the second in grey: distances are in Å. (From Loll et al., 2005. Reproduced by permission of Nature Publishing Group.)

model, known as the S-state cycle (Figure 16.7). In this model, five states of the enzyme, designated S_n, are proposed to exist, with n 0–4, where each state corresponds to a different level of oxidation of the tetra-Mn centre (Kok et al., 1970). When S_4 is generated, it reacts in less than a microsecond to release dioxygen and return to the reduced form of the enzyme, S_0. The stable state of the enzyme in the dark is S_1, which corresponds to $Mn(III)_2Mn(IV)_2$, so that only three photochemical turnovers are required before O_2 is released.

The structure of this centre, which contains the Mn_4Ca cluster and the redox-active Tyr_Z residue, has been recently solved (Figure 16.8) at 3 Å resolution, and both confirms and confounds earlier structures—one of the major problems seems to be reduction of Mn^{3+} and Mn^{4+} to Mn^{2+} by X-ray generated radicals in the course of data collection from the crystals. However, what this most recent structure shows clearly is that the fourth Mn(4) caps a distorted cubane $CaMn_3O_4$ cluster, as suggested from the earlier studies. Combining crystallographic refinement with EXAFS data, the metal metal distances are: Mn1–Mn2 and Mn2–Mn3, 2.7 Å apart (probably connected by di-μ-oxo bridges), Mn1–Mn3 and Mn3–Mn4, 3.3 Å apart (suggestive of mono-μ-oxo bridges), and Ca^{2+} forming the vertex of a trigonal pyramid, equidistant (~3.4 Å) from Mn1, Mn2 and Mn3. Combining spectroscopic studies (notably FTIR) with the assumption that the S_1 state involves an oxidation distribution of $Mn(III)_2Mn(IV)_2$, the authors suggest some of the possible oxidation states of the four Mn ions in the course of the catalytic cycle. It is proposed that Mn1 and Mn3 can either be in oxidation state III or IV in the S_1 state, that Mn4, which is not oxidized during the transitions from S_0 to S_3 is present as Mn(IV), while Mn2 probably changes from Mn(III) to Mn(IV) in the S_1–S_2 transition.

Clearly, the S_0 state must have one Mn(II), one Mn(III) and two Mn(IV) while the S_4 must have three Mn(IV) and one Mn(V)—but where they are located in the tetra-Mn centre and how their four-electron reduction is coupled with water splitting and dioxygen evolution remains to be established. (For recent reviews see Goussias et al., 2002; Ferreira et al., 2004; Rutherford and Boussac, 2004; Iverson, 2006.)

REFERENCES

Ash, D.E. (2004) Arginine metabolism: enzymology, nutrition and clinical significance, *J. Nutr.*, **134**, 2760S–2764S.

Barynin, V.V., Whittaker, M.M., Antonyuk, S.V., Lamzin, V.S., Harrison, P.M., Artymiuk, P.J. and Whittaker, J.W. (2001) Crystal structure of manganese catalase from *Lactobacillus plantarum*, *Structure*, **9**, 725–738.

Ferreira, K.N., Iverson, T.M., Maghlaoui, K., Barber, J. and Iwata, S. (2004) Architecture of the photosynthetic oxygen-evolving center, *Science*, **303**, 1831–1838.

Goussias, C., Boussac, A. and Rutherford, A.W. (2002) Photosystem II and photosynthetic oxidation of water: an overview, *Phil. Trans. R. Soc. Lond. B*, **357**, 1369–1381.

Iverson, T.M. (2006) Evolution and unique bioenergetic mechanisms in oxygenic photosynthesis, *Curr. Opin. Chem. Biol.*, **10**, 91–100.

Kok, B., Forbush, B. and McGloin, M.P. (1970) Cooperation of changes in photosynthetic O_2-evolution-I: a linear four-step mechanism, *Photochem. Photobiol.*, **11**, 457–475.

Loll, B., Kern, J., Saenger, W., Zouni, A. and Biesiadka, J. (2005) Towards complete cofactor arrangement in the 3.0 Å resolution structure of photosystem II, *Nature*, **438**, 1040–1044.

Miller, A.-F. (2004) Superoxide dismutases: active sites that save, but a protein that kills, *Curr. Opin. Chem. Biol.*, **8**, 162–168.

Rutherford, A.W. and Boussac, A. (2004) Water photolysis in biology, *Science*, **303**, 1782–1784.

Voet, D. and Voet, J.G. (2004) *Biochemistry*, 3rd edition, Wiley, Hoboken, NJ, 1591 pp.

Yocum, C.F. and Pecoraro, V.L. (1999) Recent advances in the understanding of the biological chemistry of manganese, *Curr. Opin. Chem. Biol.*, **3**, 182–187.

– 17 –

Molybdenum, Tungsten, Vanadium and Chromium

INTRODUCTION

In this last chapter on particular metal ions, we have regrouped four metals that together with manganese and nickel are used as alloys to produce specialist steels[1]. However, it is not due to their capacity to confer particular properties on steel, but rather for their biological chemistry that we have grouped them together. Molybdenum is the only second row transition element that is essential for most living organisms, and the few species that do not need molybdenum use tungsten, which is molybdenum's third row homologue. In the case of vanadium, the close similarity between its chemical properties and those of molybdenum, has led to its replacement of molybdenum in the FeMo cofactor of some bacterial nitrogenases. However, it is also involved in the activity of haloperoxidases, and four-coordinate vanadate can mimic cellular metabolites via its analogy with phosphate. Vanadium compounds have also been found to have an insulin-like effect. While chromium is an element that appears to be biologically necessary, we have yet to find its precise function, although it appears to be required for proper carbohydrate and lipid metabolism in mammals.

MOLYBDENUM AND TUNGSTEN

While it is relatively rare in the earth's crust, Mo is the most abundant transition metal in seawater. When we consider that the oceans are the closest we get today to the primordial soup in which life first arose, it is not surprising that Mo has been widely incorporated into biological systems. Indeed, the only organisms that do not require Mo, use W, which lies immediately below Mo in the Periodic Table, instead. The biological versatility of Mo and

---

[1] Steel is an alloy of iron and carbon, containing typically up to 2% carbon. The addition of other metals in alloys can give special properties such as superior strength, hardness, durability or corrosion resistance.

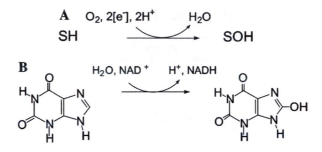

Figure 17.1 The contrasting reactions catalysed by monooxygenases (a) and molybdenum hydroxylases (b).

W results not only from their redox-active ranging through oxidation states VI–IV, but because the intermediate V valence state is also accessible, they can act as interfaces between one- and two-electron redox systems, which allows them to catalyse hydroxylation of carbon atoms using water as the ultimate source of oxygen, (Figure 17.1) rather than molecular oxygen, as in the flavin-, haem- or Cu-dependent oxygenases, some of which we have encountered previously. For reviews see Hille, 2002; Brondino et al., 2006; Mendel and Bittner, 2006.

If we assume that the early conditions on our planet were not only anaerobic, but also hot, tungsten would have been much better adapted than molybdenum; since low-valent tungsten sulfides would have been more soluble in aqueous solutions, their tungsten–sulfur bonds are more stable and their reduction potentials lower than their molybdenum equivalents. As the earth's crust cooled, and cyanobacterial photosynthesis transformed the atmosphere from anaerobic to aerobic, the oxygen sensitivity of tungsten compounds, together with the greater water solubility of high-valence molybdenum oxides, and the dramatically different redox balance, would have pushed the scales in favour of molybdenum. This hypothesis is supported by the distribution of the two metals, molybdenum enzymes present in all aerobes and tungsten enzymes only in obligate anaerobes (often thermophiles). A few anaerobes can use either metal, depending on availability.

Another factor that characterizes molybdenum and tungsten enzymes is that instead of using the metal itself, directly coordinated to amino acid side-chains of the protein, an unusual pterin cofactor, Moco, is involved in both molybdenum- and tungsten-containing enzymes. The cofactor (pyranopterin–dithiolate) coordinates the metal ion via a dithiolate side-chain (Figure 17.2). In eukaryotes, the pterin side-chain has a terminal phosphate group, whereas in prokaryotes, the cofactor (R in Figure 17.2) is often a dinucleotide.

With the exception of bacterial nitrogenase, whose Fe–Mo-cofactor will be discussed later, all molybdenum enzymes contain the molybdenum pyranopterindithiolate cofactor (MoCo), which is the active component of their catalytic site (and of tungsten enzymes, in organisms that do not use molybdenum). The biosynthetic pathway for this pterin cofactor appears to be universally conserved in biology, underlining its importance. Interestingly however, baker's yeast, a much-used 'model' eukaryote, is the only organism known that does not contain Mo-enzymes (it is also one of the few organisms which does not contain ferritin). The MoCo cofactor can exist in the fully oxidized (Mo^{VI}) and in the fully reduced (Mo^{IV}) forms, with some enzymes generating the (Mo^{V}) form as a catalytic intermediate.

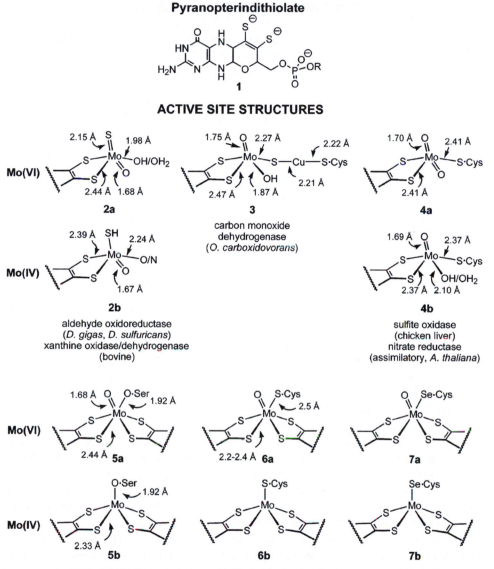

Figure 17.2 The structure of the pterin cofactor (**1**) which is common to most molybdenum- and tungsten-containing enzymes and schematic active site structures for members of the xanthine oxidase (**2,3**), sulfite oxidase (**4**) and DMSO reductase (**5–7**) enzyme families. (From Enemark et al., 2004. Copyright (2004) American Chemical Society.)

MOLYBDENUM ENZYME FAMILIES

Molybdenum-containing enzymes can be divided into three families, the xanthine oxidase (**2,3**), sulfite oxidase (**4**) and the DMSO reductase (**5–7**) families, the active sites of which are represented in the oxidized (Mo^{VI}) and reduced (Mo^{IV}) states (Figure 17.2). They each catalyse a particular type of reaction, and have a characteristic active site structure. Members of the xanthine oxidase family have one Moco entity, which closes a five-membered ene-1,2-dithiolate chelate ring together with a *cis*-$Mo^{VI}OS$ group. The family members aldehyde oxidoreductase and xanthine oxidase/dehydrogenase (**2ab**) catalyse the hydroxylation of carbon centres, whereas the third family member (**3**), the CO dehydrogenase from *Oligotropha carboxidovorans*, which oxidizes CO to CO_2 is an exception, and has the sulfido ligand coordinated by a Cu^I centre. The second, sulfite oxidase family (isolated from avian or mammalian liver) also includes nitrate reductases, from plants that assimilate nitrate (**4ab**). Like the first, it contains one equivalent of the cofactor, this time with a cysteine ligand provided by the protein. Finally, the third, so-called DMSO reductase family, catalyses diverse reactions, including DMSO reductase (**5ab**), dissimilatory nitrate reductase (**6ab**) and formate dehydrogenase (**7ab**). All members of this third family have two cofactor molecules bound to a single molybdenum ion. The coordination sphere is completed by a single M=O group and a sixth ligand, which is a serine residue in DMSO reductase (in other members of the family it can be a cysteine, a selenocysteine or a hydroxide). As we will see later in the chapter, all tungsten enzymes, which have been structurally characterized, contain two pyranopterindithiolate ligands per tungsten atom.

Xanthine oxidase (or xanthine oxidoreductase) inserts hydroxyl groups into derivatives of the purine bases adenine and guanine, finally converting xanthine to the more water soluble uric acid (Figure 17.1b), the final product of purine degradation, which is excreted in the urine. The $LMo^{VI}OS(OH)$ core of the enzyme in the oxidized state (the pterin cofactor is designated L) has a distorted square pyramidal coordination geometry (Figures 17.2 and 17.3) with the Mo=O group in the apical position. The bidentate enedithiolate ligand of the cofactor lies in the equatorial plane together with the Mo=S and the Mo–OH groups. The structure of the bovine xanthine oxidoreductase is presented in Figure 17.3. It consists of four domains, two Fe/S domains (I and II) in the N-terminal portion of the molecule (shown in green and blue, respectively), followed by the FAD domain (in grey) and the molybdenum-binding domain in the C-terminal part of the molecule.

The mechanism by which xanthine oxidase brings about hydroxylation is considered here, and must take account of the fact that water, rather than O_2, is the ultimate source of oxygen incorporated into the product. In a single-turnover experiment using $H_2^{18}O$, the radioisotope is not incorporated into the product; whereas, when the enzyme from that experiment is incubated with substrate in unlabelled water, $8-^{18}O$-uric acid is produced. It follows that a catalytically labile site on the enzyme is the proximal oxygen donor, and all the evidence points to the M–OH group of the molybdenum centre. A mechanism, involving base-assisted nucleophilic attack on the substrate by the Mo–OH group (the catalytically labile oxygen) with concomitant hydride transfer to the Mo=S group is presented in Figure 17.4. The product (P) is bound in an end-on manner to the molybdenum, via the newly introduced hydroxyl group. Glu1261 is conserved in all xanthine oxidases and aldehyde oxidases, and is thought to act as a general base catalyst, initiating the reaction.

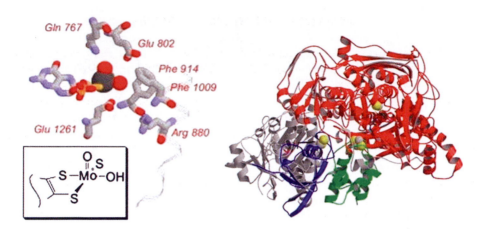

Figure 17.3 The active site structure of xanthine oxidoreductase and the structure of the bovine enzyme with the two Fe–S domains (green and blue), the FAD domain (grey) and the molybdenum-binding domain (red). (From Hille, 2005. Copyright 2005, with permission from Elsevier.)

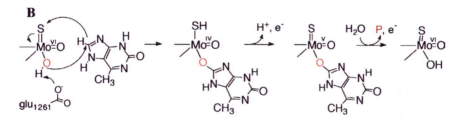

Figure 17.4 Reaction mechanism proposed for xanthine oxidase. (From Hille, 2005. Copyright 2005, with permission from Elsevier.)

The oxo-transfer chemistry of molybdenum in sulfite oxidase is probably the best characterized, in terms of synthetic models, structural and mechanistic data, of all the elements we have described up till now. The reaction cycle (Figure 17.5) involves binding of sulfite to the oxidized Mo^{VI}, two-electron reduction of the Mo centre and release of sulfate. The Mo^{VI} centre is restored by successive one-electron transfers from a cytochrome (b_5 in mammals). The primary oxo-transfer reaction:

$$Mo^{VI}O_2 + X \Leftrightarrow Mo^{IV}O + XO$$

involves direct atom transfer between the substrate and the metal centre via a covalently bound intermediate. The metal centre is then reduced in two successive one-electron steps. One of the major difficulties of synthetic model compounds was to prevent the usually irreversible μ-oxo dimerization reaction between $Mo^{VI}O_2$ and $Mo^{IV}O$ centres to form the stable dinuclear $Mo^{V}O$ species with an $[Mo_2O_3]^{4+}$ core (which can be easily avoided in the enzymatic systems).

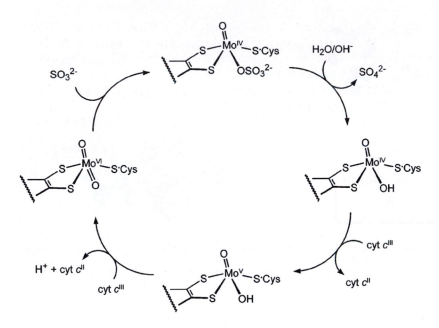

Figure 17.5 Proposed reaction cycle of sulfite oxidase. (From Enemark et al., 2004. Copyright (2004) American Chemical Society.)

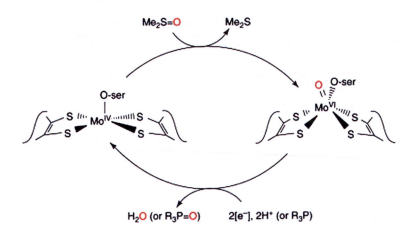

Figure 17.6 The catalytic cycle of dimethylsulfoxide reductase. (From Hille, 2002. Copyright 2002, with permission from Elsevier.)

In the case of the DMSO reductase family, as pointed out above, the metal centre is bound to two molecules of the cofactor. DMSO reductase itself catalyses the reduction of dimethylsulfoxide to dimethylsulfide with incorporation of the oxygen atom of DMSO into water. The active site of the oxidized enzyme is an $L_2Mo^{VI}O$(O-Ser) centre, which, upon reduction, loses the M=O ligand to give a L_2Mo^{IV}(O-Ser) centre. In the catalytic mechanism (Figure 17.6), the reduced enzyme L_2Mo^{IV}(O-Ser) reacts with DMSO to give the oxidized form $L_2Mo^{VI}O$(O-Ser) of the enzyme with the oxygen of the DMSO now incorporated into the M=O group of the oxidized enzyme. In the second phase of the catalytic mechanism, the oxidized enzyme is reduced by a cytochrome to release the oxygen derived from DMSO as water (this reaction can also be carried out by a water-soluble phosphine R_3P to give the phosphine oxide $R_3P=O$).

TUNGSTEN ENZYMES

As mentioned above, tungsten enzymes are found in place of molybdenum enzymes in thermophilic bacteria and hyperthermophilic archaebacteria. Like the molybdenum enzymes, they can be classified into three broad families—all containing two pterin cofactor molecules per W. In this respect, they are similar to the DMSO reductase family of Mo enzymes. Members of the first two catalyse redox reactions. The first, the aldehyde oxidoreductase family, catalyse the oxidation of aldehydes to carboxylic acids. The reducing equivalents are transferred to a [4Fe–4S] centre. While there is still some ambiguity, it is likely that, in addition to the four ligands from the cofactor, the oxidized enzyme contains the group $W^{VI}O\cdot(OH)$ (Figure 17.7): the reduced form probably has a single W^{IV}–OH. As in the xanthine oxidase family of Mo enzymes, there appear to be no ligands contributed by the protein. The second family consists of enzymes that function to reductively fix CO_2. They have amino acid sequence homologies with the DMSO reductase family of Mo enzymes, with cysteine or selenocysteine as a ligand from the protein coordinating the metal in the oxidized enzyme as $L_2W^{VI}OX$ (Figure 17.7). The third family, with only one member, is made up by acetylene hydratase, which adds water to the double bond of acetylene forming acetaldehyde. Although it contains a [4Fe–4S] centre, this does not appear to participate directly in the catalysis, unlike the situation in aconitase (see Chapter 13). Instead, it seems from model studies and from the observation that the enzyme as isolated requires activation by a strong reductant, that the catalysis of acetylene hydration involves the participation of a W^{IV} site.

A number of organisms appear to be able to use either molybdenum or tungsten, as a function of their bioavailability.

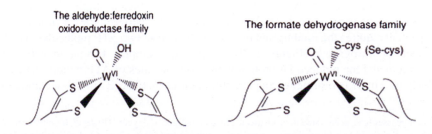

Figure 17.7 Active site structures of tungsten-containing enzymes (from Hille, 2002).

NITROGENASES

In the biological nitrogen cycle, an important role is played by a relatively limited number of anaerobic microorganisms capable of converting about one-third of atmospheric dinitrogen into ammonia, which can subsequently be incorporated into glutamate and glutamine, and from there into other nitrogen containing molecules. This represents $\sim 10^8$ tons per year, the same as is produced by the Haber–Bosch industrial process—albeit that the latter functions at both high pressures (150–350 atm) and high temperatures (350–550 °C). The microorganisms that fix nitrogen include the bacterium *Rhizobium*, involved in the symbiotic fixation of nitrogen in the root nodules of leguminous plants. Since this enzyme is extremely sensitive to oxygen[2], the plant roots produce haemoglobin with a high affinity for oxygen, leghaemoglobin (like haemoglobins from insect larvae or lamprey, it has the classic 'globin fold' found in mammalian haemoglobins and myoglobins—see Chapter 3), which maintains an anaerobic environment around the enzyme (Downie, 2005).

All nitrogenases consist of two types of subunit, one of which contains a special Fe–S cluster, known as the P-cluster, and a second, which contains an iron and sulfur-containing cofactor, includes a heterometal. The heterometal is usually molybdenum, hence the cofactor is known as FeMoCo. However, in some species, and under conditions of particular metal bioavailability, molybdenum can be replaced by vanadium or iron. These 'alternative' nitrogenases contain vanadium instead of molybdenum (when Mo levels are low and V is available), and the other contains only iron (when both Mo and V levels are low). However, by far the greatest advances in our understanding of the structure and mechanism of nitrogenases have come from studies on the MoFe-nitrogenases from free-living nitrogen-fixing bacteria such as *Azotobacter*, *Clostridium* and *Klebsiella* (for a review see Peters and Szilagyi, 2006).

The overall reaction catalysed by nitrogenases is:

$$N_2 + 8H^+ + 8e^- + 16ATP + 16H_2O \rightarrow 2NH_3 + H_2 + 16ADP + 16P_i$$

As we can see, the process of nitrogen fixation is extremely energy-intensive, requiring both large amounts of ATP and of reducing equivalents. The nitrogenase is made up of two proteins (Figure 17.8), termed the MoFe protein and the Fe protein. The $\alpha_2\beta_2$ heterotetrameric MoFe protein contains both the FeMo-cofactor and the so-called P-cluster, with the functional unit constituted by an $\alpha\beta$ dimer, containing one FeMo-cofactor and one P cluster. In contrast, the Fe protein is a homodimer, which binds a single [4Fe–4S] cluster at the interface between the two subunits. Unlike many other multiple-electron transfer reactions in biochemistry, each individual-electron transfer between the Fe-protein and the MoFe-protein requires the binding and hydrolysis of at least two ATP molecules. The basic mechanism of nitrogenase (Figure 17.9) involves (i) complex formation between the MoFe-protein and the reduced Fe-protein with two molecules of ATP bound, (ii) electron transfer between the two proteins coupled with hydrolysis of ATP, (iii) dissociation of the

---

[2] Some nitrogenases have been found to be unaffected by the presence of oxygen. However, if you do cut into the root nodules of a common legume like pea or bean, you will see that it has a blood-red colour due to the high levels of leghaemoglobin.

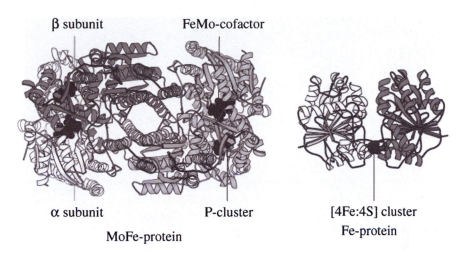

Figure 17.8 Ribbon diagrams of the heterodimeric MoFe-protein (left) and homodimeric Fe-proteins of nitrogenase (right). The α- and β-subunits to the left of the MoFe-protein are shown in light and dark shading, respectively, while the metalloclusters are shown as dark space-filling models on the right side. The two subunits of the Fe-protein are shown in light and dark shading, with the 4Fe–4S cluster at the dimer interface as a space-filling model. (From Rees et al., 2005. Reproduced with permission of the Royal Society.)

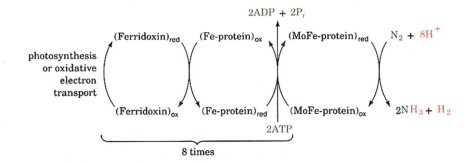

Figure 17.9 The role of components 1 and 2 in the fixation of nitrogen.

oxidized Fe-protein from the complex accompanied by its re-reduction and exchange of the 2ADPs for ATPs, (iv) repetition of this cycle of association, reduction, ATP hydrolysis and dissociation to transfer one electron at a time to the MoFe-protein. Once a sufficient number of electrons and protons have been accumulated, available substrates can be reduced. Usually when 8 reducing equivalents have been accumulated, and 16 molecules of ATP hydrolysed, the enzyme can bind and reduce the very stable triple bond of a dinitrogen molecule to two molecules of ammonia. Concomitantly, two protons and two electrons are converted to gaseous hydrogen. Electrons derived from photosynthesis or from the mitochondrial-electron transport chain are transferred to the Fe-protein.

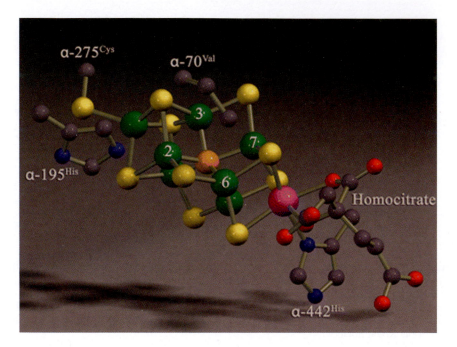

Figure 17.10 Structure of the MoFe-cofactor highlighting Fe atoms 2, 3, 6 and 7 with the MoFe-protein ligands Cysα273 and Hisα442 and the side-chains of Hisα195 and Valα70. (From Barney et al., 2006. Copyright (2006) National Academy of Sciences, USA.)

The structure of both the MoFe-cofactor (Figure 17.10) and of the P-cluster (Figure 17.11) became apparent when the structures of nitrogenases were determined by X-ray crystallography. The MoFe-cofactor has 1 Mo, 7 Fe and 9 S organized in an elongated structure, linked to the protein by only two residues, Cysα273 and Hisα442, which coordinate Fe1 and the Mo atom, respectively, at opposite ends of the extended cluster. This is in marked contrast to other iron–sulfur clusters, which typically have a protein side-chain ligand per metal ion. In order to complete the coordination sphere of the eight metal centres, there are a number of additional inorganic sulfides together with bidentate coordination of the Mo atom to a molecule of homocitrate[3], completing its octahedral coordination. Evidence has been presented from higher resolution X-ray studies of a previously undetected central atom (possibly a nitrogen) in the cofactor core, whose role remains uncertain.

In the dithionite-reduced state (Figure 17.11a), P^N, the P-cluster can be considered as two [4Fe–3S] clusters bridged by a hexa-coordinate sulfur. In the P^{OX} state, which is oxidized by two electrons relative to P^N, two of the iron atoms Fe5 and Fe6 have moved away from the central sulfur atom, and are now coordinated by the amide nitrogen of Cysα87 and the hydroxyl of Serα186, maintaining the irons in a four-coordinate state.

---

[3] A homologue of citrate (see Chapter 5) with an additional CH_2 group.

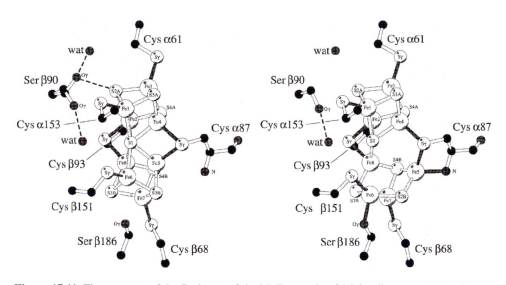

Figure 17.11 The structure of the P-cluster of the MoFe-protein of *Klebsiella pneumoniae* nitrogenase in the P^N (a) and P^{OX} (b) states. (From Mayer et al., 1999. Copyright 1999, with permission from Elsevier.)

The Fe-protein has the protein fold and nucleotide-binding domain of the G-protein family of nucleotide-dependent switch proteins, which are able to change their conformation dependent on whether a nucleoside diphosphate (such as GDP or ADP) is bound instead of the corresponding triphosphate (GTP or ATP). However, nucleotide analogues, which induce the conformational switch of the Fe-protein, do not allow substrate reduction by the MoFe-protein, nor does reduction of the MoFe-protein by other electron-transfer reagents (whether small proteins or redox dyes) drive substrate reduction. Only the Fe-protein can reduce the MoFe-protein to a level that allows it to reduce substrates such as nitrogen.

Electrons arriving at the Fe-protein are transferred to the P-cluster and from there to the MoFe-protein, which is the site of interaction with dinitrogen or any of the other subtrates that are reduced by nitrogenase. The redox chemistry of nitrogen reduction, on the basis of model reactions first proposed by Chatt, involves nitrogenous species at the level of diazene (N_2H_2) and hydrazine (N_2H_4) before the final release of two molecules of ammonia. Recent evidence for a diazene-derived species bound to the FeMo-cofactor supports this view (Barney et al., 2006) as does evidence that hydrazine (N_2H_4) is a substrate for nitrogenase. A binding site for N_2 and for alkyne substrates has been localized on the iron–sulfur face of the FeMo cofactor defined by the Fe atoms 2, 3, 6 and 7 (highlighted in green in Figure 17.10), and ENDOR spectroscopy has shown that the alkene product of alkyne reduction is probably bound end-on to a single Fe atom of the FeMo cofactor.

A series of model studies initiated in the early 1960s by the groups of Chatt and Hidai demonstrated that dinitrogen could be bound and reduced to ammonia at a single metal centre by Mo and W complexes (Chatt et al., 1978; Hidai, 1999). However, although examples of virtually all the proposed intermediates in a 'Chatt' cycle were isolated, no catalytic reduction of N_2 to NH_3 was ever achieved. Catalytic reduction of dinitrogen to ammonia

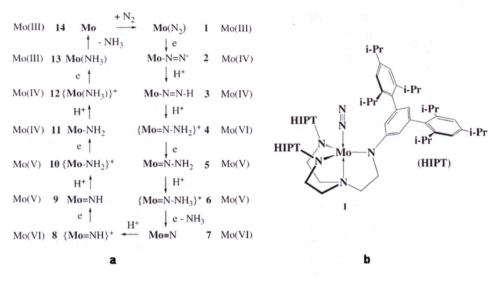

a   b

Figure 17.12 (a) Proposed intermediates in the reduction of dinitrogen at a [HITPN$_3$N]Mo designated (Mo) centre through stepwise addition of protons and electrons. (b) Structure of [HITPN$_3$N]Mo(N$_2$), designated as **Mo(N$_2$)** in (a). (From Schrock, 2005. Copyright (2005) American Chemical Society.)

at a single molybdenum centre has been achieved recently by the group of Richard R. Schrock[4] using the HITP [3,5-(2,4,6-i-Pr$_3$C$_6$H$_2$)$_2$C$_6$H$_3$] ligand. Eight of the proposed intermediates in a hypothetical 'Chatt-like' reduction of dinitrogen (Figure 17.12a) were prepared and characterized, including paramagnetic **Mo(N$_2$)** (**1**, Figure 17.12b). All these intermediates, except **7**, were extremely sensitive to oxygen, and several of them were subsequently used successfully to reduce dinitrogen to ammonia with protons and electrons (Yandulov and Schrock, 2003).

The catalytic reduction of the dinitrogen triple-bond by single-site metal nitrogen interme- diates raises the question of why Nature goes to the trouble of using the much more complex 7Fe:9S:Mo:homocitrate cluster of the FeMo cofactor in biological nitrogen fixation. One would expect that a simpler one- or two-metal centre for nitrogen fixation would have been dominant in the evolution if it had been biologically functional. Yet over more than a billion years evolutionary pressures have retained this cofactor-based nitrogenase system, despite the requirement for the unusual metabolite, homocitrate and at least 20 additional proteins for its assembly and insertion. Indeed, even the 'alternative' nitrogenases are thought to be minor variations on the cofactor with V or Fe replacing Mo. As Howard and Rees (2006) point out at the end of their overview of biological nitrogen fixation, entitled 'How many metals does it take to fix N$_2$?', the number of metal atoms required is 20, corresponding to the metal com- position of the FeMo cofactor, the P cluster and the Fe protein. They all seem to be required, and to date no one has found a way to simplify the system. Perhaps, after all, this simply underlines the affirmation[5] 'enzyme catalysis—not different, just better!'.

---

[4] Who shared the 2005 Nobel Prize in Chemistry with R.H. Grubbs and Y. Chauvin.
[5] The title of a Nature paper by the Amory Houghton Professor of Chemistry and Biochemistry at Harvard University, Jeremy R. Knowles (Knowles, 1991).

VANADIUM BIOCHEMISTRY

Vanadium is beneficial and possibly essential for humans. It is certainly essential for a number of organisms. Vanadate (oxidation state V) and its derivatives are phosphate analogues, showing both ground state and transition state analogy (both structural and electronic) with phosphorus compounds. The analogy of five-coordinate vanadium compounds with the transition state of phosphate ester hydrolysis is well documented, and explains why so many vanadium compounds are potent inhibitors of phosphatases, ribonucleases and ATPases.

Haloperoxidases represent the first, and best-characterized class of vanadium enzymes, capable of catalysing the two-electron oxidation of a halide by hydrogen peroxide. The chloroperoxidases, found in many algae, seaweed, lichens and fungi, can oxidize both Cl^- and Br^-, whereas bromoperoxidases, found in many marine extracts, can only oxidize Br^-. The X-ray structures of a number of vanadate-dependent haloperoxidases have been reported (Figure 17.13). On the basis of spectroscopic evidence, it is now thought that the oxidation state of the vanadium remains at V throughout catalysis, and that the mechanism for both types of vanadium haloperoxidases are the same, as indicated in Figure 17.14. The reaction proceeds by initial binding of H_2O_2 followed by protonation of bound peroxide and addition of the halide. NMR spectroscopy confirms the presence of VO_2–O_2, and there is no evidence for direct binding of halide to the vanadium ion. The rate-limiting step in the catalysis is the nucleophilic attack of the halide on the protonated protein–peroxide complex, generating an 'X^+' species, which reacts directly with organic substrates (RH) to halogenate them (RX). In the absence of RH this step will generate singlet oxygen.

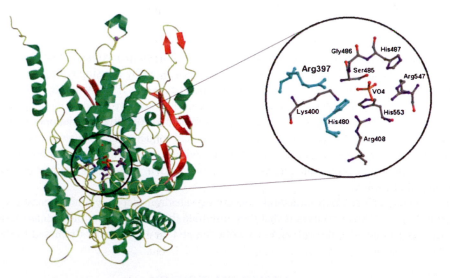

Figure 17.13 The structure and active site of the bromoperoxidase subunit from *C. pilulifera*. Residues conserved in all vanadium bromo- and chloroperoxidases are in grey, those that vary in cyan. (From Ohshiro et al., 2004. Copyright 2004 The Protein Society.)

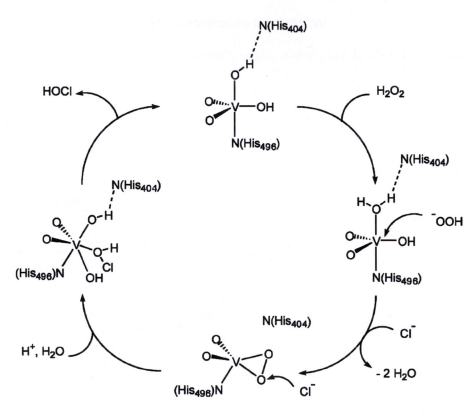

Figure 17.14 Proposed mechanism for the vanadium chloroperoxidase oxidation of chloride by hydrogen peroxide. (From Ligtenbarg et al., 2003. Copyright 2003, with permission from Elsevier.)

The halide specificity of the vanadium-dependent bromoperoxidase from the marine algae *Corallina pilulifera* (Figure 17.13) has been changed by the single amino acid substitution of Arg379 by either Trp or Phe. Both mutant enzymes R379W and R379F showed significant chloroperoxidase, as well as bromoperoxidase activity, supporting the existence of a specific halogen-binding site within the catalytic cleft of vanadium haloperoxidases.

It is also of interest to point out that the amino acid sequence and structure of the active site of vanadium haloperoxidases is conserved within several families of phosphatases, with conservation of the amino acids involved in vanadate binding in one and phosphate binding in the other.

Information, particularly structural, concerning vanadium-dependent nitrogenases, is relatively limited. The consensus is that they resemble the molybdenum nitrogenase in most aspects except for the presence of a FeV cofactor, and they will not be discussed further.

VANADIUM BIOLOGY

High levels of vanadium are found in the mushroom *Amanita muscaria* and in marine tunicates (sea squirts). In the former organism, a siderophore-like ligand that binds vanadium(IV)

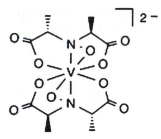

Figure 17.15 Structure of amavadine.

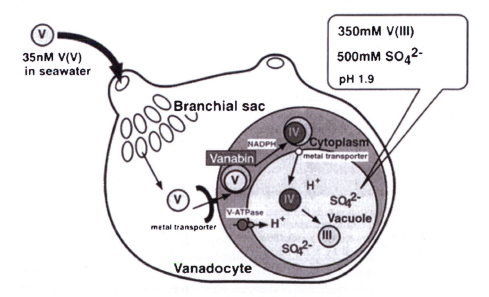

Figure 17.16 Model of the pathway for reduction and accumulation of vanadium in ascidian vanadocytes. (From Crans et al., 2004. Copyright (2004) American Chemical Society.)

called amavadine is found. Amavadine is a metal complex containing one equivalent of vanadium and two equivalents of the ligand S,S-2,2′-hydroxyiminopropionic acid (Figure 17.15). The complex is stable to hydrolysis, and has reversible one-electron redox properties, suggestive of a possible role in biology as a one-electron redox mediator.

Vanadium, as $VOSO_4$, has been found to interfere with siderophore-mediated iron transport in bacteria and plants. This seems to imply that vanadium can be transported by siderophores, and a number of studies focussing on applications of hydroxamate V-complexes in biology have been initiated.

Tunicates (ascidians or sea-squirts) are invertebrate marine organisms, which can accumulate vanadium at concentrations approaching 350 mM (the concentration of vanadium in sea water is $\sim 10^{-8}$ M). This vanadium is taken up as V(V) from seawater (Figure 17.16), reduced to oxidation state III or IV and stored in a soluble form in the blood cells within acidic vacuoles at concentrations a million fold higher than in their external surroundings.

Vanadium seems to be bound in the cytoplasm to vanadium binding proteins (vanabins, of molecular weights 12–16 kDa). However, the precise role of vanadium in these marine organisms remains unknown.

Finally, we briefly consider the insulin-like effect of vanadium compounds. As was pointed out in Chapter 5, the regulation of intermediary metabolism is a very complex phenomenon, and there are few examples less complicated than the action of insulin, which through interaction with its receptor in a large number of target tissues initiates a series of signalling cascades that not only affect carbohydrate and lipid metabolism, but also have many other metabolic repercussions. Vanadium compounds have been shown to enhance the effect of insulin by stimulating the phosphorylation of the insulin receptor (*in vitro*) and inhibiting protein phosphatases. They also seem, particularly in the case of the vanadyl cation, to bind to transferrin, thereby facilitating their entry into cells via the transferrin/transferrin receptor pathway (although, how they dissociate from the receptor inside the cell remains unclear). They may also influence the redox balance of cells, interacting with the glutathione system. However, despite their potential beneficial effects, failure to bring use of vanadium salts into therapeutic practice is due on the one hand to their toxicity, and on the other to their limited window of therapeutic action—while they may target some parts of the complex insulin signalling cascade, they cannot exert the exquisite specificity of the natural hormone both to activate and to ensure the extinction of its activation cascade, once its objectives have been achieved.

CHROMIUM IN BIOLOGY

As was pointed out in Chapter 1, chromium has become immensely popular as a nutritional supplement, for promotion of muscle development and as a weight-loss agent (Vincent, 2003), second only to calcium as a mineral supplement (Nielsen, 1996). However, while there are indications that Cr administration may be useful as an adjuvant therapy in Type 2 diabetes as well as in the regulation of diabetes during pregnancy, the precise biochemical mode of action of chromium remains unclear (Vincent, 2000a, 2004). The biologically relevant form, the trivalent Cr^{3+} ion, seems to be required for proper carbohydrate and lipid metabolism in mammals. However, chromium deficiency is difficult to achieve. Although no Cr-dependent enzymes or proteins have been identified, there are indications that chromium is excreted in the form of the oligopeptide, chromodulin. Chromodulin has a molecular weight of 1438, containing Gly, Cys, Glu and Asp with the latter two making up more than half the total amino acid residues. Chromodulin tightly binds four atoms of Cr^{3+} (k_d 10^{-21} M^{-4}) in a highly cooperative manner. It has been proposed that the chromium-loaded chromodulin (holochromodulin) may function in the amplification system for insulin signalling (Vincent, 2000b). Following insulin binding to its receptor, the receptor acquires tyrosine kinase activity, which is involved in transmitting the signal from insulin inside the cell. It is suggested that holochromodulin binds to the receptor in its active conformation, thereby amplifying the kinase activity of the receptor. When signalling is turned off, the receptor reverts to its original conformation, releasing the holochromodulin, which is ultimately excreted in the urine. Transferrin, the Fe-transport protein in serum, has been shown to be responsible for transporting Cr^{3+} to tissues in an insulin-responsive manner.

While there are obvious similarities between both the name and the proposed mechanism of action of chromodulin and the Ca^{2+}-binding protein calmodulin, much remains to be done to establish unequivocally the mechanism of chromodulin action at the molecular level.

REFERENCES

Barney, B.M., Lukoyanov, D., Yang, T.-C., Dean, D.R., Hoffman, B.M. and Seefeldt, L.C. (2006) A methyldiazene ($NH=N–CH_3$)-derived species bound to the nitrogenase active-site FeMo cofactor: implications for mechanism, *Proc. Natl. Acad. Sci. U.S.A.*, **103**, 17113–17118.

Brondino, C.D., Romao, M.J., Moura, I. and Moura, J.J.G. (2006) Molybdenum and tungsten enzymes: the xanthine oxidase family, *Curr. Opin. Chem. Biol.*, **10**, 109–114.

Chatt, J., Dilworth, J.R. and Richards, R.L. (1978) Recent advances in the chemistry of nitrogen fixation, *Chem. Rev.*, **78**, 589–625.

Crans, D.C., Smee, J.J., Gaidamauskas, E. and Yang, L. (2004) The chemistry and biochemistry of vanadium and the biological activities exerted by vanadium compounds, *Chem. Rev.*, **104**, 849–902.

Downie, J.A. (2005) Legume haemoglobins: symbiotic nitrogen fixation needs bloody nodules, *Curr. Biol.*, **15**, R196–R198.

Enemark, J.H., Cooney, J.J.A., Wang, J.-J. and Holm, R.H. (2004) Synthetic analogues and reaction systems relevant to the molybdenum and tungsten oxotransferases, *Chem. Rev.*, **104**, 1175–1200.

Hidai, M. (1999) Chemical nitrogen fixation by molybdenum and tungsten complexes, *Coord. Chem. Rev.*, **185–186**, 99–108.

Hille, R. (2002) Molybdenum and tungsten in biology, *TIBS*, **27**, 360–367.

Hille, R. (2005) Molybdenum-containing hydroxylases, *Arch. Biochem. Biophys.*, **433**, 107–116.

Howard, J.B. and Rees, D.C. (2006) How many metals does it take to fix N_2? A mechanistic overview of biological nitrogen fixation, *Proc. Natl. Acad. Sci. U.S.A.*, **103**, 17088–17093.

Knowles, J.R. (1991) Enzyme catalysis: not different, just better, *Nature*, **350**, 121–124.

Ligtenbarg, A.G.J., Hage, R. and Feringa, B.L. (2003) Catalytic oxidations by vanadium complexes, *Coord. Chem. Rev.*, **237**, 87–101.

Mayer, S.M., Lawson, D.M., Gormal, C.A., Roe, S.M. and Smith B.E. (1999) New insights into structure-function relationships in nitrogenase: A 1.6 A resolution X-ray crystallographic study of Klebsiella pneumoniae MoFe-protein, *J. Mol Biol.*, **292**, 871–891.

Mendel, R.R. and Bittner, F. (2006) Cell biology of molybdenum, *Biochim. Biophys. Acta*, **1763**, 621–635.

Nielsen, F. (1996) Controversial chromium: does the superstar mineral of the mountebanks receive appropriate attention from clinicians and nutritionists?, *Nutr. Today*, **31**, 226–233.

Ohshiro, T., Littlechild, J., Garcia-Rodriguez, E., Isupov, M.N., Iida, Y., Kobayashi, T. and Izumi, Y. (2004) Modification of halogen specificity of a vanadium-dependent bromoperoxidase, *Protein Sci.*, **13**, 1566–1571.

Peters, J.W. and Szilagyi, R.K. (2006) Exploring new frontiers of nitrogenase structure and function, *Curr. Opin. Chem. Biol.*, **10**, 101–108.

Rees, D.C., Tezcan, F.A., Haynes, C.A., Walton, M.Y., Andrade, S., Einsle, O. and Howard, J.B. (2005) Structural basis of biological nitrogen fixation, *Phil. Trans. R. Soc.*, **363**, 971–984.

Schrock, R.R. (2005) Catalytic reduction of dinitrogen to ammonia at a single molybdenum center, *Acc. Chem. Res.*, **38**, 955–962.

Vincent, J.B. (2000a) The biochemistry of chromium, *J. Nutr.*, **130**, 715–718.

Vincent, J.B. (2000b) Elucidating a biological role for chromium at a molecular level, *Acc. Chem. Res.*, **33**, 503–510.

Vincent, J.B. (2003) The potential value and potential toxicity of chromium picolinate as a nutritional supplement, weight-loss agent and muscle development agent, *Sports Med.*, **33**, 213–230.

Vincent, J.B. (2004) Recent advances in the nutritional biochemistry of trivalent chromium, *Proc. Nutr. Soc.*, **63**, 41–47.

Yandulov, D.V. and Schrock, R.R. (2003) Catalytic reduction of dinitrogen to ammonia at a single molybdenum center, *Science*, **301**, 76–78.

– 18 –

Metals in Brain and Their Role in Various Neurodegenerative Diseases

INTRODUCTION: METALS IN BRAIN

Metal ions are absolutely essential to fulfil a series of important biological functions in the brain, such as nerve transmission and the synthesis of neurotransmitters. They include spectroscopically silent metal ions such as potassium, sodium, calcium, magnesium and zinc together with the more spectroscopically accessible iron, copper, manganese and a few others. The role of some of these metal ions in brain function is particularly important. As we saw in Chapter 9, the alkali metal ions Na^+ and K^+ are involved in the opening and closing of ion channels, which generate electrochemical gradients across the plasma membranes of neurons. This plays a crucial role in the transmission of nervous impulses not only within the brain but also in the transmission of signals from the brain to other parts of the body. For more on metal-based neurodegeneration see Crichton and Ward, 2006.

CALCIUM

Within cells, including nerve cells, fluxes of Ca^{2+} ions play an important role in signal transduction (see Chapter 11). Most eukaryotic cells export calcium across the plasma membrane or deposit it in membrane-enclosed storage sites in order to maintain free cytosolic Ca^{2+} levels at 100–200 nM, roughly 10,000 times less than in the extracellular space. This allows calcium to function as a second messenger and also as a carrier of biological signals that guide cells from their origin to their ultimate death.

When intracellular Ca^{2+} increases, the ubiquitous eukaryotic Ca^{2+}-binding protein calmodulin (CaM) binds Ca^{2+} ions (see Chapter 11). This causes a major conformation change, exposing a previously buried hydrophobic patch on the CaM molecule, which can bind to a large number of target enzymes, modifying their activity. The structure of the $(Ca^{2+})_4$–CaM bound to a target polypeptide is shown in Figure 18.1.

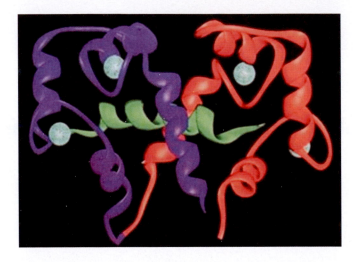

Figure 18.1 Ribbon diagram showing the NMR structure of $(Ca^{2+})_4$–calmodulin bound to a 26 residue target peptide. The two domains of calmodulin are in purple and red, the Ca^{2+} ions as spheres and the target peptide in green. (From Voet and Voet, 2004. Reproduced with permission from John Wiley & Sons., Inc.)

Ca^{2+} is also involved in signalling from neuronal synapses to the cell nucleus, resulting in neuronal activity-dependent control of neuronal gene expression[1]. This synapse-to-nucleus signalling plays a key role in circadian rhythms, long-term memory and neuronal survival. The transient rise in free Ca^{2+} concentration after neuronal excitation can be transmitted from the cytoplasm to the nucleus in several different ways (Figure 18.2). Following rises in intracellular Ca^{2+}, the nuclear transcription factor downstream regulatory element antagonistic modulator (DREAM) is activated. DREAM is abundant in the nucleus and has three Ca^{2+}-binding motifs, the E-F hands described in Chapter 11. It is proposed that DREAM remains bound to a downstream regulatory element (DRE), which acts as a gene silencer when nuclear Ca^{2+} is low. When the nuclear Ca^{2+} levels rise, DREAM dissociates from the DNA, causing derepression of DRE, and activation of downstream genes such as that which codes for dynorphins (which may act as an antidote to the pleasurable effects of cocaine) and attenuation of pain signalling *in vivo*. Within seconds of Ca^{2+} entry into the cytoplasm, through both *N*-methyl-D-aspartate (NMDA) receptors and L-type voltage-gated Ca^{2+} channels (VGCC), CaM is activated and translocates to the nucleus where it participates in the activation of Ca^{2+}/cAMP responsive element binding protein (CREB)-dependent gene expression. CaM also mediates CREB phosphorylation via the adenyl cyclase/phosphokinase A (AC/PKA) and the MAP kinase (MAPK) pathways, which begin to exert their influence subsequently. Almost as rapidly, CaM activates another target protein in mammalian brain, calcineurin (CaN), and a heterodimeric phosphatase, which dephosphorylates a member of the nuclear factor of activated T-cells (NFAT) family of transcription factors, NFATc4. The NFATc group of transcription factors play a key role

---

[1] Synapses are the local sites of communication between neurones.

Early stages of activity-dependent signaling to the nucleus

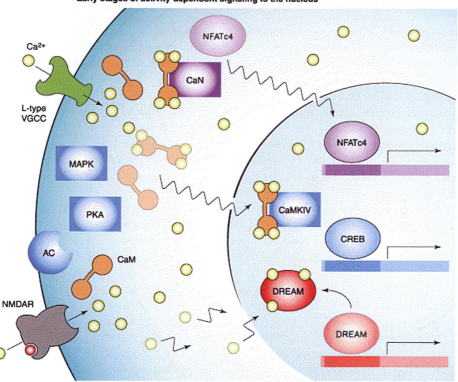

Figure 18.2 Signal transduction cascades set in motion by a Ca^{2+} rise near the cell membrane. (From Deisseroth et al., 2003. Copyright 2003, with permission from Elsevier.)

in neuronal plasticity as well as vascular development and muscular hypertrophy. NFATc4 is expressed in neurons of the hippocampus, the memory and learning centre of the brain. Upon dephosphorylation, NFATc4 undergoes translocation from the cytosol to the nucleus.

Synaptotagmins are yet another family of Ca^{2+}-binding proteins, localized on the membranes of synaptic vesicles, where they seem to be involved in the release of neurotransmitters. While the mechanism by which they are involved in Ca^{2+}-mediated synaptic transmission is unclear, it seems likely that the neurotoxicity of heavy metals such as Pb is due to a higher affinity of synaptotagmins for Pb^{2+} than for Ca^{2+}.

The Ca^{2+}–CaM-dependent protein kinase CaMKII plays a central role in Ca^{2+} signal transduction and is highly enriched in brain tissue, accounting for ~2% of total hippocampal protein and around 0.25% of total brain protein. It is the most abundant protein in the post-synaptic density, the region of the post-synaptic membrane, which is physically connected to the ion channels that mediate synaptic transmission. The structural modification of synaptic proteins is thought to be the molecular event, which is involved in the memory storage process. The substrates phosphorylated by CaMKII are implicated in homeostatic regulation of the cell, as well as in activity-dependent changes in neuronal function that appear to underlie complex cognitive and behavioural responses, including learning and memory.

ZINC

Another metal ion that has been implicated in brain function is Zn^{2+}. The brain barrier systems, i.e. the blood-brain and blood-cerebrospinal fluid barriers, ensure that there are adequate zinc supplies for brain function and prevention of neurological diseases. A large portion of the zinc is present as zinc metalloproteins in neurons and glial cells. Approximately 10% of the total zinc in the brain, probably ionic zinc, exists within vesicles, which are responsive to dietary zinc deprivation. Such vesicular zinc is released into the synaptic cleft during neurotransmission and modulates NMDA-specific post-synaptic receptors in a rapid dose-dependent response that is reversible. Zinc enhances GABA release via potentiation of α-amino-3-hydroxy-5-methyl-4-isoxalolepropionate (AMPA)/kainate receptors in the CA3 region of the hippocampus, followed by a decrease in pre-synaptic glutamate release in the same region.

Zinc plays an important role in regulating brain development particularly during foetal and early post-natal life. Zinc deficiency adversely affects the autonomic nervous system regulation as well as hippocampal and cerebellar development leading to learning impairment and olfactory dysfunction. Furthermore, the susceptibility to epileptic seizures (which may decrease vesicular zinc) is also enhanced by zinc deficiency in vulnerable individuals. Zinc deficiency may lower the body's adaptability to stress. In this situation, intracellular free calcium concentration may be altered prior to the decrease in zinc concentration in the extracellular fluid. Anxiety-like behaviour, which is observed in animal models of induced zinc deficiency, can be corrected to some extent, by the administration of zinc. This is due to the fact that zinc, which is an antagonist of NMDA receptors, exhibits anti-depressant-type activity, by inducing brain-derived neurotrophic factor (BDNF) gene expression, which increases the synaptic zinc levels in the hippocampus. Preliminary clinical studies have also demonstrated the benefit of zinc supplementation in anti-depressant therapy.

On the other hand excessive synaptic release of zinc followed by entry into vulnerable neurons contributes to severe neuronal cell death. This is caused by the sequential activation of Akt and GSK-3beta, which play an important role in directing hippocampal neural precursor cell death.

Changes in the expression of metallothioneins, MTs, induced by changes in zinc status as well as other factors will have adverse effects on brain function. Stress will increase the expressions of MTs and their mRNAs in the hippocampus.

Mutations that cause reduced expression of the full-length survival motor neuron (SMN) protein are a major cause of spinal muscular atrophy (SMA), a disease characterized by degeneration of the α-motor neurons in the anterior horn of the spinal cord. The severity of SMA may be influenced by the actions of modifier genes. One potential modifier gene is *ZPR1*, an essential protein with two zinc fingers, present in the nucleus of growing cells that relocates to the cytoplasm in starved cells. ZPR1p is down-regulated in patients with SMA and it interacts with complexes formed by SMN. The functional significance of *ZPR1* gene down-regulation, was examined in a mouse model with targeted ablation of the *Zpr1* gene. Such ZPR1-deficient mice exhibit axonal pathology and neurodegeneration, thereby confirming that ZPR1 deficiency is a contributing factor in neurodegenerative disorders.

The expression of ZPR1 is suppressed in humans with severe SMA although the mechanism of its suppression remains an outstanding question. Sequence analysis of genomic DNA has not identified mutations in the *ZPR1* gene that correlate with SMA. It is therefore suggested that epigenetic changes in the *ZPR1* gene or an uncharacterized mutation in a regulatory region of the *ZPR1* gene may contribute to the changes in *ZPR1* gene expression.

COPPER

Genetic and nutritional studies have illustrated the essential nature of copper for normal brain function. Deficiency of copper during the foetal or neonatal period, will have adverse effects both on the formation and the maintenance of myelin. In addition, various brain lesions will occur in many brain regions including the cerebral cortex, olfactory bulb and corpus striatum. Vascular changes have also been observed. It is also of paramount importance that excessive amounts of copper do not occur in cells, due to redox-mediated reactions such that its level within cells must be carefully controlled by regulated transport mechanisms. Copper serves as an essential cofactor for a variety of proteins involved in neurotransmitter synthesis, e.g. dopamine β-hydroxylase, which transforms dopamine to nor-adrenaline, as well as in neuroprotection via the Cu–Zn superoxide dismutase present in the cytosol. A transporter of the SLC31 or Ctr family of proteins, Ctr1, mediates cellular copper uptake into cells where it rapidly binds to intracellular copper chaperone proteins (see Chapter 8). The copper chaperone Atox1 delivers copper to the secretory pathway and docks with either the copper-transporting ATPase ATP7B in the liver or ATP7A in other cells. ATP7B directs copper to plasma ceruloplasmin or to biliary excretion in concert with a newly discovered chaperone, Murr1. ATP7A directs copper within the *trans*-Golgi network (TGN) to the proteins dopamine β-monooxygenase, peptidylglycine monooxygenase, lysyl oxidase and tyrosinase, depending on the cell type. The ATP7A and ATP7B P-type ATPases are, respectively, affected in the genetic disorders of copper metabolism, Menkes' disease and Wilson's disease (see also Chapter 14).

DISORDERS OF COPPER METABOLISM: WILSON'S AND MENKES DISEASES

The neurological diseases caused by disorders of copper metabolism described in 1912 by Samuel Wilson, a young London registrar, and by the Columbia University paediatrician John Menkes 50 years later could hardly be more different. Wilson described a familial nervous disorder, which he called progressive lenticular[2] degradation, associated with cirrhosis of the liver. Large amounts of copper were present in the brain leading to progressive neurological dysfunction. The disease is characterized by progressive copper accumulation in the brain, liver, kidneys and cornea. In contrast, in Menkes disease, first described as an

---

[2] Lenticular—pertaining to the lens of the eye.

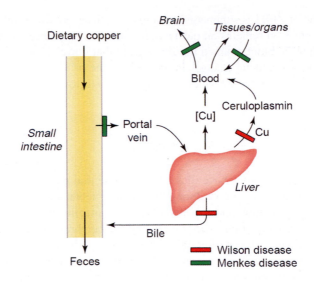

Figure 18.3 Pathways of copper that are blocked in Menkes and Wilson's disease. (From Crichton and Ward, 2006. Reproduced with permission from John Wiley & Sons., Inc.)

X-chromosome-linked fatal neurodegenerative disorder, the transport of copper across the intestinal tract is blocked, resulting in overall copper deficiency (Figure 18.3).

Yet in both diseases the mutations are found in copper-transporting P-type ATPases. As we saw in Chapter 8, copper is transported by the chaperone ATOX1 to the P-ATPase ATP7B, located predominantly in the trans-Golgi network (TGN), and this is the Wilson's disease protein. The principal function of ATP7B is to transport copper into the secretory pathway both for excretion in the bile and for incorporation into ceruloplasmin (Figure 18.4). In Wilson's disease, lack of functional ATP7B results in secretion of Cu-free apoceruloplasmin, which is rapidly degraded. In normal subjects, when hepatocyte copper content increases, ATP7B cycles to a cytoplasmic compartment near to the bile canicular membrane, where copper is accumulated in vesicles prior to its biliary excretion, most likely associated with Murr1. Murr1 is the product of a gene deficient in canine copper toxicosis, a condition in which affected dogs suffer from defective biliary copper excretion and accumulate massive amounts of lysosomal copper in the liver. In Wilson's disease, mutations in ATP7b result in cytosolic copper accumulation within the hepatocyte, provoking oxidative damage, leakage of copper into the plasma and finally copper overload in most tissues.

Menkes disease is an X-chromosome-linked neurodegenerative disorder of childhood characterized by massive copper deficiency. The boys affected generally die in early childhood, usually in the first decade with abnormalities that can all be related to deficiencies in copper-containing enzymes. The Menkes protein is also a P-type ATPase, known as ATP7A, which appears to play a major role in copper absorption in the gut and in copper reabsorption in the kidney. It has six copper binding motifs in the amino terminal region which also interact with copper chaperones, while eight transmembrane domains form a channel through which copper is pumped, driven by hydrolysis of ATP. In normal circumstances the protein is located primarily at the TGN, and relocalizes to the plasma membrane

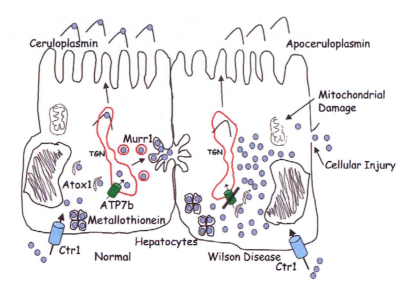

Figure 18.4 Proteins involved in copper uptake, incorporation into ceruloplasmin and biliary excretion in normal and Wilson's disease hepatocytes. (From Crichton and Ward, 2006. Reproduced with permission from John Wiley & Sons., Inc.)

in elevated copper conditions to expel the excess copper from the cell. Copper transport in the brain of Menkes patients is also blocked. Whereas hippocampal neurones release copper in response to activation of NMDA receptors, neurones from animals lacking functional ATP7a do not.

ACERULOPLASMINAEMIA

Ceruloplasmin has long been thought to be a ferroxidase and it has been proposed that ceruloplasmin has a custodial role *in vivo*, ensuring that Fe^{2+} released from cells is oxidized to the potentially less toxic Fe^{3+} prior to its incorporation into apotransferrin. Aceruloplasminaemia is a neurodegenerative disease associated with the absence of functional ceruloplasmin due to the presence of inherited mutations within the ceruloplasmin gene. This condition results in disruption of iron homeostasis, with extensive iron accumulation in a number of tissues such as brain and liver. However, in these patients, as in aceruloplasminaemic mice, both copper transport and metabolism are normal, providing strong evidence against the role of ceruloplasmin as a major copper transporter.

CREUTZFELDT–JAKOB AND OTHER PRION DISEASES

Creutzfeldt–Jakob and other prion diseases have been associated with disorders of copper metabolism. The first cases of Creuzfeldt–Jakob disease in humans were described by Creuzfeldt and Jakob over 80 years ago. Although scrapie was known as a fatal neurological disorder of sheep as early as the 1700s, its transmissibility was first demonstrated in 1939. During the 1980s, a bovine spongiform encephalitis (BSE) epidemic occurred in cattle, which was attributed to the feeding of BSE-prion-contaminated bone and meat to cattle. More recently, cases of vCJD in vulnerable individuals were thought to be due to their consumption of such BSE-contaminated beef.

The causative agent is known to be the prion protein. There is a considerable body of evidence which indicates that the progression of mammalian prion diseases involves a process in which the normal cellular Prp (PrP^C) predominantly α-helical normal form is converted into PrP^{Sc} through a post-translational process during which it acquires a high β-sheet content. This initial event is then followed by the spontaneous formation of a self-propagating aggregate. Finally, the newly formed prion must replicate itself in a process, which involves two separate steps: growth of the infectious particle by addition of the aggregate and amplification of the number of infectious particles. Two models have been proposed for the conversion of PsP^C to PsP^{Sc}, which are illustrated in Figure 18.5. The "Refolding" or heterodimer model proposes that the conformational change involving direct conversion of PsP^C to PsP^{Sc} is kinetically controlled, with a high activation energy barrier preventing spontaneous conversion at detectable rates. It is possible that an enzyme or a chaperone could facilitate this reaction. The "Seeding" or nucleation model postulates that PsP^C and PsP^{Sc} (or a PsP^{Sc}-like molecule, light coloured spheres) are in equilibrium, with the equilibrium largely in favour of PsP^C. PsP^{Sc} is only stable when it forms a multimer, hence PsP^{Sc} is stabilized when it adds onto a crystal-like seed or aggregate of PsP^{Sc} (dark spheres). Seed formation is an extremely rare event, however, once a seed is present, monomer addition ensues rapidly.

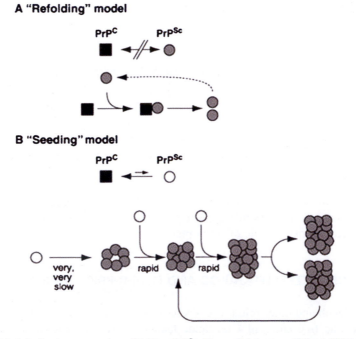

A "Refolding" model

PrP^C PrP^{Sc}

B "Seeding" model

PrP^C PrP^{Sc}

very,
very
slow   rapid   rapid

Figure 18.5 Models for the conversion of PrP^c to PrP^{Sc}. (From Crichton and Ward, 2006. Reproduced with permission from John Wiley & Sons., Inc.)

It has been well documented that the prion protein binds copper(II) ions and that, among divalent metal ions, PsP^C selectively binds Cu(II). The major Cu(II)-binding site has been identified as being within the unstructured amino terminal region (encompassing residues 60–91 of human PsP^C). Specifically, copper binds to a highly conserved octapeptide repeat domain, consisting of four sequential repeats of the sequence ProHisGlyGlyGlyTrpGlyAsn. The Cu(II) to octapeptide-binding stoichiometry is 1:1, i.e. the octapeptide repeat region binds four Cu(II) ions, and copper binding is most favoured at physiological pH, falling off sharply under mildly acidic conditions. While most studies of copper binding have focused on the octarepeat region, evidence has been found for a fifth preferential Cu(II) coordination site, between residues His96 and His111, outside of the octarepeat domain, with a nanomolar dissociation constant. Interestingly, circular dichroism studies show that copper coordination is associated with a loss of irregular structure and an increase in β-sheet conformation.

Evidence is also growing that PsP^C plays an important role in copper homeostasis, in particular at the pre-synaptic membrane; that it may be involved in triggering intracellular calcium signals; and that it may play a neuroprotective role in response to copper and oxidative stress (Figure 18.6). Exposure of neuroblastoma cells to high Cu(II) concentrations stimulated endocytosis of PsP^C, whereas deletion of the four octarepeats or mutation of the histidine residues in the central two repeats abolished endocytosis of PsP^C (see Chapter 8).

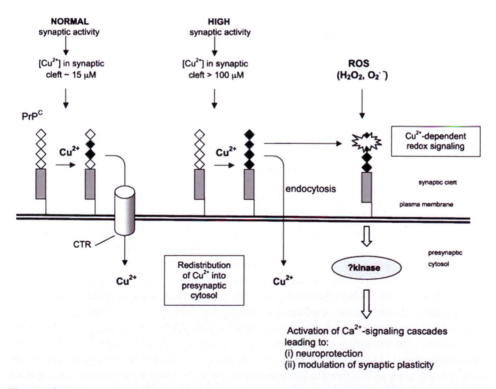

Figure 18.6 Schematic representation of the physiological role of prion protein (Prp^c) in copper homeostasis and redox signalling. (From Crichton and Ward, 2006. Reproduced with permission from John Wiley & Sons., Inc.)

However, studies at physiological concentrations of copper, led to the conclusion that PsP^C does not participate in the uptake of extracellular Cu(II). It has been suggested (Figure 18.6) that in the unique setting of the synapse, PsP^C acts to buffer Cu(II) levels in the synaptic cleft, following the release of copper ions as a result of synaptic vesicle fusion. Cu^{2+} ions released during neurotransmitter vesicle exocytosis are buffered by PsP^C, and subsequently returned to the pre-synaptic cytosol. This can occur either by transfer of copper to copper transport proteins (CTR) within the membrane, or in the case of higher copper concentrations, by PsP^C-mediated endocytosis. Copper-loaded PsP^C may interact with ROS, triggering redox signalling and subsequently activation of Ca^{2+}-dependent signalling cascades. These changes in intracellular Ca^{2+} levels lead to modulation of synaptic activity and to neuroprotection.

IRON

Enormous advances have been made in the last 10 years in understanding iron homeostasis, including the identification of genes and proteins involve in its uptake and transfer, e.g. DcytB, DMT1, Ireg1 and the modulation of their translation as well as of proteins involved in iron transport and storage by IRP-1 and IRP-2 (see Chapter 13). However, exactly how the brain regulates fluxes and storage of iron into neurons, oligodentrites, astrocytes and glial cells remains an enigma. Brain iron uptake into the brain is regulated by the expression of transferrin receptor 1 on endothelial cells of the blood-brain barrier (BBB). Transferrin-bound iron in the systemic circulation is endocytosed by brain endothelial cells, and iron is released to brain intestinal fluid, possibly by the iron exporter, ferroportin. Transferrin synthesized by the oligodentrites in the brain will bind the majority of the iron that traverses the BBB after the oxidation of the iron, possibly by a glycophosphoinositide-linked ceruloplasmin found in astrocytic foot processes that surround brain endothelial cells. Neurons acquire iron from diferric transferrin. However, the source of iron within microglia cells is unclear—other phagocytic cells such as macrophages, take up iron via transferrin receptors and release iron via ferroportin. In normal circumstances, the iron content within the brain varies greatly from one region to another. Significantly greater iron concentrations, as micrograms per gram of protein, are found in the substantia nigra and the globus palidus than in liver (Götz et al., 2004), as well as other brain regions, dentate gyrus, interpeduncular nucleus, thalamus, ventral pallidus, nucleus basilis and red nucleus (Figure 18.7). Regions of the brain associated with motor functions tend to have more iron than non-motor-related regions. This may explain why movement disorders are often associated with brain iron loading. The form in which this iron is incorporated into various proteins remains unclear. In oligodendrocytes it is bound to both H- and L-chain ferritin, in microglia to L-ferritin, while neurons contain mostly neuromelanin. In contrast astrocytes contain hardly any ferritin. Since H- and L-ferritin have different functions, their specific locations might indicate specific biological roles. H-ferritin is associated with stress responses as well as catalysing the oxidation of Fe^{2+} to Fe^{3+} via the ferroxidase centre. As to the movement of iron between different brain regions, this in the main remains unclear. It is thought that transferrin and ferritin may be important, since mRNA receptors for these iron proteins are detectable in grey matter and white matter, respectively. The fate of non-transferrin-bound iron, which may cross the BBB, remains unclear.

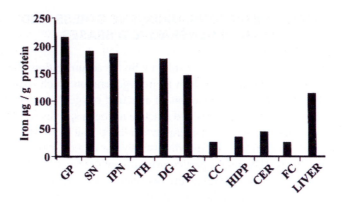

Figure 18.7 Distribution of iron in human brain. GP, globus pallidus; SN, substantia nigra; IPN, interpeduncular nucleus; TH, thalamus; DG, dentate gyrus; RN, red nucleus; CC, cerebral cortex; HIPP, hippocampus; CER, cerebellum; FC, frontal cortex. (From Crichton and Ward, 2006. Reproduced with permission from John Wiley & Sons., Inc.)

In recent years there have been several reports related to the adverse effects of iron deficiency on brain development, both pre- and post-natal. Iron is involved in many central nervous system processes that could affect infant behaviour and development. In various epidemiological studies it is reported that children with iron deficiency anaemia have poorer performances on tests of some specific cognitive function. Animal experiments have identified some of the defects of reduced iron availability on brain function, which include post-translational changes (which result in a failure of iron incorporation into protein structures which are subsequently degraded), vulnerability of the developing hippocampus (with loss of the neuronal metabolic marker cytochrome c oxidase), and altered dendritic structure. Iron deficiency will also have a direct effect on myelin, including a decrease in myelin lipids and proteins, as well as neurotransmitter systems, since iron is essential for a number of enzymes including tryptophan hydroxylase (serotonin) tyrosine hydroxylase (norepinephrine and dopamine). Long-term follow-up studies of iron deficiency in the human infant brain indicate that such alterations in myelination result in slower conduction in both the auditory and visual systems. Both of these sensory systems are rapidly myelinating during the period of iron deficiency and are critical for learning and social interaction. Together with the reduced energy, impaired glial function, altered activation of monoamine circuits, this may alter experience-dependent processes, which are critical to brain structure and function during early development.

An inevitable consequence of ageing is an elevation of brain iron in specific brain regions, e.g. in the putamen, motor cortex, pre-frontal cortex, sensory cortex and thalamus, localized within H- and L-ferritin and neuromelanin with no apparent adverse effect. However, ill-placed excessive amounts of iron in specific brain cellular constituents, such as mitochondria or in specific regions brain, e.g. in the substantia nigra and lateral globus pallidus, will lead to neurodegenerative diseases (Friedreich's ataxia and Parkinson's disease (PD), respectively). We discuss here a few of the examples of the involvement of iron in neurodegenerative diseases. From more on iron metabolism see Crichton, 2001.

REDOX METAL IONS, OXIDATIVE STRESS AND
NEURODEGENERATIVE DISEASES

A general mechanism can be proposed to explain why the accumulation of metal ions such as iron and copper in the brain can result in neurodegeneration. Such redox-active metals can generate oxidative stress by production of reactive oxygen species (ROS) such as hydrogen peroxide and hydroxyl radicals and reactive nitrogen species (RNS) such as nitric oxide and peroxynitrite. There is considerable evidence that both ROS and RNS are involved in a number of neurodegenerative diseases by causing oxidative damage to lipids, proteins and nucleic acids. The brain is particularly susceptible to oxidative damage because of its high oxygen consumption and relatively poor anti-oxidative defence mechanisms.

   An additional element in these pathophysiologies is the production of unstable, reactive aldehyde intermediates from lipid peroxidation, from glucose–protein or glucose–lipid interactions (glycation) or from oxidative modification of amino acids in proteins (amino acid oxidation). Radical species, such as the hydroxyl radical can cause peroxidative degradation of the polyunsaturated fatty acids present in membrane phospholipids, cholesterol esters and triglycerides. The major initial reaction products of lipid peroxidation are lipid hydroperoxides, which break down to give a broad array of smaller aldehyde fragments, which can be classified into three families: 2-alkenals, such as acrolein and its methyl homologue, crotonaldehyde; 4-hydroxy-2-alkenals, the most prominent of which is 4-hydroxy-2-nonenal (HNE); and ketoaldehydes, like malondialdehyde (MDA) (Figure 18.8). Direct oxidation by ROS of certain amino acid side-chains in proteins (proline, arginine, lysine and threonine) or oxidative cleavage of the protein backbone can lead to the formation of protein carbonyl derivatives (Figure 18.9). Such protein carbonyls can also be formed by Michael addition reactions of α,β-unsaturated aldehydes such as HNE, malondialdehyde

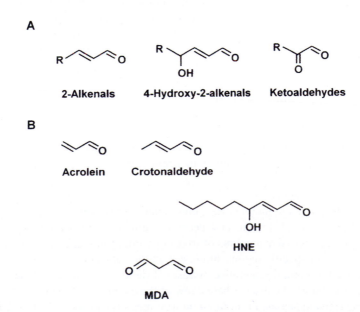

Figure 18.8 The three families of reactive aldehydes (A) and examples of each (B).

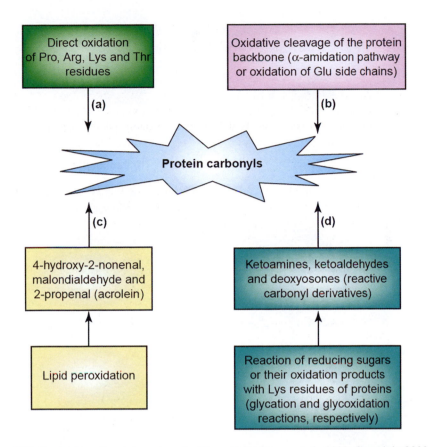

Figure 18.9 Production of protein carbonyls. (From Dalle-Donne et al., 2003. Copyright 2003, with permission from Elsevier.)

and acrolein with either the sulfhydryl group of cysteine, the imidazole group of histidine or the amino group of lysine in proteins. Carbonyl groups can also be introduced into proteins by addition of reactive carbonyl compounds (ketoamines, ketoaldehydes and deoxyosones) produced by a complex series of reactions between reducing sugars or their oxidation products with the amino groups of lysine residues in proteins, by mechanisms known as glycation and glyoxidation.

The outcome of this oxidative stress is to generate carbonyl functions, which together with other oxidative processes, cause damage to proteins, resulting in misfolding, and in many cases increased β-sheet content. The damaged, misfolded proteins aggregate, overwhelming the cytosolic ubiquitin/proteasome protein degradation system[3], and accumulate within intracellular inclusion bodies. Such intracellular inclusion bodies are found in a great many neurodegenerative diseases such as Alzheimer's disease (AD), PD, amyotrophic lateral sclerosis (ALS) and Huntington's disease (Figure 18.10).

---

[3] The proteasome is a cytoplasmic protein degradation machine that hydrolyses proteins, which have been designated for degradation by having a number of molecules of the small ubiquitous protein, ubiquitin, attached to them.

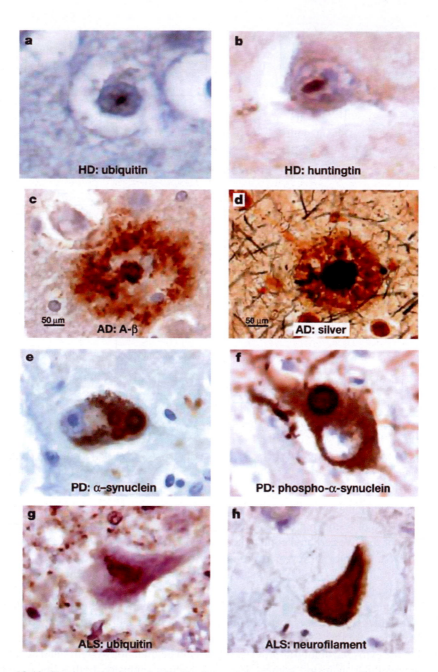

Figure 18.10 Characteristic inclusion bodies in neurodegenerative diseases, all labelled with antibodies (except (d)) as indicated. (a) and (b) HD intranuclear inclusion labelled for ubiquitin and huntingtin (cerebral cortex). (c) and (d) AD neuritic plaque labelled with Aβ (cerebral cortex) and silver stained. (e) and (f) PD, Lewy bodies labelled for α-synuclein and phosphorylated α-synuclein (substantia nigra). (g) and (h) ALS labelled with ubiquitin and neurofilaments (medulla oblongata). (From Ross and Poirier, 2004. Reproduced by permission of Nature Publishing Group.)

Parkinson's Disease, PD

In PD there is a two-fold increase in the iron content of the substantia nigra and the lateral globus pallidus. The etiology of such iron excesses are unknown although it has been suggested that changes in iron release mechanisms across the BBB, or dysregulation of iron transport across the membranes of specific brain regions may be involved. There are increased levels of iron in both Lewy bodies[4] (Figure 18.10) within cytosolic compartments as well as in dopaminergic neurons of the substantia nigra in PD patients, which will cause oxidative damage.

Despite the increased brain iron content in PD there is no corresponding up-regulation of ferritin expression. As we saw earlier, iron regulatory proteins, IRP-1 and IRP-2 act as iron sensors and regulate ferritin synthesis (see Chapter 8). IRP-2 seems to play the predominant role in post-transcriptional regulation of iron metabolism in brain. Changes in ubiquitination appear to be an important facet of PD; since the degradation of IRP-2 requires its ubiquitination, and proteasomal degradation this may be an explanation for the failure to up-regulate ferritin synthesis (which would require the inactivation of IRP-2).

Genetically engineered mice, which lack IRP-2 but have the normal complement of IRP-1, develop adult onset neurodegenerative disease associated with inappropriately high expression of ferritin in degenerating neurons. Mice that are homozygous for a targeted deletion of IRP-2 and heterozygous for a targeted deletion of IRP-1 develop severe neurodegeneration with severe axonopathy, and increased levels of ferric iron and ferritin expression as well as neuronal cell bodies degenerating in the substantia nigra.

Three proteins—α-synuclein, parkin and ubiquitin carboxy-terminal hydrolase (UCH-L1)—have been shown to be linked with PD. A putative model of the role of these proteins, is shown in Figure 18.11. In healthy neurons, the cytoplasmic concentration of α-synuclein (blue) is tightly controlled. Glycosylation (green) of α-synuclein is required for its ubiquitination (Ub) (yellow circles), probably by parkin (which is a protein-ubiquitin ligase, the enzyme family responsible for ubiquitination of condemned proteins). Proteasomal degradation of ubiquinylated α-synuclein produces peptide–ubiquitin conjugates, and subsequent recycling of ubiquitin might be controlled by UCH-L1. An increase in cytoplasmic α-synuclein concentration, as a result of increased synthesis or via inactivation of parkin, can promote its aggregation. If the gene encoding α-synuclein is mutated, the increased concentration can promote oligomerization of α-synuclein to form structured profibrils, which might be pathogenic in PD. Profibrils are morphologically heterogeneous, and spheres, chains and rings have been identified. These intermediates are eventually converted to fibrils and then to Lewy bodies, which are the pathological hallmark of PD brains.

Large amounts of iron are sequestered in substantia nigra and in locus coerulus as a neuromelanin–iron complex in dopaminergic neurons particularly in PD. Neuromelanin, a granular dark brown pigment, is produced in catecholaminergic neurons of the SN and

---

[4] Aggregates of ubiquinated α-synuclein are found in so-called Lewy bodies, intracellular inclusions present within dopaminergic neurones, axons and synapses of the substantia nigra, and are a characteristic feature of PD.

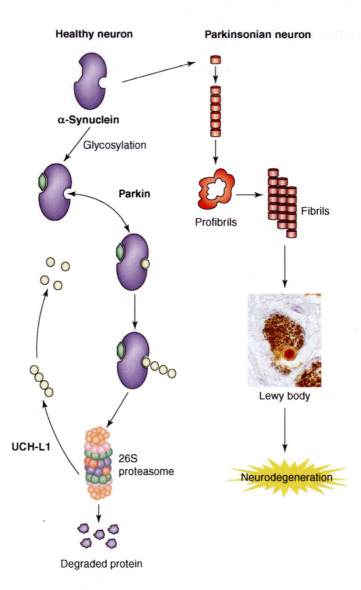

Figure 18.11 A putative model of the three proteins linked with Parkinson's disease. (From Barzilai and Melamed, 2003. Copyright 2003, with permission from Elsevier.)

locus coeruleus and is possibly the product of reactions between oxidized catechols with a variety of nucleophiles, including thiols from glutathione and proteins. It can be seen in Figure 18.10(e) and (f) as a fine granular brown label. The function of neuromelanin in the pigmented neurons is unknown but it could play a protective role via attenuation of free radical damage by binding transition metals, particularly iron. In normal individuals, the neuromelanin–iron complex is found in both the substantia nigra and locus coeruleus and increases linearly with age in the substantia nigra. Whether the ability of the neurones to synthesis neuromelanin is impaired in PD patients is unknown, although it is reported that the absolute concentration of nigral neuromelanin is less than 50% in PD with respect to age-matched controls. *In vitro* it has been shown that melanin can bind a significant amount of iron at two sites although the pigment appears to be only 50% saturated with iron in PD. Iron is bound to the catechol groups.

Alzheimer's Disease, AD

AD is one of the most common neurodegenerative maladies in Western societies. Clinical symptoms occur typically between the ages of 60 and 70 years. This disease, for which no effective treatment is currently available, initially presents with symptoms of memory loss, after which a progressive decline of both cognitive and motor function occurs. Both genetic and environmental factors are implicated in its development. Females are more susceptible than males, which may be attributable to the higher constitutive activity of the synaptic zinc transporter ZnT3. Studies showed that female mice exhibited age-dependent hyperactivity of the ZnT3 transporter which was associated with increased amyloid peptide, Aβ, deposition.

There is considerable evidence that defective homeostasis of redox-active metals, i.e. iron and copper, together with oxidative stress, contributes to the neuropathology of AD. The characteristic histology of AD is the deposition of both Aβ, as neurotic plaques (Figure 18.12a), and of the protein tau, as neurofibrillary tangles NFT (Figure 18.12b), predominantly in the cerebral cortex and hippocampus.

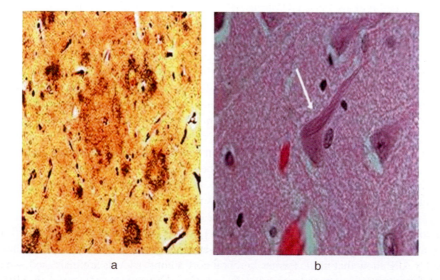

a   b

Figure 18.12 (a) Characteristic histo-pathological findings of Alzheimer's disease are senile plaques—a collection of degenerative pre-synaptic endings with astrocytes and microglia. Plaques are stained with silver stains and are of varying size. (b) Neurofibrillary tangles of Alzheimer's disease. The tangles are present as long pink filaments in the cytoplasm. Each is composed of cytoskeletal intermediate filaments.

Amyloid precursor protein, APP, a type I membrane protein, resembles a cell surface receptor and is physiologically processed by site-specific proteolysis. In normal circumstances, APP is cleaved by α-secretase to yield APPsα and the C-terminal fragment containing p3 (Figure 18.13). The membrane anchored α-carboxy terminal fragment, α-CTF, is then cleaved by γ—secretase within the membrane, releasing p3 peptide and the APP intracellular domain (AICD) (Figure 18.13). The presence of increasing amounts of iron may alter α-secretase activity; one hypothesis suggested that iron might be required as a cofactor or be an allosteric modifier of α-secretase activity. Iron may also decrease α-secretase cleavage rates.

In amyloidogenesis, the APP is cleaved sequentially by the proteolytic enzymes β- and γ-secretase, initially to yield APPβs and C99 fragments followed by γ-secretase cleavage of APPβs to produce β-amyloid peptide, Aβ, ($A\beta_{42}$ and $A\beta_{40}$)[5] and AICD. Aβ accumulation and aggregation is considered to be the initiating factor in AD pathogenesis although it is known that such deposition occurs over many years, if not decades, prior to the clinical signs of cognitive impairment. Aβ may be oxidized within the membrane, perhaps as a result of the increased Cu and Fe levels in the brain, from where it is ultimately liberated in a soluble form, to precipitate in the amyloid plaques. Aβ peptides will increase calcium influx through voltage-gated calcium channels (N- and L-types) by reducing magnesium blockade of NMDA receptors, as well as forming cation-selective ion channels after Aβ peptide incorporation into the cellular membrane, thereby increasing excitotoxicity. Aβ peptides may interfere with long-term hippocampal potentiation and cause synaptic dysfunction in AD.

Iron may modulate APP processing, by virtue of the presence of a putative iron response element in APP mRNA (based on sequence homology). The IRE was mapped within the 5′-untranslated regions (5′-UTR) of the APP transcript (+51 to +94) from the 5′cap site (Figure 18.14). The APP mRNA IRE is located immediately upstream of an interleukin-1 responsive acute box domain (+101 to +146). In response to intracellular iron chelation, translation of APP was selectively down-regulated, thereby causing a striking decrease in the production of APP_{sol}. Iron influx reversed this inhibition, by a pathway similar to iron control of the translation of the ferritin-L mRNAs by iron-responsive elements in its 5′UTRs. In addition, increase in cytokine production, namely IL-1 increased IRP binding to the APP 5′-UTR, thereby decreasing APP production. When the APP cRNA probe is mutated in the core IRE domain, IRP binding is abolished. In addition binding of the IRP to the IRE might interfere with APP translation and translocation across the endoplasmic reticulum membrane. This interference could be significant since α-secretase activity has been shown to require membrane bound APP. The role played by IRP-2 in AD remains undefined but may be more important than was previously thought.

Aβ has a very effective binding domain for copper in its N-terminal domain and can bind copper in nanomolar amounts (Figure 18.15). It is unclear whether APP or Aβ when associated with copper, are in fact neuronal metallochaperones. Knock out and knock in mice for APP show that in the former, cerebral cortex copper levels are increased, whereas in the latter reduced copper levels were found. Copper was also influential in APP processing in the cell; copper will reduce levels of Aβ and cause an increase in the secretion of the APP ectodomain.

---

[5] The subscript refers to the number of residues in the peptide.

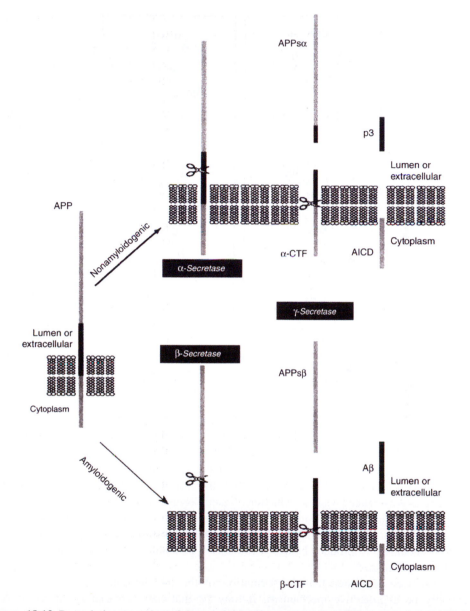

Figure 18.13 Proteolytic processing of the amyloid precursor protein (APP) by secretases. APP is a type 1 transmembrane glycoprotein. The majority of APP is processed by the non-amyloidogenic pathway (thick arrow) APP is first cleaved by α-secretase within the amyloid-β protein (Aβ) domain leading to APPsα secretion and precludes Aβ generation. Membrane anchored α-carboxy terminal fragment (CTF) is then cleaved by γ-secretase within the membrane releasing the p3 peptide and the APP intracellular domain (AICD). Alternatively amyloidogenesis takes place when APP is first cleaved by β-secretase, producing APPsα. Aβ and AICD are generated upon cleavage by γ-secretase of the β-CTF fragment retained in the membrane. (From Wilquet and De Strooper, 2004. Copyright 2004, with permission from Elsevier.)

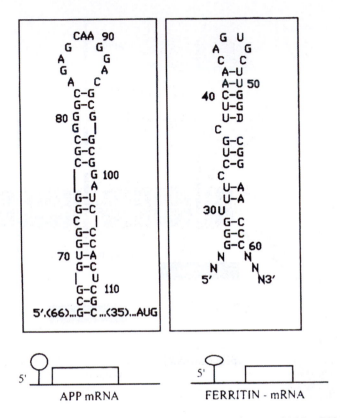

Figure 18.14 Evidence for an iron-responsive element in the 5′-UTR of APP mRNA. APP 5′-UTR sequences were computer-folded to generate the predicted RNA stem. (From Rogers et al., 2002. Reproduced by permission of the Journal of Biological Chemistry.)

Neurofibrillar tangles, NFTs, contain redox-active iron. Accumulation of tau in neurofibrillary tangles is associated with the induction of haem oxygenase 1, HO-1 a potent anti-oxidant, which plays an important role in metabolizing haem released from damaged mitochondria. HO-1 will reduce oxidative damage but Fe^{2+} will be released which may participate in Fenton chemistry to produce hydroxyl radicals. Tau within the neurofibrillary tangles is oxidatively damaged.

It has been suggested that the formation of the β-sheet configuration of Aβ may actually be a protective mechanism. It may be that the increased synthesis of APP and Aβ is an attempt by the brain cells to detoxify the elevated levels of redox-active metals, copper and iron; other studies suggest that zinc and copper are inhibitory and prevent β-sheet formation. Membrane-bound Aβ may be damaged by metal-induced ROS prior to their liberation from the membrane, and consequently precipitated by zinc which is released from synaptic vesicles (Figure 18.16). *In vitro*, zinc rapidly accelerates Aβ aggregation, the zinc being associated with the N-terminal region of Aβ which has an autonomous zinc binding domain. Zinc induces conformational change of the 1–16 N-terminal region of AP3.

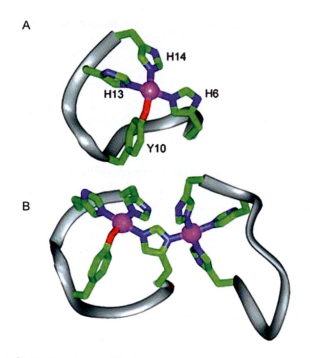

Figure 18.15 (a) Cu^{2+} binding site in APP. (b) Proposed aggregation via a His bridge between two copper atoms to form a dimer. (From Curtain et al., 2001. Reproduced by permission of the Journal of Biological Chemistry.)

Huntington's Disease

This is one of a family of diseases in which the expansion of CAG codon repeats resulting in extended polyglutamine (polyQ) tracts in the expressed protein. HD has a frequency of 4 in 10^5 among European populations (less than 1 in 10^6 in Japanese and African populations), and is the most common of the polyQ diseases. It causes movement disorders, cognitive deterioration and psychiatric disturbances. Symptoms begin appearing insidiously, typically between the ages of 35–50: the disease is progressive and fatal some 15–20 years after onset. Motor disturbances include choreiform[6] involuntary movements of proximal and distal muscles and progressive impairment of voluntary movements. In patients with juvenile onset HD the symptoms include bradykinesia (slowness of voluntary movements and of speech), rigidity and dystonia (intense irregular muscle spasms); the involuntary movements of the children often take the form of tremor, and they often suffer from epileptic seizures. HD is characterized by a remarkable specificity of neuronal loss. The most sensitive region is the striatum, with the caudate nucleus and the putamen particularly affected; in advanced cases there is also loss of neurons in the thalamus, substantia nigra and the subthalamic nucleus. The age of onset of the disease as well as the severity of the symptoms

---

[6] Choreiform movements are purposeless, involuntary movements such as flexing and extending of fingers, raising and lowering of shoulders or grimacing.

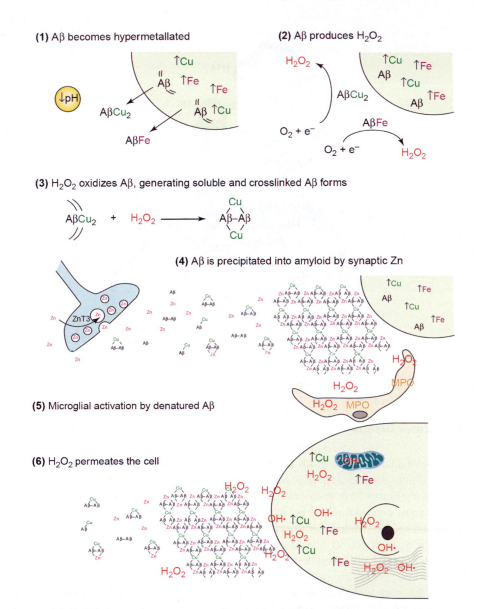

(1) Aβ becomes hypermetallated

(2) Aβ produces H_2O_2

(3) H_2O_2 oxidizes Aβ, generating soluble and crosslinked Aβ forms

(4) Aβ is precipitated into amyloid by synaptic Zn

(5) Microglial activation by denatured Aβ

(6) H_2O_2 permeates the cell

Figure 18.16 Hypothetical model for the metallobiology of Aβ in Alzheimer's disease. (From Bush, 2003. Copyright 2003, with permission from Elsevier.) The proposed sequence of events: (1) concentration of iron and copper increase in the cortex with aging. There is an overproduction of APP and Aβ in an attempt to suppress cellular metal-ion levels. (2) Hyper-metallation of Aβ occurs which may facilitate H_2O_2 production. (3) Hyper-metallated Aβ reacts with H_2O_2 to generate oxidized and cross-linked forms, which are liberated from the membrane. (4) Soluble Aβ is released from the membrane and is precipitated by zinc which is released from the synaptic vesicles. Oxidized Aβ is the major component of the plaque deposits. (5) Oxidized Aβ initiates microglia activation. (6) H_2O_2 crosses cellular membranes to react with Cu and Fe, and generate hydroxyl radicals which oxidize a variety of proteins and lipids.

are a function of the length of the glutamine stretches—individuals with 35 CAG repeats or less do not develop the disease; those with 35–39 have an increased risk; while repeats of 40 or over will always lead to the disease in the course of a normal lifespan.

The gene for the protein involved, huntingtin of unknown function, consists of 67 exons, extending over 180 kb of DNA and codes for a protein of 3144 residues (one of the longest polypeptide chains known). The polyQ domain is close to the N-terminus, with the CAG repeat in the first exon. In common with other polyQ diseases, huntingtin with its polyQ repeat is found within aggregates or inclusions within the nuclei of neurones (Figure 18.10). These inclusions in several cases are formed by insoluble amyloid-like fibres, reminiscent of the aggregated forms of proteins found in other neurological diseases. A conformational transition from random coil to β-sheet, which shares most of the features typical of amyloids, takes place during the process of fibre formation.

Alterations in brain iron metabolism have been reported, resulting in increased iron accumulation in Huntington's disease. This was particularly the case in basal ganglia from patients with HD compared to normal controls. In studies in embryonic stem cells, huntingtin was found to be iron-regulated, essential for the function of normal nuclear and perinuclear organelles and to be involved in the regulation of iron homeostasis.

Friedreich's Ataxia

Friedreich's ataxia is the most common hereditary ataxia[7] and is the most prevalent cerebellar ataxia among children and adults in Europe. It was first described in 1863 by Nikolaus Friedreich. His clinical observations described the essential characteristics of the disease as an adolescent-onset ataxia, particularly associated with clumsiness in walking, accompanied by sensory loss, lateral curvature of the spine, foot deformity and heart disease. Detailed neuropathological examination showed cerebrospinal degeneration. Friedreich's ataxia (FRDA) is yet another of the 15 neurological diseases in man, which are known to be caused by the anomalous expansion of unstable trinucleotide repeats. However, unlike Huntington's disease, the trinucleotide expansion occurs in a non-coding region of the gene. The FRDA gene is composed of seven exons spread throughout 95 kb of DNA (Figure 18.17), and encodes for the 210 residue protein, frataxin. Frataxin protein levels are severely decreased in FRDA patients, and most FRDA patients are homozygous for the GAA expansion in intron 1. Since the mutation is in the non-coding intron, the consequence of the GAA expansion is to decrease the amount of frataxin mRNA which is synthesized, thereby accounting for the decreased amount of protein. The characteristic pathogenesis of FRDA includes abnormal iron accumulation in mitochondria, hypersensitivity to oxidative stress, deficiency of Fe–S enzymes and respiratory chain electron transporters, reflecting the role of frataxin in FeS cluster assembly (see Chapter 6).

---

[7] Inability to coordinate voluntary bodily movements, particularly muscular movements.

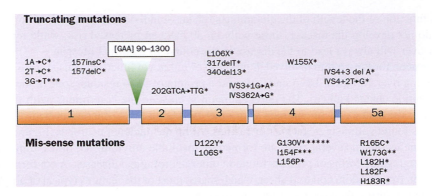

Figure 18.17 Frataxin mutations. The commonest mutation is the GAA expansion in the first intron of the frataxin gene (98%). Boxes represent exons and blue bars introns of the frataxin gene. Asterisks indicate the number of families reported with each mutation. From Dürr, 2002. Copyright 2003, with permission from Elsevier.

REFERENCES

Barzilai, A. and Melamed, E. (2003) Molecular mechanisms of selective dopaminergic neuronal death in Parkinson's disease, *Trends Mol. Med.*, **9**, 126–132.

Bush, A.I. (2003) The metallobiology of Alzheimer's disease, *Trends Neurosci.*, **26**, 207–214.

Crichton, R.R. (2001) *Inorganic Biochemistry of Iron Metabolism: From Molecular Mechanisms to Clinical Consequences,* Wiley, Chichester, 326 pp.

Crichton, R.R. and Ward, R.J. (2006) *Metal Based Neurodegeneration: From Molecular Mechanisms to Therapeutic Strategies*, Wiley, Chichester, 227 pp.

Curtain, C.C., Ali, F., Volitakis, I., Cherny, R.A., et al. (2001) Alzheimer's disease amyloid-beta binds copper and zinc to generate an allosterically ordered membrane-penetrating structure containing superoxide dismutase-like subunits, *J. Biol. Chem.*, **276**, 20466–20473.

Dalle-Donne, I., Giustarini, D., Colombo, R., Rossi, R. and Milzani, A. (2003) Protein carbonylation in human diseases, *Trends Mol. Med.*, **9**, 164–176.

Deisseroth, K., Mermelstein, P.G., Xia, H. and Tsien, R.W. (2003) Signalling from synapse to nucleus: the logic behind the mechanisms, *Curr. Opin. Neurobiol.*, **13**, 354–365.

Dürr, A. (2002) Friedreich's ataxia: treatment within reach, *Lancet Neurol.*, **1**, 370–374.

Götz, M., Double, K., Gerlach, M., Youdim, M.B.H. and Riederer, P. (2004) The relevance of iron in the pathogenesis of Parkinson's disease, *Ann. N. Y. Acad. Sci.*, **1012**, 193–208.

Rogers, J.T., Randall, J.D., Cahill, C.M., Eder, P.S., et al. (2002) An iron-responsive element type II in the 5′-untranslated region of the Alzheimer's amyloid precursor protein transcript, *J. Biol. Chem.*, **277**, 45518–45528.

Ross, C.A. and Poirier, M.A. (2004) Protein aggregation and neurodegenerative disease, *Nat. Med.*, **10**, S10–S17.

Voet, D. and Voet, J.G. (2004) Biochemistry, 3rd edition, Wiley, Hoboken, NJ, 1591 pp.

Wilquet, V. and De Strooper, B. (2004) Amyloid-beta precursor protein processing in neurodegeneration, *Curr. Opin. Neurobiol.*, **14**, 582–588.

– 19 –

Biomineralization

"For the harmony of the world is made manifest in Form and Number, and the heart and soul and all poetry of Natural Philosophy are embodied in the concept of mathematical beauty."[1]

INTRODUCTION

When we survey the living world around us we can only wonder at the diversity of its shapes and forms. The Scottish polymath[2], D'Arcy Thompson addressed this subject in his classic book *On Growth and Form*, first published in 1917, and in a revised edition in 1942 (the latter, a mere 1116 pages long). The central thesis of his book was that biologists placed too much emphasis on the role of evolution, and not enough on the roles of physical laws and mechanics, as determinants of the form and structure of living organisms. He decided that the laws concerned with static and dynamic forces of tension, compression and shear occurred in all living structures and influenced both growth, function and form. The bones of a skeleton in a museum would be a limp heap on the floor without the clamps and rods that pull them together, and Thompson argued that in living animals tension plays as important role in holding the skeleton together as does weight. In the same way, tension holds together the arches of medieval cathedrals, and steel cables provide the tensile strength on which suspension bridges are hung (Thompson, 1942).

We now recognize that, while much of biology relies on inorganic structures, biominerals, to supply the tensile strength and the other material properties that we associate with, for example bone, the diversity of form and shape depends on the organic matrix in which the biomineral is allowed to form. It is a little like the construction of buildings with reinforced concrete—the mould determines in what shape and form the concrete will set. And it is

---

[1] D'Arcy Wentworth Thompson "On Growth and Form" (1917).
[2] Thompson was offered the chair of Classics, Mathematics or Zoology at the University of Saint Andrews—he chose the latter.

just so in biomineralization—the organic mould is the organic matrix in which the process of selective precipitation of the inorganic mineral to be formed is directed, indeed, one might even say orchestrated, by the organic component.

Calcium is probably the most widely distributed element in biominerals, particularly in the 'hard parts' of organisms, such as teeth and bones. With the recognition that numerous minerals based on a great number of cations (among which figure Ba, Ca, Cu, Fe, K, Mg, Mn, Na, Ni, Pb, Sr and Zn) as hydroxides, oxides, sulphides, sulphates, carbonates and phosphates, the more restrictive term 'calcification' has given way to the more global 'biomineralization'. Since it is not our intention to cover the whole field of biomineralization, we restrict ourselves to three examples. First, we consider the formation of the mineral iron core in the storage protein ferritin and related proteins. Then we examine the calcification process involving calcium carbonate in ascidians and molluscs and calcium phosphate in bone and dental enamel.

IRON DEPOSITION IN FERRITIN

Ferritins, as was described in Chapter 8, are widely distributed throughout living organisms. They have a highly conserved structure made up of 24 protein subunits (apoferritin) which assemble to form a roughly spherical hollow shell, with an external diameter of ~120 Å and an internal diameter of ~80 Å (Figure 19.1). The subunits are a classical four helical bundle, with a short fifth helix at the C-terminal end of the protein. Mammalian ferritins are heteropolymers made up of variable proportions of two types of subunit, H and L: tissues involved predominantly in iron storage, such as liver and spleen have a high content of L-chains. In contrast, tissues like heart and brain that require protection from the potential toxic effects of 'free' iron in the generation of hydroxyl radicals, have a preponderance of H chains. While both types of subunit have closely similar tertiary structures they have only around 50% homology of sequence. There is another important difference between H- and L-chain subunits. H-subunits have a dinuclear iron site, the so-called ferroxidase site, which catalyses the oxidation of Fe^{2+}, whereas L-subunits have nucleation sites, at which the iron core of the ferritin molecule begins to be formed. As we will see shortly, both types of chains are required for optimal iron assimilation.

Typically, mammalian ferritins can store up to 4500 atoms of iron in a water-soluble, non-toxic, bioavailable form as a hydrated ferric oxide mineral core with variable amounts of phosphate. The iron cores of mammalian ferritins are ferrihydrite-like ($5Fe_2O_3 \cdot 9H_2O$) with varying degrees of crystallinity, whereas those from bacterioferritins are amorphous due to their high phosphate content. The Fe/phosphate ratio in bacterioferritins can range from 1:1 to 1:2, while the corresponding ratio in mammalian ferritins is approximately 1:0.1.

Access of iron to the interior of the protein could be through channels, which traverse the shell along the three- and four-fold axes of symmetry of the protein. The three-fold channels are predominantly hydrophilic, with three glutamate and three aspartate residues at each end of the funnel-shaped channel. In contrast, the four-fold channels are essentially lined with hydrophobic residues.

Iron incorporation into mammalian ferritins is thought to involve the following steps (Crichton, 2001): (1) Uptake of Fe^{2+} into the protein shell, most probably through the hydrophilic three-fold channels. (2) Oxidation of ferrous iron by the dinuclear ferroxidase

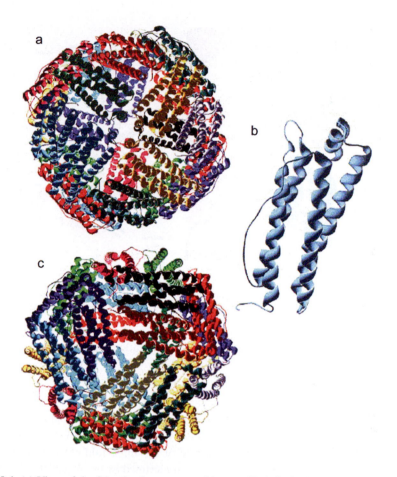

Figure 19.1 (a) View of the 24-subunit structure of human H-chain ferritin (rHuHF) viewed down the four-fold symmetry axis. (b) The subunit of rHuHF, with the short E-helix at the top of the four-helix bundle. (c) rHuHF viewed down the three-fold symmetry axis. (From Lewin et al., 2005. Copyright with permission from The Royal Society of Chemistry, 2005.)

sites situated within the four-helix bundle of H-chain subunits by molecular O_2, producing H_2O_2. (3) Migration of Fe^{3+} from the ferroxidase sites to sites of nucleation on the interior surface of the L-chain subunits of the protein shell, which facilitates mineralization. (4) Growth of the ferrihydrite mineral core via iron oxidation and mineralization on the surface of the growing crystallite. It is this final phase that will concern us principally here.

Iron pathways into ferritin

We assume that all substances involved in ferritin iron deposition (Fe^{2+}, Fe^{3+}, O_2) need to gain access to the interior of the apoferritin protein shell. The most likely pathway is via the three-fold channels, which would involve passing through the 12 Å long channel, and then traverse a further distance of ~8 Å along a hydrophilic pathway from the inside of the

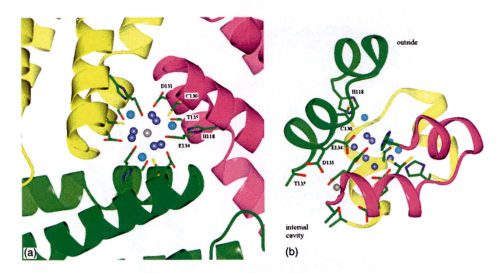

Figure 19.2 Metal-binding sites in the channel aligned on the three-fold symmetry axis shows binding to three zinc atoms and their symmetrically related subunits; the first is in the entrance of the funnel-shaped channel (in cyan), the second is in an alternative position (in blue) and the third is aligned on the three-fold axis (in grey). The overall stoichiometry is seven zinc cations per channel, i.e. 56 zinc cations per molecule. The two representations are with two different orientations; part a is aligned on the three-fold axis and part b is perpendicular to the axis. (From Toussaint et al., 2006. Copyright 2006, with permission from Elsevier.)

channel to the ferroxidase site. Calculations of electrostatic potential in HuHF show that the negative outer entrance is surrounded by patches of positive potential and that this attracts cations towards the channel entrance. The role of the three-fold channel in the entry of divalent cations into the interior of the protein is underlined by our recent studies on recombinant human H-chain ferritin (rHuH) show the way in which Zn^{2+} ions can transit through this channel, using the configurational flexibility of a key cysteine residue to move the ion through the channel, in a manner reminiscent of transit through the potassium channel described in Chapter 9 (Figures 19.2–4). Zn(II) was used as a redox-stable alternative for Fe(II), to allow the characterization of the different metal-binding sites (Figure 19.3).

Iron oxidation at dinuclear centres

H- and L-homopolymers incorporate iron atoms, but incorporation into the H form occurs at rates several times faster than L ferritins. H-chain ferritins are members of a larger class of (μ-carboxylato)diiron proteins (see Chapter 13). The diiron centre is located in the central region of the four-helix subunit bundle (Figure 19.3), and was identified by a large number of kinetic experiments as the catalytic ferroxidase centre. Substitution of residues at the ferroxidase centre of HuHF, such as E27A, E62K + H65G, H107A and Y34F, results in greatly decreased ferroxidase activity. Rapid oxidation, which is complete in less than 10 s for HuHF with 48 Fe^{2+}/protein, requires binding of iron (II) at both sites A and B. The stoichiometries of both oxygen consumption and of proton release subsequent to

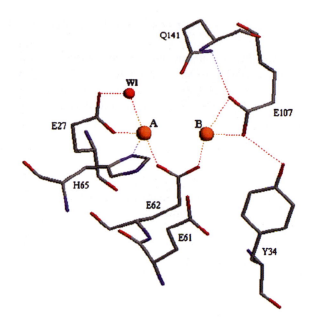

Figure 19.3 Ferroxidase centre environment from the Zn derivative crystal structure of rHuHF. (From Toussaint et al., 2006. Copyright 2006, with permission from Elsevier.)

Fe^{3+} hydrolysis have been determined by using a combined oximeter and pH stat. The overall reaction at the ferroxidase centre is postulated to be:

$$\text{Protein} + 2Fe^{2+} + O_2 + 3H_2O \rightarrow \text{Protein-}[Fe_2O(OH)_2] + H_2O_2 + 2H^+ \quad (19.1)$$
(ferroxidase complex)

The stoichiometry of $2Fe(II)/O_2$, and the structure of the ferroxidase iron site suggest that the first step after iron (II) binding would be transfer of two electrons, one from each Fe(II), to a dioxygen molecule bound at the same site, to give a formal peroxodiferric intermediate: formally this represents dioxygen binding, followed by Fe(II) oxidation (Equations (19.2) and (19.3)):

$$\text{Protein-}[Fe_2]^{4+} + O_2 \rightarrow \text{Protein-}[Fe_2^{4+}O_2]^{4+} \quad (19.2)$$
(dioxygen complex)

$$\text{Protein-}[Fe_2^{4+}O_2]^{4+} \rightarrow \text{Protein-}[Fe_2^{6+}O_2^{2-}]^{4+} \quad (19.3)$$
(peroxointermediate)

The peroxo intermediate would then undergo iron (III) hydrolysis (Equations (19.4) and (19.5)) to give first the μ-oxobridged Fe(III) dimer and then upon addition of another two molecules of H_2O, a protein-$[Fe_2O(OH)_2]$ species at the ferroxidase centre:

$$\text{Protein-}[Fe_2^{6+}O_2^{2-}]^{4+} + H_2O \rightarrow \text{Protein-}[Fe_2O]^{4+} + H_2O_2 \quad (19.4)$$
(μ-oxobridged complex)

$$\text{Protein-}[Fe_2O]^{4+} + 2H_2O \rightarrow \text{Protein-}[Fe_2O(OH)_2] + 2H^+ \qquad (19.5)$$

The detection of a peroxodiferric intermediate in the ferritin ferroxidase reaction establishes the ferritin ferroxidase site as being very similar to the sites in the O_2-activating (μ-carboxylato)diiron enzymes. However, in ferritins, the peroxodiferric intermediate forms diferric oxo or hydroxo precursors, which are transferred to biomineralization sites with release of hydrogen peroxide.

Ferrihydrite nucleation sites

The initial stages of iron incorporation requires the ferroxidase sites of the protein. Thereafter, the inner surface of the protein shell provides a surface which supplies ligands that can partially coordinate iron but which leave some coordination spheres available for mineral phase anions, thereby enabling the biomineralization process to proceed, with formation of one or more polynuclear ferrihydrite crystallites. Iron is transferred from the ferroxidase sites to the core nucleation sites by the net reaction:

$$\text{Protein-}[Fe_2O(OH)_2] + H_2O \rightarrow \text{Protein} + 2FeOOH_{core} + 2H^+ \qquad (19.6)$$

Two metal-binding sites were identified on the inner surface of the B helix at the subunit dimer interface of rHuHF that bind Zn^{2+} ions, and constitute the putative nucleation centres. One site involves residues shared by two symmetrically related subunits His60, Glu64 and Glu67, while His57 and Glu61 from the same subunit are implicated in the second metal-binding site (Figure 19.4). L-chains are more efficient in promoting nucleation than H-chains, and typically have Glu residues in place of His at positions 57 and 60. Amino acid substitutions or chemical modification at these nucleation site residues lead to diminished core formation. Core formation could be initiated by proximal binding of two Fe^{3+} at the putative ferrihydrite heteronucleation centre. It has been proposed that Glu61 could alternately act as a ligand to the ferroxidase site and to the nucleation site, and hence serve as a go-between to move iron from one site to another. We can find no evidence for this in any of the rHuH ferritin structures that we have determined. It is clear that modification of both the ferroxidase centre and the nucleation centre leads to ferritins that do not oxidize or incorporate iron.

Hence, the overall reaction for iron oxidation and hydrolysis at the ferroxidase centre, followed by further hydrolysis and migration to the core nucleation sites consists of two reactions, the protein-catalysed ferroxidase reaction itself and the Fe(II) plus H_2O_2 detoxification reaction (Equations (19.7) and (19.8), respectively):

$$2Fe^{2+} + O_2 + 4H_2O \rightarrow 2FeOOH_{core} + H_2O_2 + 4H^+ \qquad (19.7)$$

$$2Fe^{2+} + H_2O_2 + 2H_2O \rightarrow 2FeOOH_{core} + 4H^+ \qquad (19.8)$$

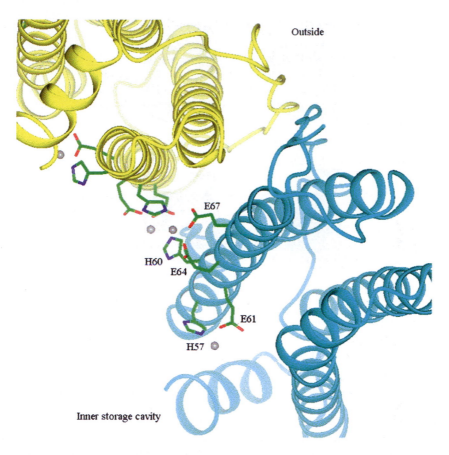

Figure 19.4 The two sites of metal binding at the surface of the inner shell of the ferritin constitute some putative nucleation centres. One site involves residues shared by two symmetrically related subunits His60, Glu64 and Glu67, while His57 and Glu61 from the same subunit are implicated in the second metal-binding site. (From Toussaint et al., 2006. Copyright 2006, with permission from Elsevier.)

Biomineralization

As we pointed out earlier, the H subunit catalyses the ferroxidase reaction, which occurs at all levels of iron loading, but decreases with increasing amounts of iron added (48–800 Fe/protein). Reaction (19.8) catalysed by both H- and L-chain ferritins, occurs largely at intermediate iron loadings of 100–500 Fe/protein. Once nucleation has taken place, the role of the protein is to maintain the growing ferrihydrite core within the confines of the protein shell, thus maintaining the insoluble ferric oxyhydroxide in a water-soluble form.

We now consider how the biomineralization chamber which is constituted by the interior of the apoferritin protein shell influences the growth of the core. Once sufficient core has

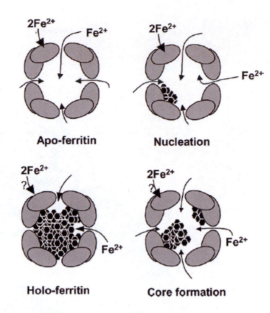

Figure 19.5 Representation of the crystal growth mechanism for ferritin core formation. (From Lewin et al., 2005. Copyright with permission from The Royal Society of Chemistry, 2005.)

been deposited ($>$100 Fe atoms), Fe(II) oxidation can proceed on the mineral surface of the growing core with the stoichiometry reported in reaction (19.9):

$$4Fe^{2+} + O_2 + 6H_2O \rightarrow 4FeOOH_{core} + 8H^+ \quad\quad (19.9)$$

This is the reaction which dominates at $>$800 Fe/protein. The model which best describes the global biomineralization process in mammalian ferritins is shown in Figure 19.5, and is called the crystal growth mechanism. The core surface is considered to be the key factor in determining the rate of mineral formation. As described above, in the initial stages iron can be oxidized by the ferroxidase centres of the protein and transferred to the nucleation sites where they progressively form small polynuclear iron clusters that can then act as nucleation centres for mineral growth. Most probably, one of these clusters will become the dominant nucleation centre and growth of the mineral would then occur from this centre. At some stage, oxidation by the protein would cease to be important with essentially all of the oxidation taking place on the surface of the mineral.

While core formation during hydrolysis of Fe(III) produces electrically neutral ferrihydrite, it also produces protons: two per Fe(II) oxidized and hydrolysed, whether due to iron oxidation and hydrolysis at the ferroxidase centre, followed by further hydrolysis and migration to the core nucleation sites or by direct Fe(II) oxidation and hydrolysis on the mineral surface of the growing core. These protons must either be evacuated from the cavity or else their charges must be neutralized by incoming anions, and it

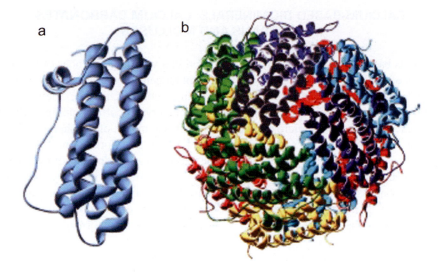

Figure 19.6 The three-dimensional structure of the *Listeria innocua* ferritin subunit (a) and the 12-mer (dodecameric) structure viewed down a three-fold symmetry axis. (From Lewin et al., 2005. Copyright with permission from The Royal Society of Chemistry, 2005.)

seems likely that both mechanisms are employed. In most ferritin molecules, some hydroxyl ions of the core (mostly on the core surface) are replaced by phosphate ions, while electrostatic calculations suggest that expulsion of protons (as well as Fe^{3+} or Fe^{2+} ions) or uptake of orthophosphate (or other anions such as chloride) would be facilitated by the electrostatic field gradient through the four-fold channels in human H-chain ferritin.

Why mammalian ferritin cores contain ferrihydrite-like structures rather than some other mineral phase is less easy to understand, and presumably reflects the way in which the biomineral is built up within the interior of the protein shell together with the geometry of the presumed nucleation sites. The phosphate content in the intracellular milieu can readily be invoked to explain the amorphous nature of the iron core of bacterioferritins and plants. Indeed, when the iron cores of bacterioferritins are reconstituted in the absence of phosphate, they are found to be more highly ordered than their native counterparts, and give electron diffraction lines typical of the ferrihydrite structure. Recently it has been reported that the 12 subunit ferritin-like Dps protein (Figure 19.6), discussed in Chapter 8, forms a ferrihydrite-like mineral core, which would seem to imply that deposition of ferric oxyhydroxides within a hollow protein cavity (albeit smaller) leads to the production of this particular mineral form (Su et al., 2005;Kauko et al., 2006).

Ferrihydrite is not the only mineral form of iron that is found in nature. Magnetite, the Fe^{2+}/Fe^{3+} mineral (Fe_3O_4) is found in magnetic bacteria (Matsunaga and Okamura, 2003) as well as in bees, birds and fish, where it is believed to function as a navigational magnetic sensor. Some bacteria can form pseudo-single crystals of akaganeite (β-FeOOH), many millimetres long in a polysaccharide-template directed process, and goethite (α-FeOOH) is found in the radular teeth of limpets and in some human haemosiderins.

CALCIUM-BASED BIOMINERALS: CALCIUM CARBONATES
IN ASCIDIANS AND MOLLUSCS

Biominerals based on calcium carbonate are found in a very large number of organisms. The shells of molluscs are among the most abundant biogenic minerals, and are composed of 95–99% calcium carbonate crystal and less than 5% organic matrix. Shell layers are formed from calcium carbonate as aragonite or calcite crystal corresponding to different mineral textures (also called microstructures). The organic matrix of the shell is composed of proteins and polysaccharides, which are thought to direct the formation of the calcium carbonate crystal and thus are responsible for the extraordinary properties of the shell. For example, nacre, the aragonite layer of the shell, exhibits a fracture resistance 3000 times higher than that of abiotic aragonite. The different crystal polymorphisms and microstructures of the layers are controlled by proteins secreted from outer epithelial cells in different regions of the mantle. Figure 19.7 shows the molecular correspondence at the

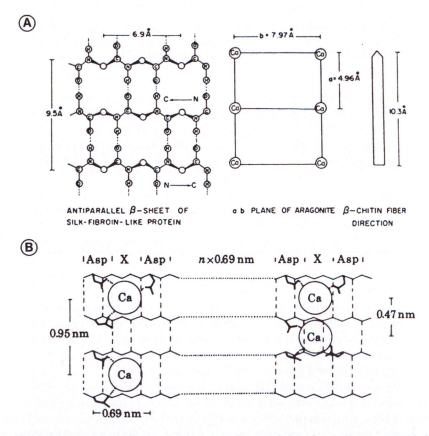

Figure 19.7 Molecular correspondence of the inorganic–organic interface in the nacreous shell layer of *Nautilus repertus*. (a) Structural relationships between protein sheets, aragonite crystals and chitin fibres. (b) Possible complementarity of Ca binding. (From Mann et al., 1989. Reproduced with permission from John Wiley & Sons., Inc.)

inorganic–organic interface in the nacre shell layer of *Nautilus repertus*. There is a close geometric match between the periodicity of the protein β-sheet and the lattice spacings of the aragonite, and it has been suggested that there is molecular complementarity between Ca atoms in the aragonite and aspartic acid residues organized in the sequence As-X-Asp (where X = a neutral residue) along the β-sheet.

Ascidians (Tunicates) are a class of marine organisms (including sea squirts), which have a characteristic tough covering, called a tunic. The tunic is composed of cellulose and contains small calcium-containing 'spicules' in a wide variety of shapes (Figure 19.8):

(a)

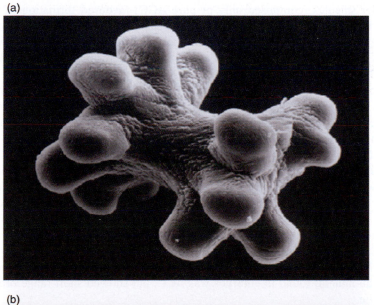

(b)

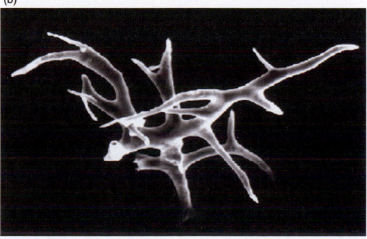

Figure 19.8 Scanning electron micrographs of spicules from the tunicate *P.pachydermatina* (a) Dogbone spicule from the tunic. (b) Antler spicule from the branchial sac. (From Wilt, 2005. Copyright 2005, with permission from Elsevier).

they are thought to contribute to the stiffness of the tissues. They are formed in close association with cells called sclerocytes, which secrete a very tough enveloping organic layer as well as in forming the mineralized spicule. The two spicules shown in Figure 19.8 from the tunicate *Pyura pachydermatina* both contain an unusual, and unstable form of calcium carbonate called amorphous calcium carbonate (ACC) together with the organic envelope. The knobbed or 'dogbone' tunic spicules also contain an overlay of calcite. Addition of macromolecules extracted from the calcite layer of these spicules speeded up formation of calcite crystals, whereas the corresponding extract from the ACC layer inhibited crystal formation. In contrast, macromolecules from antler spicules (composed only of pure ACC) favoured formation of the stable hydrated form of ACC. This clearly shows that the proteins and other macromolecules occluded with the mineral phase play an important role both in stabilization of unstable forms like ACC and in selection of the polymorphic form which is selected (in this case ACC or calcite).

The shells of molluscs, such as clam, oyster abalone, scallop and fresh water snail, use $CaCO_3$ as the principal constituent for an extraordinary array of diverse structures. Several approaches have led, and will lead further in the future, to a greater understanding of how these complex forms are generated.

One interesting method is to examine the development of the shell from the larval form of the organism all the way to the adult. ACC referred to above has been found together with aragonite in larval shells, and that the ACC is gradually transformed into aragonite over a period of hours or days. This novel suggestion that stable crystalline forms of calcium carbonate are formed from an amorphous precursor form, which transforms to the crystal in a slow, regulated way is becoming increasingly supported by experimental evidence. This is underlined by the observation that the aragonite found in larvae is less crystalline than aragonite obtained from non-biogenic sources, and that the mineral found in embryos was ACC, but showed short range order similar to aragonite.

Shell formation continues after morphogenesis, and the adult shell may contain either aragonite alone, or a mixture of layers of aragonite and calcite. In species that have a mixture, the layer of calcite, called the prismatic layer, is deposited close to the periostracum, which covers the external surface of the shell (facing the sea water). The inner nacre layer of aragonite is then deposited (Figure 19.9). Shells that have both a prismatic layer of calcite and nacre of aragonite must control the form of $CaCO_3$, which is deposited. It has been found that soluble molecules of the matrix extracted from the nacre favour aragonite formation *in vitro*, whereas extracts from prismatic layers favour calcite formation *in vitro*. The implication is that the cells that secrete the matrix are programmed to change the composition of the matrix at precise times and places, in order to regulate the change from calcite to aragonite during the transition from prism to nacre.

The final approach is to use a combination of microarray techniques, antisense oligonucleotides, functional tests and the use of monoclonal antibodies to identify matrix proteins and families of matrix proteins involved in the biomineralization process, and ultimately to establish their function. The recent publication of the genome of the sea urchin

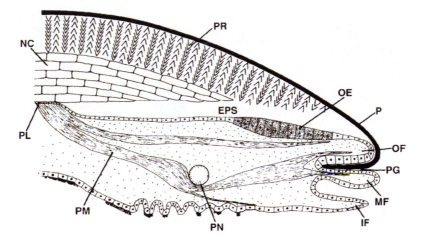

Figure 19.9 Diagram of a section through the edge of the shell and attached mantle of a bivalve mollusc. P, periosstracum; PR, prismatic shell layer; NC, nacre (from Wilt, 2005).

Strongylocentrotus purpuratus represents a promising start to the bioinformatic approach for the identification, characterization and functional analysis of particular molecules of the matrix involved in biomineralization.

BIOMINERALIZATION IN BONE AND ENAMEL FORMATION

Bone and teeth in mammals and bony fishes all rely on calcium phosphates in the form of hydroxyapatite $[Ca_5(PO_4)_3OH]_2$, usually associated with around 5% carbonate (and referred to as carbonated apatite). The bones of the endoskeleton and the dentin and enamel of teeth have a high mineral content of carbonated apatite, and represent an extraordinary variety of structures with physical and mechanical properties exquisitely adapted to their particular function in the tissue where they are produced. We begin by discussing the formation of bone and then examine the biomineralization process leading to the hardest mineralized tissue known, the enamel of mammalian teeth.

Bone is organized at five hierarchical levels of structure, shown in Figure 19.10: (1) cortical and cancellous (with an open porous structure) bone; (2) osteons (also called Haversian systems), the chief structural component of bone, consisting of lamellae (layer) of bone surrounding a long narrow passage, (the Haversian canal), which are clearly visible in a cross section of a long bone (e.g. the femur); (3) at the micron (μm) level, the lamellae and canals which constitute the osteons; (4) at the submicron scale, the mineral matrix embedded in the collagen fibres; and (5) at the nanoscale, mineral crystals, collagen and water molecules. We will be concerned in what follows by a consideration of the organic matrix, the apatite mineral and the water, which constitute the predominant constituents of bone.

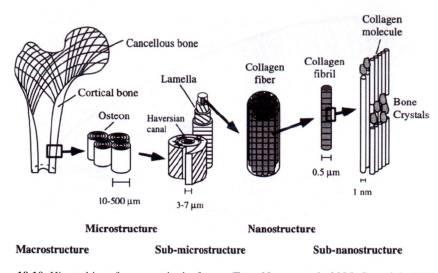

Figure 19.10 Hierarchies of structure in the femur. (From Nyman et al., 2005. Copyright 2003, with permission from Elsevier.)

THE ORGANIC MATRIX, MINERAL PHASE AND BONE MINERALIZATION

The organic matrix, which constitutes ~32% of the volume of bone, is made up of 90% type I collagen, with smaller amounts of non-collagenous proteins such as osteocalcin and osteonectin, which assist in the biomineralization process. Collagen is secreted by bone-forming cells, osteoblasts, as a precursor, procollagen, a helical rod made up of three intertwining polypeptide chains (Figure 19.11). It has a typical amino acid sequence, with glycine residues in almost every third position, and frequent occupancy of the second and third positions by proline and hydroxyproline, respectively (the latter introduced by the action of the Fe^{3+}- and ascorbate-dependent prolyl hydroxylase). After the amino- and carboxyl-terminal ends have undergone proteolytic cleavage, collagen begins to assemble into a collagen fibril, some 300 nm long and 1.2 nm in diameter, stabilized by cross-links between hydroxylysine residues (introduced by the copper dependent enzyme lysyl hydroxylase). Other types of cross-link can also be formed.

   The mineral phase constitutes ~43% of the volume of bone and mostly contains calcium and phosphate, with small, but highly significant amounts of carbonate (and a few other impurities). The bone mineral is not hydroxyapatite, but rather can be classified as carbonated apatite $[Ca_5(PO_4CO_3)_3]$. Within the mammalian skeleton, bone mineralization depends on the organization of the cross-linked collagen network. Initially, water fills the void space within the organic framework of the collagen matrix of the osteoid. Crystal nucleation occurs first at multiple independent sites within the collagen fibril as well as on the surface of the fibrils. It continues into the zones between collagen molecules, displacing water molecules as it goes, to leave final water content in bone of up to 25% of the volume. This residual water may help to stabilize collagen by forming inter- and intra-hydrogen bonds with hydrophilic residues, like the hydroxyl group of hydroxyproline and other polar side-chains (Figure 19.12).

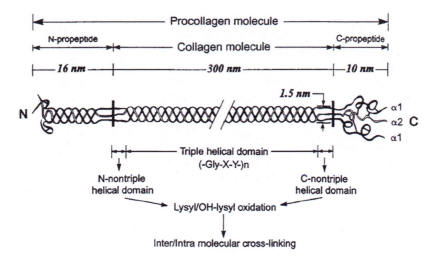

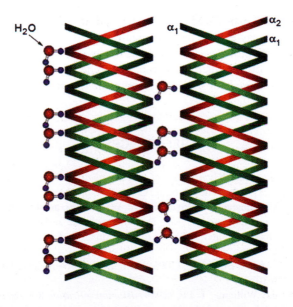

Figure 19.11 After cleavage of the terminal ends, the collagen molecule begins to assemble into a fibril via enzymatic cross-links. (From Nyman et al., 2005. Copyright 2003, with permission from Elsevier.)

Figure 19.12 Water helps to stabilize collagen by forming inter- and intra-hydrogen bonds with hydrophilic residues. (From Nyman et al., 2005. Copyright 2003, with permission from Elsevier.)

The enamel of mammalian teeth is much more heavily mineralized than bone, which makes it much harder. In addition, it does not contain collagen, although in its final mature state it does contain small amounts of specialized matrix proteins. Early tooth development is a classical illustration of the interaction between two tissue types (epithelial cells and mesenchymal cells[3]), whereby a number of signalling molecules are involved in orchestrating reciprocal interactions between the two types of tissue.

Enamel formation is though to involve the following steps in its assembly: (1) stimulation of cells called ameloblasts, derived from the epithelium, which are responsible for the secretion of enamel matrix proteins and the carbonated apatite of the enamel; (2) self-assembly of matrix proteins, notably amelogenin, which assembles into nanospheres to form 'ribbons' of matrix; (3) secretion of saturating levels of Ca^{2+} and of PO_4^-; (4) nucleation of crystal formation; (5) regulation of crystallite growth by the matrix; and (6) proteolytic degradation of the matrix and rapid filling in with carbonated apatite crystallite. The principal water-soluble protein of enamel is amelogenin, which is degraded as the enamel matures—transient degradation intermediates are found in developing enamel. This 180 residue hydrophobic protein self-assembles into nanospheres *in vitro*, which resemble the matrix ribbons found *in vivo*, and it interacts with carbonated apatite *in vitro* to limit crystallite growth just as it is thought to act *in vivo* to channel accumulation of the mineral crystallite into rods. However, there are other enamel-specific proteins that have been isolated, studied and in some cases cloned. And to make matters more complicated, experiments in which there is not complete lack of biomineralization after particular gene knock-outs of single bone proteins suggest that there is likely to be a redundancy of function, and that many of the actors in this particularly complicated process still need to be identified.

Finally, it is intriguing that in terms of biomineralization, invertebrates have based their reliance on calcium carbonates, while vertebrates appear to have used almost exclusively calcium phosphate. We say almost, because, while the use of calcium phosphates for bio-mineralization is an invention of some vertebrates, they still use calcium carbonate for the formation of otoliths[4] of the inner ear. It remains to be established if the equivalent of the gene *starmaker* required for otolith formation in zebrafish has homologues among invertebrates.

REFERENCES

Crichton, R.R. (2001) *Inorganic Biochemistry of Iron Metabolism: From Molecular Mechanisms to Clinical Consequences*. John Wiley & Sons, Chichester, pp. 326.

Kauko, A., Pullianen, A.T., Haataja, S., Mayer-Klauke, W., Finne, J. and Papageorghiou, A.C. (2006) Iron incorporation in *Streptococcus suis* Dps-like peroxide resistance protein Dpr requires mobility in the ferroxidase center and leads to the formation of a ferrihydrite-like core, *J. Mol. Biol.*, **364**, 97–109.

Lewin, A., Moore, G.R. and Le Brun, N.E. (2005) Formation of protein-coated iron minerals, *Dalton Trans.*, **22**, 3579–3610.

Mann, S., Webb, J. and Williams, R.J.P. (1989) *Biomineralization Chemical and Biochemical Perspectives*, VCH, Wienheim, 541 pp.

Matsunaga, T. and Okamura, Y. (2003) Genes and proteins involved in bacterial magnetic particle formation, *Trends Microbiol.*, **11**, 536–541.

---

[3] Mesenchymal tissue is immature, unspecialized tissue, found in the early embryo of animals, whereas epithelial cells are parenchymal cells that line an internal cavity or tube.

[4] Otolith—a tiny bony structure in the inner ear of lower vertebrates.

Nyman, J.S., Reyes, M. and Wang, X. (2005) Effect of ultrastructural changes on the toughness of bone, *Micron*, **36**, 566–582.

Su, M., Cavallo, S., Stefanini, S., Chiancone, E. and Chasteen, N.D. (2005) The so-called *Listeria inocua* ferritin is a Dps protein. Iron incorporation, detoxification and DNA protection properties. *Biochemistry*, **44**, 5572–5578.

Thompson, D.W. (1942) *On Growth and Form*. Cambridge University Press, Cambridge, 1116 pp.

Toussaint, L., Bertrand, L., Hue, L., Crichton, R.R. and Declercq, J.-P. (2006) High-resolution X-ray structures of human apoferritin H-chain mutants correlated with their activity and metal-binding sites, *J. Mol. Biol.*, **365**, 440–452.

Wilt, F.H. (2005) Developmental biology meets materials science: morphogenesis of biomineralized structures, *Dev. Biol.*, **280**, 15–25.

– 20 –

Metals in Medicine and the Environment

INTRODUCTION

In many crucial biological processes, such as oxygen transport, electron transport, intermediary metabolism, metals play an important part. Therefore, disorders of metal homeostasis, metal bioavailability or toxicity caused by metal excess, are responsible for a large number of human diseases. We have already mentioned disorders of iron metabolism (see Chapter 7) and of copper metabolism (see Chapter 14). The important role, particularly of redox metals such as copper and iron, and also of zinc, in neurodegenerative diseases, such as Parkinson's disease, Alzheimer's disease, etc. has also been discussed (see Chapter 18). We will not further discuss them here.

Metals are also found to play an astonishing number and variety of roles in modern medicine both as therapeutic and as diagnostic agents. Metals, such as arsenic, gold and iron have been used to treat different human diseases since antiquity. Sometimes, the medical applications use the metal itself, and we will illustrate this by the therapeutic use of lithium in bipolar disorder (manic-depression). However, there are a constantly growing number of metal-based drugs, involving a broad spectrum of metals. Among these, arguably, the most prominent is *cis*-platinum, and we will illustrate how this widely used anti-cancer drug functions by binding to DNA and disrupting DNA replication.

One of the prominent trends in contemporary medicine is the recognition of the increasingly important role of preventive medicine[1], such as early recognition and treatment of hypertension and elevated blood cholesterol in the prevention of coronary artery disease. A major tool in prevention is based on the use of non-invasive techniques, such as magnetic resonance imaging (MRI). The magnetic resonance image can be enhanced by the administration of suitable MRI contrast agents. The unique properties of the cationic form of some metals (such as the lanthanide gadolinium), namely their high paramagnetism, can be exploited to this end; and we will discuss the development of new, and potentially targeted, metal-based MRI contrast agents. For more on metal-based drugs see Sigel and Sigel, 2004; Gielen and Tiekink, 2005.

Finally, the introduction of metal ions into the environment, often through the activities of man himself, can constitute a further health hazard. The example of the toxic effects of

---

[1] Chinese proverb: poor doctor heals, good doctor prevents.

environmental cadmium will be reviewed. The consequences of acid rain, notably in making soil aluminium more abundant in ground water have been mentioned in Chapter 1. Here we will discuss the toxicity of aluminium in more detail.

METALLOTHERAPEUTICS WITH LITHIUM

Lithium is the simplest therapeutic agent for the treatment of depression and has been used for over 100 years—lithium carbonate and citrate were described in the British Pharmacopoeia of 1885. Lithium therapy went through periods when it was in common use, and periods when it was discouraged. Finally, in 1949, J.J.F. Cade reported that lithium carbonate could reverse the symptoms of patients with bipolar disorder (manic-depression), a chronic disorder that affects between 1% and 2% of the population. The disease is characterized by episodic periods of elevated or depressed mood, severely reduces the patients quality of life and dramatically increases their likelihood of committing suicide. Today, it is the standard treatment, often combined with other drugs, for bipolar disorder and is prescribed in over 50% of bipolar disorder patients. It has clearly been shown to reduce the risk of suicide in mood disorder patients, and its socioeconomic impact is considerable—it is estimated to have saved around $9 billion in the USA alone in 1881.

The molecular basis of mood disorder diseases and their relationship to the effects of lithium remain unknown. How does lithium function? To begin with, we should note that although the hydration shell of Li^+ is similar in size to Na^+, its ionic radius is much closer to that of Mg^{2+}. This led to the suggestion that lithium ions might exert their action by competing with Mg^{2+} for binding sites on proteins. As we have seen in Chapter 10, there are a great many magnesium-dependent enzymes involved in metabolic pathways, not to mention the extensive involvement of Mg^{2+} in nucleic acid biochemistry. It would be difficult to explain the relative specificity of the lithium effect if it interacted with all of these magnesium-binding sites. This suggests that only proteins with rather low affinities for magnesium are targeted by therapeutic concentrations of lithium. As we pointed out in Chapter 1, the concentration of lithium found in the serum of treated patients is around 1 mM. This is around the concentration of free magnesium within cells.

It is now thought that there are two general targets of lithium in the cell, both signal transduction pathways that are active in brain. These are the serine–threonine protein kinase, glycogen synthase kinase-3 (GSK-3) and the inositol(1,4,5)-trisphosphate [$Ins(1,4,5)P_3$] signalling pathway. GSK-3 is abundant in brain, where it is involved in signal transduction cascades. The phosphorylation of target protein serine and threonine residues results in the regulation of the cytoskeleton, gene expression via a number of transcription factors, apoptosis and glycogen synthesis activity. Lithium is a potent and selective inhibitor of the enzyme through competition for Mg^{2+} binding. This inhibition, with its plethora of potential consequences for cellular signalling, may explain to some extent the neuroprotective effects of lithium therapy in bipolar disorders.

A second well characterized signal transduction pathway, which is subject to lithium inhibition, is the phosphoinositol cascade (see Chapter 11, in particular Figure 11.8). A number of enzymes of this pathway contain a common amino acid sequence that constitutes a lithium-sensitive Mg^{2+} binding site, and it has been proposed that lithium exerts some of its effect by affecting phosphoinositol metabolism in as yet uncertain manner. There are known to be a number of indirect interactions between GSK-3 and phosphoinositide signalling, but clearly much more remains to be uncovered before the mode of action of lithium in the brain is understood.

CISPLATIN: AN ANTI-CANCER DRUG

Cisplatin, *cis*-[PtCl$_2$(NH$_3$)$_2$], is extensively used for the treatment of testicular and ovarian cancers and increasingly against other types of solid tumours (head–neck, lung, cervical and bladder), and gives a greater than 90% cure rate in the case of testicular cancer. It was first synthesized by Peyrone in 1845 (known as Peyrone's salt), and its structure was elucidated by Alfred Werner[2] in 1893. In the 1960s it was rediscovered serendipitously[3] when Rosenberg et al. investigated the effects of electric fields on bacterial growth. In the presence of NH$_4$Cl, Pt electrodes and sunlight, *E. coli* cultures grew up to 300 times their normal length, but the cells failed to divide. They found that the electric field was not responsible for the arrest of cell division, but that small amounts of certain platinum compounds formed during the electrolysis were responsible. Reasoning that if it inhibited cell division it might be effective as an anti-cancer drug, they then found that whereas the *trans* isomer was extremely toxic, the *cis* isomer (Figure 20.1) was active against several forms of cancer,

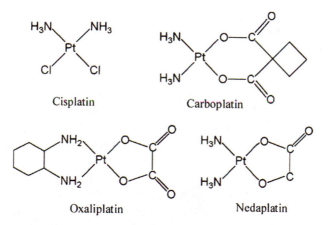

Figure 20.1 Platinum complexes currently in clinical use.

---

[2] Alfred Werner received the Nobel Prize in Chemistry in 1913 for his research into the structure of coordination compounds.

[3] Horace Walpole wrote to his friend Horace Mann in January 28, 1754: 'I once read a silly fairy tale, called *The Three Princes of Serendip*: as their highnesses travelled, they were always making discoveries by accidents, and sagacity, of things they were not in quest of. For instance, one of them discovered that a mule blind of the right eye had travelled the same road lately, because the grass was eaten only on the left side, where it was worse than on the right—now do you understand "serendipity"?' For Walter Gratzer (a regular contributor for many years to *Nature*), serendipity is not that when you drop your buttered toast on the floor that it falls, as it invariably does, buttered side down, but that when you pick it up you discover the contact lens that you lost a few days earlier.

although it too had severe side effects. Its applicability is still limited to a narrow range of tumours, and some tumours have natural resistance to the drug or develop resistance after treatment. Because of its side effects, limited solubility in aqueous solutions and intravenous mode of administration, a search for more effective and less toxic analogues has been initiated. Only a few thousands of platinum complexes that have been evaluated have achieved routine clinical use. These include carboplatin, oxaliplatin and nedaplatin (Figure 20.1). They all have at least one N-H group, which is responsible for important hydrogen-bond donor properties.

The mechanism of action of cisplatin is relatively well understood. The drug enters cells not only by passive diffusion, but also by an active transport mechanism. Ctr1, the major copper influx transporter, described in Chapter 8, has been convincingly demonstrated to transport cisplatin and its analogues, carboplatin and oxaliplatin. Evidence also suggests that the two copper efflux transporters ATP7A and ATP7B regulate the efflux of cisplatin. The precise role that copper transport proteins play in mediating cisplatin resistance remains enigmatic. The concentration of chloride ions in blood and extracellular body fluids is 100 mM, which is high enough to suppress cisplatin hydrolysis. Once inside the cell, the concentration of chloride ions is much lower (4 mM), resulting in the hydrolysis of the drug to form the mono-aqua $[PtCl(H_2O)(NH_3)_2]^+$ cation, and more slowly the di-aquo $[Pt(H_2O)_2(NH_3)_2]^{2+}$ (Figure 20.2). These positively charged species then cross the nuclear

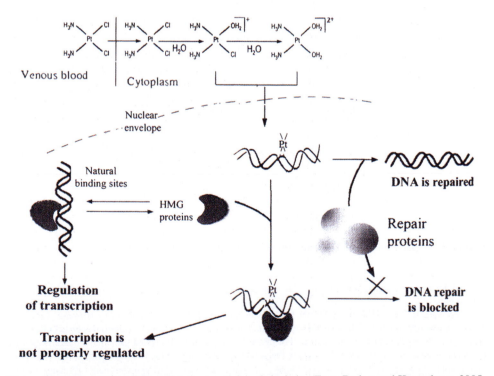

Figure 20.2 Mechanism of the anti-tumour activity of cisplatin. (From Brabec and Kasparkova, 2005. Reproduced with permission from John Wiley & Sons., Inc.)

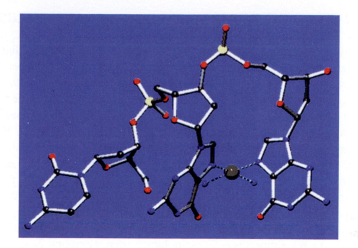

Figure 20.3 X-ray structure of d(CGG) chelated to a cisplatin unit. (From Reedijk, 2003. Copyright (2003) National Academy of Sciences, USA.)

membrane and bind to DNA, although they can also bind to RNA and to sulfhydryl groups in proteins. Bifunctional cisplatin binds to DNA, first forming monofunctional adducts, preferentially at guanine residues, which subsequently form major intra-strand cross-links between adjacent purine residues. In all adducts, the cisplatin is coordinated to the N7 atom of the purine. These cross-links inhibit DNA replication, block transcription and ultimately trigger programmed cell death (apoptosis). The X-ray structure of one of the adducts of Pt(II) with a d(CGG) fragment is shown in Figure 20.3. The chelation of the two N7 sites of the guanine residues in *cis* conformation produces a kink in double-stranded DNA, affecting its secondary structure.

It seems to be generally accepted that the anti-tumour activity of cisplatin is mediated by the recognition of the platinated adducts by cellular proteins. In particular, high-mobility group (HMG) proteins, which are believed to play a key role in the regulation of gene expression (transcription), have been found to specifically recognize and bind to cisplatin-modified DNA. It has been suggested that HMG proteins play a role in the mechanism of cisplatin toxicity, either by hijacking the proteins away from their natural binding sites or alternatively by protecting the cisplatin adducts from DNA repair mechanisms. A number of other proteins have been found to bind to platinated DNA.

In excess of 3000 cisplatin analogues have been synthesized in the search for platinum anti-cancer drugs with broader spectrum of action against different tumours, with fewer side effects and activity against cisplatin-resistant tumours. Attention has focused on compounds that only form monofunctional DNA adducts, on *trans*-platinum complexes, poly-platinum compounds and platinum (IV) analogues, and the search continues. Statistically, for one new clinically active compound to be discovered, 10,000 new compounds need to be synthesized and screened, so high-throughput methodologies are being developed and the search for new cisplatin analogues continues apace.

CONTRAST AGENTS FOR MAGNETIC RESONANCE IMAGING

While many, often spectacular, advances have been made in our understanding of the progression of a great number of diseases at the molecular level, the development of molecular imaging, which allows *in vivo* visualization of molecular events at the cellular level is having a revolutionary role in medical diagnosis. MRI is an important tool for the diagnosis of disease, and a non-invasive method of acquiring 3-D images of human soft tissue (unlike X-rays that can locate the electron-dense hard tissue, such as bone). MRI scanners generate multiple 2-D cross-sections (slices) of tissue and 3-D reconstructions. MRI utilizes the same longer wavelength radiowaves that are used in nuclear magnetic resonance (NMR) spectroscopy. The samples (or patients) are placed in a powerful magnetic field and exposed to radiofrequency pulses, and the relaxation times of excited nuclei (typically protons from water in the tissue) are detected. The contrast in an MRI image is the result of the interplay of numerous factors including the proton density of the tissues being imaged, the relative relaxation times T_1 (spin–lattice relaxation) and T_2 (spin–spin relaxation) and the instrumental parameters. For a review see Dzik-Jurasz, 2003.

The diagnostic power of the technique is illustrated by the MRI scans of patients with two different neurological conditions. Figure 20.4a shows the characteristic 'eye of the tiger' sign observed in patients with Hallervorden–Spatz syndrome (due to a mutation in the gene encoding pantothenate kinase 2). The MRI picture is a T_2-weighted image that shows diffuse bilateral low-signal intensity of the globus pallidus (due to iron deposition) with a region of hyperintensity in the internal segment (the high signal is thought to represent tissue oedema). Neuroferritinopathy is another neurological disorder, in which the insertion of an adenine residue in the gene for the ferritin light chain results in an altered carboxy terminal sequence of the protein. The T_2-weighted MRI image in Figure 20.4b is quite characteristic, with symmetrical degeneration of the globus pallidus and putamen and low-signal intensity in the internal capsule.

The use of MRI allows accurate, non-invasive diagnosis of many pathological conditions. However, the use of contrast agents, often in the form of paramagnetic metallochelates,

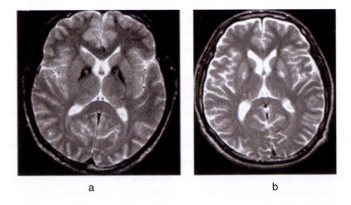

a   b

Figure 20.4 MRI scans of the brain of a patient with (a) Hallervorden–Spatz syndrome and (b) neuroferritinopathy. (From Crichton and Ward, 2006. Reproduced with permission from John Wiley & Sons., Inc.)

makes the method even more sensitive and the diagnosis more specific. MRI contrast agents are not directly visualized in the resulting image, only their effects are observed. They enhance image contrast as a result of their influence on the relaxation times of nearby water protons, and as a consequence on the NMR signal. Paramagnetic molecules, because of their unpaired electrons, are potent MRI contrast agents, decreasing the T_1 and T_2 relaxation times of nearby proton spins, and enhancing the signal observed. The most extensively studied paramagnetic metal ions are transition metal ions (high-spin Mn(II) and Fe(III), each with five unpaired electrons) and lanthanides (essentially Gd(III) with its seven unpaired electrons). Since free metal ions are toxic to biological systems, they have to be administered in a non-toxic form bound to suitable ligands or chelates. The first contrast agent to be approved for clinical use, Gd-DTPA (the structure is described in Chapter 1), has been administered to more than 20 million patients over a period of 10 years of clinical use.

While the first generation of MRI contrast agents is relatively unspecific, they nonetheless allow the evaluation of physiological parameters such as the status of the blood-brain barrier or renal function. Gd-DTPA-enhanced MRI imaging can assist in the diagnosis and treatment follow-up of many types of cancer. However, in order to improve the diagnostic efficacy of contrast agents, they need to be made target specific so that they accumulate in specific biological locations. Approaches that have been tried include coupling contrast agents to antibodies against membrane receptors, to transferrin in order to image tumour cells, which over-express the transferrin receptor, or to annexin V, a protein that binds to phosphatidylserine, as a marker of apoptosis. Phosphatidylserine moves from the interior to the exterior of cell membranes when a cell undergoes apoptosis. Several 'smart' sensor probes have been designed, which are activated only in the presence of their intended target. As one example, a gadolinium-based smart contrast agent has been developed to demonstrate gene transfection (Figure 20.5). When the enzyme β-galactosidase is expressed in engineered cells, the β-galactose ring protecting the Gd^{3+} is cleaved allowing bulk water access to the paramagnetic gadolinium ion. As we have seen in Chapter 11, changes in

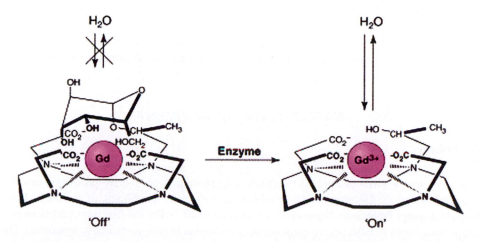

Figure 20.5 MRI contrast agent activated by β-galactosidase. (From Meade et al., 2003. Copyright 2003, with permission from Elsevier.)

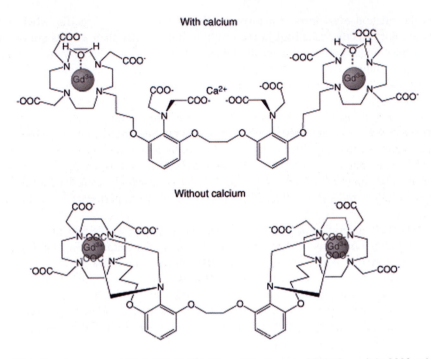

Figure 20.6 Contrast agent activated by Ca(II). (From Meade et al., 2003. Copyright 2003, with permission from Elsevier.)

intracellular Ca^{2+} are important in cell signalling. A gadolinium-based contrast agent, which can specifically detect Ca^{2+}, has been designed, which has two distinct conformations as a function of Ca^{2+} concentration (Figure 20.6). In the absence of Ca^{2+}, the aromatic iminoacetates of the ligand interact with the two Gd^{3+} ions. In the presence of Ca^{2+}, the iminoacetates rearrange to bind Ca^{2+}, thereby allowing water to bind directly to Gd^{3+}. This shortens the relaxation time of water protons and results in signal enhancement on T_1 imaging. It should be pointed out that most of the 'smart' contrast agents to date have been tested in animal or cellular models, but have not made the ultimate step to clinical applications.

METALS IN THE ENVIRONMENT

Cadmium

The most severe form of chronic cadmium (Cd) poisoning caused by prolonged oral Cd ingestion is Itai-itai disease, which developed in numerous inhabitants of the Jinzu River basin in Toyama Prefecture, Japan in the 1950s (Figure 20.7). For the first time, cadmium pollution was shown to have severe consequences on human health, particularly in women. The most important effects were softening of the bones and kidney failure. The name of the disease is derived from the painful screams (Japanese: 痛い *itai*) caused by the severe pain in the

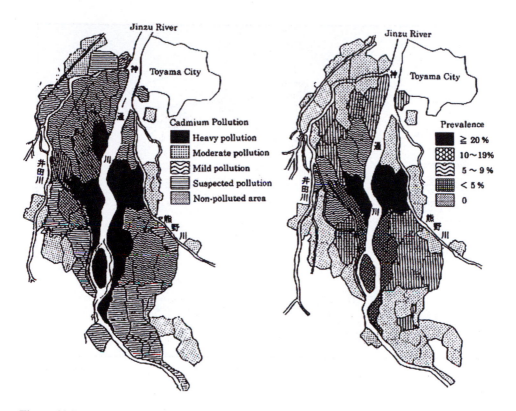

Figure 20.7 Itai-itai disease: (left) the degree of Cd pollution and (right) the presence of the disease in women over 50 years of age.

joints and the spine. Its cause was found to be environmental Cd pollution originating from the effluent from a zinc mine located in the upper reaches of the river, with the inhabitants, thereby exposed, developing severe chronic Cd poisoning. In the Cd-polluted areas, 50–70% of the amount of Cd ingested is orally derived from rice, and in practice a close association was reported between the prevalence of Itai-itai disease and the Cd concentration in rice.

Occupational and environmental pollution with cadmium can result from heavy metal mining, metallurgy and industrial use, manufacturing of nickel–cadmium batteries, pigments, plastic stabilizers and anti-corrosive products. Important sources of human intoxication are cigarette smoke due to high concentrations of cadmium in cigarettes (smokers on a packet a day can easily double their cadmium intake), as well as food, water and air contamination. Chronic intoxication is associated with obstructive airway disease, emphysema, irreversible renal failure, bone disorders and immuno-suppression. In humans, cadmium exposure has been associated with cancers of the prostate, lungs and testes (it is classified as a carcinogen). At the cellular level, cadmium affects proliferation, differentiation and causes apoptosis. However, since Cd^{2+} is not redox-active, the generation of reactive oxygen species (ROS) and DNA damage must be due to indirect effects. Cadmium also modulates gene expression and signal transduction, and reduces the activities of proteins involved in anti-oxidant defences. It has also been shown to interfere with DNA repair.

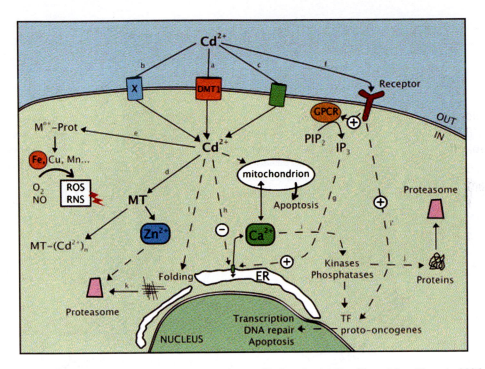

Figure 20.8 Schematic representation of cadmium traffic in animal cells. (From Martelli et al., 2006. Copyright 2006, with permission from Elsevier.)

The large and easily polarized Cd^{2+} ion is a soft Lewis acid, with a preference for easily oxidized soft ligands, particularly sulfur, so it would be expected to displace Zn^{2+} from proteins where the zinc coordination environment is sulfur dominated. The close similarities in the ionic radii of Cd^{2+} and Ca^{2+} (0.95 and 1.00 Å, respectively) favours exchange of the two metals in calcium-binding proteins. Cadmium can also interfere with iron. A schematic representation of cadmium traffic in animal cells is presented in Figure 20.8. As might be expected, by mimicking other essential metal ions, Cd^{2+} adopts a Trojan horse strategy[4] to be assimilated by cells. In the digestive tract, Cd^{2+} is transported by the broad specificity divalent metal transporter, DMT1 (Figure 20.8a), which is the intestinal transporter for non-haem iron (see Chapter 7). Since there is an iron-responsive element (IRE)-containing form of DMT1, which is targeted to the apical membrane of enterocytes and is regulated by iron regulatory protein (IRP) binding, dietary cadmium uptake and therefore its toxicity will depend on the iron status of the individual. Since women have a higher dietary iron intake than men, because of menstrual blood losses, they are more at risk, and the risk is even greater during pregnancy when DMT1 expression is greatly increased. Cadmium export through the basolateral membrane may not be as efficient as uptake, since cadmium accumulates in enterocytes on a high-cadmium diet. The iron transporter at the

---

[4] In order to break the 10-year long siege of Troy, the Greeks constructed an enormous wooden horse, hiding their best warriors in the stomach of the animal.

basolateral membrane, ferroportin, may be involved in the transport of cadmium into the blood stream, as may calcium-ATPases and zinc exporters.

Cadmium enters neurons via voltage-gated calcium channels even in the presence of external calcium, suggesting that these channels are the main cadmium entry pathways in nerve cells (Figure 20.8c). Since Cd^{2+} is able to cross even when these channels are blocked, other calcium channels such as ligand-gated, for example N-methyl-D-aspartate (NMDA) receptors or store-operated calcium channels (see Chapter 11), may also participate in cellular cadmium uptake. Although there is little evidence that cadmium can enter cells via zinc transporters, recent studies suggest that the zinc transporter ZIP8 is involved in cadmium uptake by mouse testicular cells. Other alternative pathways for cadmium uptake probably also exist (Figure 20.8.b).

Once inside the cell the small, cysteine-rich protein, metallothionein (MT, Figure 20.8d), which serves to bind intracellular zinc and copper (see Chapter 8), is a major target for cadmium binding. Genes coding for MT are strongly induced by both Zn^{2+} and Cd^{2+} by activation of the cadmium- (and zinc-)sensitive transcription factor MTF-1. The importance of MT in cadmium toxicity is underlined by the observation that mice lacking MT are more sensitive to cadmium exposure than wild-type mice, while MT-overexpressing cells are more resistant. It appears that the abundant intracellular thiol-containing tripeptide, glutathione, may also be involved in the detoxification and excretion of cadmium.

The chemical similarities between cadmium and zinc imply that cadmium could probably exchange with these metals in zinc-binding proteins (Figure 20.8e), although very little concrete evidence has been adduced in animal cells, with the exception of metallothionein. Cadmium can alter the intracellular concentration of calcium, which is an important and universal intracellular signal messenger (see Chapter 11). Acute exposure to cadmium can increase intracellular Ca^{2+} concentration via a poorly characterized cell surface G-protein-coupled 'metal-binding receptor', GPCR (Figure 20.8f), resulting in activation of phospholipase C and IP_3 production by hydrolysis of phosphatidylinositol. This triggers calcium release from intracellular stores, probably from IP_3-gated Ca^{2+} channels in endoplasmic reticulum (Figure 20.8g). This cadmium-dependent up-regulation of the internal concentration of calcium may have consequences for cellular proliferation, differentiation and apoptosis. Within the cell, cadmium has the opposite effect (Figure 20.8h). It blocks release of stored Ca^{2+} by inhibiting the activity of IP_3 and ryanodine receptors (see Chapter 11). Cadmium can also increase intracellular calcium concentration in muscle cells by promoting calcium release from the sarcoplasmic reticulum[5].

A considerable number of transcription factors have reactive cysteine residues, which enable them to respond to the redox conditions in the cell. Since cadmium perturbs redox homeostasis, it can affect this class of transcription factors. If cadmium can displace the tetra-coordinate zinc atoms in zinc finger-containing transcription factors, it will affect them as well. Many of the pathways involving activation and inactivation of transcription factors involve kinases and phosphatases, themselves under the intricate control of calcium fluxes. It is therefore no surprise that cadmium will exert effects on the activity of transcription factors, the activation of proto-oncogenes, and thereby on gene expression (Figure 20.8i and i').

---

[5] The sarcoplasmic reticulum is a fine reticular network of membrane-limited elements, which pervades the sarcoplasm of muscle cells.

Cadmium also appears to be involved in the ubiquitin-proteasome pathway. Ubiquitin binding to proteins is often signalled by post-transcriptional modifications such as phosphorylation (Figure 20.8j). Cadmium has also been shown to decrease the solubility of specific proteins, and, as we saw in Chapter 18, high concentrations of aggregated proteins can impede the proteasome (Figure 20.8j and k). It can also affect protein folding, again with deficient proteasomal action (Figure 20.8l).

Although cadmium is not strongly mutagenic, it is known that it causes increased oxidative DNA damage and that it inhibits the DNA repair systems. It has also been found to induce cell death both by necrosis and apoptosis. Since the latter is extremely calcium-dependent, it seems likely that the pro-apoptotic effects of cadmium are due to its interference with calcium homeostasis.

Aluminium

We pointed out in Chapter 1 the disastrous effects of burning brown coal, associated with the generation of sulfuric and nitric acids aerosols, which result in the subsequent acidification of soil. Acid soils, with a pH of 5.5 or lower, significantly limit crop production worldwide: approximately 50% of the world's potentially arable soils are acidic, and the production of staple food crops is particularly affected—20% of the worldwide production of maize and 13% of rice is on acid soils, while 60% of the acid soils in the world are in the tropics and subtropics. Thus, acid soils limit crop yields in many developing countries, while in developed countries, such as the United States, the extensive use of ammonia fertilizers causes further acidification of agricultural soils. The primary limitations of acid soils are toxic levels of aluminium and manganese, and sub-optimal levels of phosphorous. Acidification of soils leads to acidification of rivers and streams, increasing the solubility of aluminium, with direct consequences on fish populations, and eventually of water supplies to the general population.

The chemistry of aluminium combines features in common with two other groups of elements, namely (i) divalent magnesium and calcium, and (ii) trivalent chromium and iron. The toxic effects of aluminium are more related to its interference with calcium-dependent processes, whereas its access to tissues is probably a function of its similarity to ferric iron. Al^{3+} is a much stronger acid than Mg^{2+} and Ca^{2+}, which makes it a powerful competitor for oxygen donor ligands for these two important biological cations, and leads to profound interference with their metabolism. In contrast, Al^{3+} is a weaker acid than Fe^{3+} and therefore causes less interference with iron metabolism. It has a preference for oxygen ligands, particularly if they are negatively charged, such as carboxylate, catecholate and phosphate groups. Sulfhydryl groups do not bind Al^{3+}, and amines do not bind strongly except as part of multidentate ligand systems such as NTA and EDTA. The rate of ligand exchange, into and out of the metal coordination sphere, is extremely slow for Al^{3+}. This means that essential metabolic processes involving rapid Ca^{2+} exchange, which is 10^8-fold slower for Al^{3+}, would be seriously affected by substitution, as would Mg^{2+}-dependent enzymes (10^5-fold slower with Al^{3+}).

In the extracellular fluids of mammals, the iron-binding protein transferrin is usually only 30% saturated, so that it has 70% of its metal-binding capacity free. If Al^{3+} could get into the circulation, it could certainly occupy those binding sites. The stability constants for the two-binding sites on transferrin are similar, log K_a of 12.9 and 12.3 (Martin, 1994), substantially lower than for iron.

The daily dietary intake of aluminium in man is 2–3 mg/day, but it will not accumulate in the body if the kidneys are functioning normally. In the gastrointestinal tract less than 0.5% of an ingested dose of aluminium will be absorbed, while the skin and the lungs have also evolved as barriers to aluminium entry into the body. The problem arises when these natural barriers are bypassed by the intravenous administration of aluminium-containing solutions. An example is the use of aluminium hydroxide to treat the high levels of serum phosphate found in end-stage renal disease. Another source is the administration of intravenous solutions used for parenteral nutrition[6], which might contain aluminium. The risks from the latter are particularly important in pre-term infants and patients with kidney failure, and have led to the FDA acting to limit the amount of aluminium in such preparations.

Once in the serum, aluminium can be transported bound to transferrin, and also to albumin and low-molecular ligands such as citrate. However, the transferrrin–aluminium complex will be able to enter cells via the transferrin-transferrin-receptor pathway (see Chapter 8). Within the acidic environment of the endosome, we assume that aluminium would be released from transferrin, but how it exits from this compartment remains unknown. Once in the cytosol of the cell, aluminium is unlikely to be readily incorporated into the iron storage protein ferritin, since this requires redox cycling between Fe^{2+} and Fe^{3+} (see Chapter 19). Studies of the subcellular distribution of aluminium in various cell lines and animal models have shown that the majority accumulates in the mitochondria, where it can interfere with calcium homeostasis. Once in the circulation, there seems little doubt that aluminium can cross the blood-brain barrier.

Aluminium toxicity is the likely cause of three human disorders arising from long-term dialysis, bone disease, anaemia and dementia. The first of these conditions is consistent with interference with calcium deposition into bone, and the accumulation of aluminium in the bone matrix. It is also the likely cause of the high frequency of dementia among the natives of certain regions of South East Asia, where the soils are high in Al(III) and low in Mg^{2+} and Ca^{2+}. Elevated levels of Al(III) have been reported in the brains of Alzheimer's patients and the hypothesis that aluminium might exert its toxic effects by interfering with iron homeostasis has been advanced. However, this remains a subject of considerable controversy. Chronic exposure to aluminium results in selective cognitive impairment in rats, with altered calcium homeostasis. Taken together with the demonstration that aluminium causes dementia in dialysis patients, there can be no doubt that aluminium is neurotoxic. The potential therapeutic benefit of long-term chelation therapy to prevent accumulation of both aluminium and iron in brain, as a function of ageing, could complement a greater understanding of the role of both metal ions in the aetiology and pathogenesis of many neurological diseases.

---

[6] Parenteral—administration of a substance by any other route than via the gastrointestinal tract, particularly by injection.

REFERENCES

Brabec, V. and Kasparkova, J. (2005) $_{78}$Pt Platinum-Based Drugs, in Gielen, M. and Tiekink, E.R. eds., *Metallo-therapeutic Drugs and Metal-based Diagnostic Agents. The Use of Metals in Medicine*, John Wiley and Sons, Chichester, 489–506 pp.

Crichton, R.R. and Ward, R.J. (2005) *Metal based Neurodegeneration: From Molecular Mechanisms to Therapeutic Strategies*, John Wiley & Sons, Chichester, 227 pp.

Dzik-Jurasz, A.S.K. (2003) Molecular imaging *in vivo*: an introduction, *Br. J. Radiol.*, **76**, S98–S109.

Gielen, M. and Tiekink, E.R. (2005) *Metallotherapeutic Drugs and Metal-Based Diagnostic Agents. The Use of Metals in Medicine*, Wiley, Chichester, 598 pp.

Martelli, A., Rousselet, E., Dycke, C., Bouron, A. and Moulis, J.-M. (2006) Cadmium toxicity in animal cells by interference with essential metals, *Biochimie*, **88**, 1807–1814.

Martin, R.B. (1994) Aluminum: A Neurotoxic Product of Acid Rain, *Acc. Chem.Res.*, **27**, 204–210.

Meade, T.J., Taylor, A.K. and Bull, S.R. (2003) New magnetic resonance contrast agents as biochemical reporters, *Curr. Opin. Neurobiol.*, **13**, 597–602.

Reedijk, J. (2003) New clues for platinum antitumor chemistry: kinetically controlled meta binding to DNA, *Proc. Natl. Acad. Sci. U.S.A.*, **100**, 3611–3616.

Sigel, A. and Sigel, H. (2004) Metal ions and their complexes in medication, *Metal Ions Biol. Syst.*, **41**, 519 pp.

Index

A

ABC transporter, 121
abscissic acid, 139
absorption, 111
absorption bands α, β, γ, δ, 113
N-acetylglucosamine, 59–60
adenosine triphosphate (see ATP)
adenyl cyclase/phosphokinase A (AC/PKA),
 298
Ace1, 142
aceruloplasminaemia, 303
acetaldehyde, 78
acetyl CoA, 86, 92
acetylcholine, 151
acetylCoA synthase (ACS), 259
acetylene hydratase, 285
acid, 16, 212
 Lewis acid, 15, 166, 178
 rain, 6
acidification, 350
aconitase, 92, 133, 228, 285
acyl carrier protein (ACP), 94, 97
acrodermatitis enteropathica, 128
acrolein, 309
acrylamide, 269
actin filament, 103
action potential, 152
activators of ferrous transport (Aft), 138
active site, 39, 47
acyl phosphate, 90
acylation site, 72
Ada DNA repair protein, 204
adaptor hypothesis, 71
adenine, 55, 214
adenosine diphosphate (ADP), 167, 289
adenosylcobalamine, 214
adenylation, 37
adipocytes, 65
aequorin, 191
Aeromonas hydrophila, 205
aerosols, 350
affinity, 11

akaganeite, 329
alanine, 71
albumin, 351
aldehyde oxidoreductase, 282, 285
aldohexose, 59
aldol
 cleavage, 88
 condensation, 85
aldose, 59
alkaline phosphatase, 205–206
aluminium, 5–6, 350
 silicate, 6
 toxicity, 351
Alzheimer's disease, 5, 309
Amanita muscaria, 8, 292
amavadine, 293
ameloblasts, 336
amelogenin, 336
amine oxidases, 244
amino acids, 27, 43
 hydroxylases, 231
 metabolism, 266
 oxidation, 308
aminoacyl-adenylate, 71
aminoacyl-tRNA, 71
aminoacyl-tRNA synthetases, 71,
 73, 197
5-aminolaevulinate dehydratase 204
aminopeptidases, 206, 268
aminotransferase, 93
ammonia, 258, 286
amorphous calcium carbonate (ACC), 332
AMPA, 300
amphiphilic molecule, 65
amplification cascade, 170
β-amyloid peptide, Aβ, 314
amyloid precursor protein, APP, 314
amyloidogenesis, 314
amyloids, 319
amylopectin, 62
α-amylose, 62
amyotrophic lateral sclerosis (ALS), 205,
 250, 309

anabolism, 77
anaemia, 253
anaerobes, 280
annexin V, 343
anomers, 60
antibodies, 53
anticodon, 59, 75
anti-ferromagnetic coupling, 107
antigens, 54
antiport, 142
antisense oligonucleotides, 332
apatite, 333
apoceruloplasmin, 149
apoproteins, 34
apoptosis, 345
apotransferrin, 126, 254
APP intracellular domain (AICD), 314
Arabidopsis, 121, 139
aragonite, 330, 332
arginase, 274
arginine, 3, 30, 274, 308
argon, 7
arsenic, 9
Ascidians (Tunicates), 331
ascorbate oxidase, 247
ascorbic acid, 59, 86
aspartate, 28, 30
aspartate transcarbamoylase, 197
aspartylphosphate, 171
astrocytes, 306
ataxia, 319
atoms, 14–15
    acceptor, 15
    donor, 15
    shell, 14
ATOX1 (HAH1), 149
ATP (adenosine triphosphate), 1, 78, 79,
    167, 289
    ABC transporter, 151
    binding cassette (ABC), 119
    hydrolysis, 158
ATPases
    ATP7A, 149, 302, 342
    ATP7B, 149, 302, 342
    Calcium ATPases, 157
    Ccc2 ATPase, 112, 140
    copper-transporting P-type ATPases, 302
    cytoplasmic ATPase ZnuC, 121

heavy metal ATPase (HMA), 143
hydrogen-ATPase, 142, 157, 159
P-type ATPases, 151, 157
plasma membrane calcium ATPase
    (PMCA) pump, 187
Sodium-potassium ATPase, 151, 157
ATP synthase, 87, 101, 103
    F_0, and F_1, 100
    proton-translocating unit, 103
ATX1, 137, 140, 149
Auranofin, 11
axon, 152
axonopathy, 311
Azotobacter, 286
azurin, 243

B

Bacillus subtilis, 248
bacteria, 131
    Haemophilus influenzae, 30
    iron-storage proteins, 131
    metal storage and homeostasis, 131
    Neisseria, 30
    photosynthetic, 179
    thermophilic, 280, 285
bacterioferritins, Bfrs, 131
β-barrel, 53, 250
base, 15, 212
bicarbonate, 94
bioenergetics, 77
biogenesis, 33
biomembranes, 65
biomineralization, 321, 333–334
biominerals, 330
biosynthesis, 36, 71
    catecholamines, 231
    cholesterol, 93
    collagen, 231
    fatty acid, 93
    isoleucine, 83
    MoCo, 36
    nucleic acids, 77
    nucleotide, 93
    proteins, 71, 77
    serotonin, 231
    valine, 83

biotin, 85
1,3-bisphosphoglycerate, 90
bipolar disorder (manic-depression), 340
blood group determinants, 59
blood-brain barrier (BBB), 306
blood-clotting cascade, 28
Bohr magnetons, 107
bonds, 13, 18, 43, 220
    coordinate bonds, 15
    covalent, 3, 14, 21
    iron–carbon, 220
    peptide, 74, 200
boromycin, 3
boron, 3
bovine carboxypeptidase A, 200
bovine spongiform encephalitis (BSE), 304
bradykinesia, 317
brain-derived neurotrophic factor (BDNF)
    gene, 300
Bromine, 9
bromoperoxidases, 291

C

cadmium, 1, 346
    toxicity, 349
    traffic in animal cells, 348
    uptake, 349
CAG codon, 317
carbonic anhydrase II, 161
calcification, 322
calcineurin (CaN), 298
calcite crystal, 330, 332
calcium, 2, 165, 183–185, 193
    ATPases, 157
    channels, 349
    cluster (CaMn$_3$O$_4$), 278
    export from cells, 185
    homeostasis, 351, 191
    'hotspots', 191
    signals, 305
    uptake pathways, 185
    voltage gated Calcium channel (VGCC), 298
calcium carbonate, 330, 332, 336
calcium phosphate, 336
calmodulin (CaM), 187, 297
    CaM kinases, 194, 298–299

Calvin cycle, 77
5'-cap, 69
capping domain, 173
carbamate, 258
carbanions, 83
carbapenems, 205
carbohydrates, 43, 59–63
carbon, 2–3
    carbon dioxide, 2
    carbon monoxide, 22, 27, 29, 257
    carbon monoxide dehydrogenase
        (CODH)/acetylCoA synthases
        (ACS), 260
carbonate, 27, 29
carbonated apatite [Ca5(PO4CO3]
    3, 334
carbonic acid, 258
carbonic anhydrase, 197–198
carboplatin, 342
Carboxydothermus hydrogenoformans, 261
γ-carboxyglutamate residues, 16
γ-carboxyglutamic acid, 28
carboxylates, 39
carboxypeptidase A, 197–198, 200, 206
cardiac glycosides, 158
carotenoids, 179
carrier proteins, 34
carrier protein Δ^9 desaturases, 236
cascade, 16, 167
catabolism, 77
    glucose, 88
    oxidative, 92, 97
catalases, 220, 273–274
catalysis, 75
catechol bridges, 247
catechol dioxygenase, 233
catecholate, 39–40
cations, 22
    cation diffusion facilitator (CDF), 125,
        142–143
    CzcD, 136
cadmium, 135
cellobiose, 62
cellular proliferation, 349
cellular signalling, 183
cellulose, 62, 331
cephalosporins, 205
cerebrospinal degeneration, 319

ceruloplasmin (CP), 122, 127, 149, 243, 247
channels, 151–152
    chloride, 7
    proton translocation channel, 100
    potassium, 153
    P-loops, 154
    store-operated channels (SOCs), 185
chaperone ATOX1, 302
'Chatt' cycle, 289
chelatase, 30
chelate effect, 16–18
    coordination, 18
    stereochemistry, 18
chelating, 16
chelation therapy, 351
chelation, 17
chemistry, 2
    Inorganic, 2
    Organic, 2
Chironomus thummi, 54
chitin, 60
chlorine, 7
chloroperoxidases, 291
chlorophyll, 5, 17, 30, 178
cholesterol, 339
choreiform, 317
Chromatium vinosum, 113
chromium deficiency, 294
chromium, 8, 279
chromodulin, 294
chromosomes, 161
circadian rhythms, 298
circular dichroism, 113
citrate, 93, 137, 228
Citric acid cycle, 87
Claisen ester condensation, 84, 92
clathrin, 127
clavaminate, 233
clavaminate synthase, 234
Clostridium, 286
clusters,
    P-cluster, 36, 288
cobalamin, 30
cobalamine cofactors, 257
cobalamine, 263
Cobalt proteins, 263
Cobalt, 2, 118, 257
coding strand, 69
codons, 71
Coenzyme A, 94

coenzyme, 30
cofactor, 10, 27, 36, 233
    corrins, 27
    clusters, 27
    CuZ, 27, 36
    FeMoCo, 27, 36
    H-clusters, 36
    iron–sulfur, 27
    molybdenum cofactor, MoCo, 27, 36
    P-clusters, 27, 36
    porphyrins, 27
cognitive impairment, 314
collagen, 46, 198, 334
complex, 1, 15, 21
    linear, 19
    octahedral, 21
    planar, 19
    pyramidal, 19
    tetrahedral, 21
condensation reaction, 97
conformation (L and O), 100
conformational angles, ϕ ψ, 47–49
conformational change, 242
contrast agent, 10
coordination, 199
    chemistry, 13
    compound, 15
    geometry, 15, 18
    mononuclear zinc enzymes, 199
    number, 15, 19
cop operon, 135
    CopA, 120
    CopB, 120
    copY, 135
    copZ, 135
    CorA, 120
Corey, Robert, 47
corrinoid iron-sulfur protein (CfeSP), 262, 264
cysteine desulfurase (Nfs1), 35
copper, 1–3, 30, 35, 120, 124, 212,
    241–242, 253
    accumulation, 301
    age, 241
    centres, 39, 212, 221, 249
    chaperones, 27, 35, 112, 135, 140, 142
    clusters, 39, 252
    deficiency, 142, 149, 253
    dinuclear Type 3 copper proteins, 245
    electron-transport, 242
    enzymes, 251

genes *hCTR1* and *hCTR2*, 128
homeostasis, 135
in fungi and plants, 142
in mammals, 148
in iron metabolism, 253
insertion, 34–35
Mac1, 142
metabolism, 242
overload, 149
transport, 35, 303
transport and storage in fungi and plants, 139
transport and storage in mammals, 148
uptake, 124, 135
copper dioxygen complexes, 244
copper transport proteins (CTR), 306
coenzyme Q: cytochrome c oxidoreductase, 99
Corallina pilulifera, 292
coronary artery disease, 339
corrin, 17, 263
CotA laccase, 247
coupling, 82
covalency, 23
Creuzfeldt–Jakob disease, 303
crystal field stabilization, 107
crystal field theory 19
cupredoxins, 252
cyanide, 27, 29
cyanobacteria, 135, 241
cysteine, 27, 228
cysteine desulfurase, 35
cysteine residue, 46
cystic fibrosis transmembrane conductance regulator, 7
cystic fibrosis, 7
cystine, 46
cytochrome c oxidase (CcO), 86, 220, 241, 248
cytochrome P-450s, 220
cytochromes, 1, 222
cytokine, 314
cytosine, 55, 214

D

Desulfomicrobium norvegium, 260
decarboxylation, 85, 97, 231
DEDD family, 178
dehydratases, 228
dehydration, 88, 97

dehydrogenases, 53
alcohol, 78, 197, 202
CO dehydrogenase (CODH), 259
α-ketoacid, 79
dehydrogenation, 202
delocalization, 23
dementia, 351
denitrification, 252
denticity, 16
deoxyadenosine, 55
deoxyhaemoglobin, 110, 219
deoxynucleotides, 214
deoxyribose, 55
dephosphorylation, 158, 171, 299
depression, 340
deprotonation, 17, 202
desaturation, 231
Desferal, 17
desferrioxamine B, 39
desferrioxamine, 17
Desulfomicrobium norvegium, 260
Desulfovibrio desulfuricans, 38
detoxification of cadmium, 349
detoxification, 268
diacylglycerol (DG), 192–193
diabetes, 294
diatoms, 6
diazene (N_2H_2), 289
diazotrophs, 36
differentiation, 349
digitalis, 158
dihydrouridine, 69
dihydroxyacetone ketose, 59
dihydroxyacetone, 59
dihydroxyacetone-phosphate, 88
dihydroxybenzoate, 120
dihydroxybenzoylserine, 120
2,3-dihydroxyphenyl (DHBP), 233
dimethylsulfoxide reductase, 284
diphtheria toxin regulator (DtxR) protein, 132
dipolar charge, 20
disulfide, 36
bridges, 46
formation, 46
isomerization, 36
dimethylsulfoxide (DMSO), 284
reductases, 281–282
divalent cation transporter (DMT1), 126

DNA, 43, 56, 175
   G-quadruplex, 161
   G-quartets, 161
   groove (major, minor), 56
   ligase, 69
   oxidative damage, 350
   polymerase, 67, 178
   recombinant DNA technology, 75
   repair, 347
   synthesis, 5
dopamine, 307
dopamine β-monooxygenase, 301
Doppler shift, 109
downstream regulatory element antagonistic
   modulator (DREAM), 298
Dps protein, 131, 329
Drosophila, 153, 185
dynorphins, 298
dystonia, 317

E

elastin, 46
electrochemical gradient, 151, 153
electrochemical proton gradient, 221
electron transfer, 78
   in intermediary metabolism, 78
electron transport proteins, 222
electron, 1, 13–14, 18, 22, 106–107
   1-bonding C-bonding, 14
   d orbitals, 22
   in the electron-transfer pathways, 1
   in the respiratory chain of mitochondria, 1, 99
   orbitals, 14
   spin, 107
   unpaired, 10, 106
electronegativity, 14
electronic configuration, 22
electronic relaxation, 10
electron–transport chain, 99
electrostatic forces, 14
electrostatic interactions, 23
elongation factor G, 74
enamel, 333, 336
endocytosis, 127, 143, 305
endonucleases, 175, 178

endoplasmic reticulum (ER), 184
endosome, 127
energy splitting, 21
energy transduction, 65
enolase, 5, 91, 167, 173
enolate, 83, 173
entatic state, 243
enterobactin, 40, 41, 214
Enterococcus hirae, 120, 135
enthalpy, 17
entropy, 17
Enzymes, 5, 53
   α-ketoglutarate-dependent, 233
   Dinuclear non-haem* iron, 235
   inuclear zinc sites, 205
   iron–sulfur, 228
   Magnesium-dependent, 166
   Mononuclear non-haem* iron, 231
   mono-zinc, 198
   Of nucleic acid metabolism, 175
enzyme–substrate complex, 232
epileptic seizures, 300
epimers, 60
EPR, 32, 105, 109
   and Fe-S cluster, 32
erythrocytes, 199
   carbonic anhydrases, 199
erythropoiesis, 145
Eschericia coli, 40, 119
ESEEM, 109
ethanol, 97
ethanolamine, choline, 65
ethylenediaminetetraacetic acid (EDTA), 350
esterases, 44
eukaryotes, 34
excitotoxicity, 314
exons, 69
   ligation, 176
exonuclease activity, 177
exopeptidases, 200
'eye of the tiger', 344

F

facilitated diffusion, 151
FADH$_2$, 86

* or heme.

fatty acids, 92–94
fermentation, 97, 264
ferredoxin reductase Arh1, 35
ferredoxins (Fd), 227
    Yah1, 35
ferric citrate, 119, 131, 214
ferric enterobactin, 41
ferric oxyhydroxide, 327
ferric reductases, 122, 126, 138, 142
ferric reductase oxidases (FRO2), 123, 139
ferric-superoxide complexes, 220
ferrichrome, 41, 120, 122
ferrihydrite, 326–329
ferrioxamine, 120
ferritin, 107, 131, 144–145
    cores, 329
    ferroxidase, 326
    mammalian, 322
    H-subunits, 145
    L-subunits, 145
    synthesis, 311
ferrochelatase, 30
ferromagnetic coupling, 107
ferroportin (IREG), 126, 145
ferroportin disease, 147
ferroxidase, 122, 137, 303
    site, 322
Fet3, 122, 138
    oxidase, 140, 254, 137
Fet4, 122, 138, 143
    permeases, 125
fibroin, 51
fit1, FIT1, 139
flavin, 80
    flavin adenine dinucleotide FAD, 79, 283
    flavin adenine nucleotide, 228
    flavin mononucleotide, FMN, 79
flavodoxin (Fld), 266
fluorescence, 180
fluoride, 5, 167
fluorine, 3
fluorodeoxythymidylate, 5
folic acid, 86
formate dehydrogenase, 282
frataxin, 35
Friedreich's ataxia (FRDA), 35, 307, 319
fructose-1, 6-bisphosphate aldolase, 197
fructose-6-phosphate, 167
fumarase, 92, 133
fumarate, 228

G

gadolinium, 11, 339, 345
galactose, 59
galactose oxidase, 244
β-galactosidase, 345
gallium, 9
gamma rays, 109
gene, 66, 133
    for iron acquisition, 133
    repression, 133
gene family SLC9, 159
gene silencer, 298
gene splicing, 69
genome, 66
gephyrin, 36
globin fold, 54, 286
gluconeogenesis, 89
α-D-glucopyranose, 61
glucose, 59, 167, 88
glucose isomerase, 268
glutamate, 28
glutathione (GSH), 148
glycation, 308
glyceraldehyde, 59
glyceraldehyde-3-phosphate dehydrogenase, 88
glyceraldehyde-3-phosphate, 88
glycerol, 86
glycerophospholipids, 65
glycine, 43
glycogen phosphorylase, 193
glycogen synthesis, 167
glycogen, 59, 62
glycogenolysis, 167
glycolipids, 65
glycolysis pathway, 53, 86, 166
glycolysis, 77, 167
glycosylation, 311
glycyltyrosine, 202
glyoxidation, 309
glyoxylate, 244
goethite, 329
gold, 11
Golgi apparatus, 184
GPCR, 349
Gram-negative bacteria, 118, 132
Gram-positive bacteria, 132
guanine, 55, 214
guanosinediphosphate (GDP), 92, 289
guanosinetriphosphate (GTP), 92, 289

H

Haber–Weiss reaction, 213
haem* enzymes, 220
haem*, 17, 30
haem*-*a*3/CuB site, 249
haem*-copper oxidases, 220
haemerythrins, 236
haemochromatosis, 146
haemocyanin, 245
haemoglobins, 47, 217
haemojuvelin (HJV), 147
Haemophilus influenzae, 118
haemophores, 118
haemoproteins, 29, 114, 217
    cytochrome c, 29
    Oxygen transport, 217
haemosiderins, 329
Hallervorden–Spatz syndrome, 344
haloacid dehalogenase (HAD), 171
Haloperoxidases, 291
HAMP, 147
'hatter's shakes', 11
Haversian canal, 333
helical structures, 49
    α-helices, 47–49, 51
    3$_{10}$ helix, 49, 57
    pyrrole, 32
helicases, 68
helium, 3
helix-loop-helix EF-hand motif, 193
heme (see haem)
hepatocytes, 145
hepcidin, 145, 147
hephaestin, 122, 127
heterodimer, 36, 304
hexokinase, 80, 167
HFE, 146–147
high-mobility group (HMG), 343
histidine, 16, 27, 218
HNE, 308
holochromodulin, 294
HOLO-TF, diferric transferrin, 128
homeostasis, 120, 148, 184
homocitrate, 38, 288
homolysis, 265
Hoogsteen base pairing, 57
Hoogsteen hydrogen bonds, 161

HuHF, 324
huntingtin, 319
Huntington's disease, 309, 317
hybridization, 19
    geometry of, 19
hydration sphere, 165
hydrazine (N$_2$H$_4$), 289
hydroformylation (OXO), 268
hydrogen bonding, 18
hydrogen peroxide, 241, 308
hydrogen, 2–3
hydrogenases, 259
    [Fe] hydrogenases, 260
hydrogenation, 202
hydrolases, 197
hydrolysis, 171
    of ADP, 80
    of AMP, 80
    of ATP, 80, 88, 157
    of GTP, 74
    of phosphatidylinositol, 349
    of phosphoesters, 206
hydrophobic effect, 65
hydrophobicity, 153
hydroxamates, 39, 293
hydroxide, 22, 282
hydroxyapatite [Ca$_5$(PO$_4$)$_3$OH]$_2$, 191, 333–334
hydroxyl radicals, 308
hydroxylases, 233, 236
hydroxylation, 231, 236, 280
    alkanes, alkenes and aromatics, 236
    of carbon atoms, 280
    of C–H bonds, 231
    of the aromatic amino acids, 231
4-hydroxy-2-nonenal (HNE), 308
hyperpolarization, 152
hypertension, 339
hyperthermophilic archaebacteria, 285
hypoxia, 147

I

IFMT motif, 156
immunoglobulins, 52–53
indole, 233
inert gas, 3

---

* or heme.

inner coordination sphere, 15
inner membrane ABC transporter FhuCD, 120
inositol(1,4,5)-trisphosphate [Ins(1,4,5)P$_3$], 340
insulin, 294
intermediary metabolism, 77, 167
intermembrane space (IMS), 140
intracellular signalling messenger, 184
introns, 69, 176
iodide, 22
iodine, 9
iodoperoxidase, 10
ions, 5, 14
    ion exchangers, 153
    ionic bonding, 14
    ionic equilibria, 7
    ionic gradients, 153
    outer sphere, 15
    transport across membranes, 151
IREs, 146
iron, 1–3, 117, 126, 211–212
    absorption, 127
    age, 241
    anaemia, 1
    and oxygen, 212
    chaperone, 35
    chelation, 314
    deficiency, 139, 307
    deposition in ferritin, 322–323
    haem*, 126
    non-haem*, 126
    homeostasis, 133, 137, 145–146, 254, 306,
        319, 351
    metabolism, 319
    nucleation, 137
    octahedral geometry, 29
    oxidation, 324
    storage and transport, 136, 144
    uptake, 121, 144, 214
    in plants, Strategy I, 123, 139
    in plants, Strategy II, 124
    iron–citrate complex, 137
    iron-only hydrogenase, 38
iron regulatory protein (IRP), 348
iron-sulfur proteins (ISP), 1, 217, 226, 231
    cytosolic iron–sulfur protein assembly
        (CIA), 33
    Reiske, 33, 225, 227, 234
    scaffold protein complex (Isu1/2), 34

iron-cluster assembly (ICA), 33
iron–copper centre, 212
iron-responsive element (IRE), 348
iron–sulfur clusters, 6, 32, 228
    FeMo-cofactor, 37, 280
    [Fe$_2$–S$_2$], 33
    [Fe$_3$–S$_4$], 33
    [Fe$_4$–S$_4$], 228
    [Fe$_4$–S$_4$], 33
    HIPIP, 33, 227
    synthesis, 127
isocitrate dehydrogenase, 191
isocitrate, 92
isomer shift, 110
isomerases, 59, 197, 264
    B$_{12}$-dependent isomerases, 264
    cis/trans prolyl isomerase, 44
    phosphomannose isomerase, 197
    xylose isomerase, 269
isomerization, 84
isoprenoids, 93
isotope, 109
Isu1/Isu2, 35, 138
Itai-itai disease, 346

J

Japanese lacquer, 247

K

KcsA, 154
α-ketoglutarate, 92–93, 231
    dehydrogenase, 191
ketose, 59
kinases, 53
Klebsiella, 286
Krebs cycle, 92

L

laccase, 212, 243, 247
β-lactamases, 205
β-lactams, 205
lactate, 97

---

* or heme.

Lactobacillii, 214, 265, 273
lactoferrin, 118, 131, 214
lactonization, 174
lagging strand, 68
lanthanides, 10, 111, 345
lead, 1, 11
leading strand, 68
leghaemoglobin, 286
leucine, 159, 268
    aminopeptidases, 268
Lewis base, 15
Lewy bodies, 311
ligand, 15, 24, 109
    bidentate, 16
    biological 17, 27
    carboxylate, 16
    environments, 19
    exchange, 31, 241, 350
    field, 22
    field theory, 23
    groups, 27
    metal ions, 27
    monodentate, 16
    multi-dentate, 16
    phosphate, 16
    pyranopterindithiolate, 282
    tridentate, 16
ligand field
    octahedral, 21
    splitting, 21
    tetrahedral, 21
ligases, 197
lipids, 64
    metabolism, 294
    peroxidation, 308
lipoate, 79
lipoic acid, 86
lithium, 1, 3
    metallotherapeutics, 340
lone pair, 23, 18, 31
    s-bonds, 23
    porphyrin, 31
Lorentzian lines, 110
Lou Gehrig's disease, 250, 309
lyases, 197
lysine, 308
lysozyme, 200
lysyl hydroxylase, 334
lysyl oxidase, 244

M

macrophage, 145
magnesium, 2, 5
    blockade, 314
    enzymes, 167
    metabolism, 165
magnetic circular dichroism, 113
magnetic moment, 107
magnetic resonance imaging (MRI), 1, 339
magnetite, 329
malate, 92
malondialdehyde (MDA), 308
malonyl CoA, 94
maltose, 60
mandelate racemase (MR), 173–174
manganese, 1, 8, 271–272
manic depression, 1, 3
MAT, 97
matrix dehydrogenases, 191
matrix metalloproteinases, 198
mechanism, 200–201, 207, 216, 247, 265,
    292, 328
    carbonic anhydrase, 200
    channelling, 122
    Boyer, 100
    'Venus fly trap', 29
    for AdoCbl-dependent isomerases, 265
    for ATP synthase, 102
    for the hydrolysis of arginine, 275
    for the T to R transition in haemoglobin,
        219
    for the vanadium chloroperoxidase, 292
    for xanthine oxidase, 283
    glyceraldehydes-3- phosphate
        dehydrogenase, 90
    of alkaline phosphatase, 207
    of cisplatin, 342
    of proteinsynthesis, 74
    of ribonucleotide reductases, 216
    of urease, 259
    porphyrin metallation, 31
    'zinc-hydroxide' reaction, 201
melting, 56
membrane, 5, 151–152
    depolarization, 151
    resting potential, 152
    topography, 186
memory, 298

Menkes disease, 121, 149, 301
mercury, 11
metals, 297, 339, 346
   assimilation in bacteria, 117
   assimilation in mammals, 126, 144
   assimilation in plants and fungi, 121, 136
   binding region (MBR), 3, 251
   centres, 30
   incorporation in Fe–S clusters and
      metalloporphyrins, 36
   pathways, 117
metal ion, 3, 15, 131
   function, 3
   homeostasis, 131
   Lewis acid, 15
   Lewis base, 15
   ligand, 3
   ligand binding, 3
   mobility, 3
   storage, 131
   transport, 131
metal orbitals, 23
   p-bonds, 23
metal-free hydrogenases, 260
metallochaperone, 34, 135
metalloproteins, 27
metallothioneins (MT), 135, 148, 300
methane, 30, 251
methane monooxygenase hydroxylase,
   237, 251
Methanobacterium thermoautotrophicum, 263
Methanococcus, 173
methanogenic bacteria, 30
methionine, 29, 116, 123, 266
N-methyl-D-aspartate (NMDA), 185, 298, 349
   receptor, 185, 303
methyl coenzyme M reductase (Mcr), 263
methylmalonylCoA mutase, 264
Methylococcus capsulatus, 251
methyltetrahydrofolate, 267
methyltetrahydrofolate (CH_3H_4folate), 262
methyltransferase (MeTr), 262, 266
Michael addition reactions, 308
Michaelis complex, 202
microglia, 306
mineral phase, 334
mitochondria, 34, 188
   ISC assembly, 34
mitochondrial iron transporter (Mrs4), 138

mitochondrial respiratory chain, 92
mitogen-activated protein kinase (MAPK),
   161, 170, 298
MKKKs, 170
MMOH, 237
MMOR, 236
model,
   for reduction and accumulation of
      vanadium, 293
   for the active transport of Na^+ and K^+, 158
   for the conversion PrP^c to PrP^{Sc}, 304
   for the metallobiology of Ab in
      Alzheimer's disease, 318
   for the regulation of iron deficiency, 140
   Iron utilization, 134
   of copper homeostasis, 135, 141
   of proteins linked with Parkinson's
      disease, 312
   of the sodium-calcium exchanger
      NCX1, 188
   seeding (or nucleation) model, 304
MoFe-cofactor, 288
MoFe-nitrogenases, 286
molecular cloning, 75
molecule, 14, 107
   diamagnetic, 107
   paramagnetic, 107
   shapes, 19
molluscs, 330
   shell formation, 332
molybdenum Mo, 9, 279
   Mo cofactor, 36
   enzyme families, 282–285
   hydroxylases, 280
   molybdenum pyranopterindithiolate
      cofactor (MoCo), 280
monooxygenases, 212, 280
   bacterial multi-component
      monooxygenases (BMMs), 236
monosaccharides, 59
mood disorder diseases, 340
morphogenesis, 332
Mössbauer, Rudolf , 109
motor neurone disease 250, 309
MRI contrast agents, 339
mRNA, 139
Msc2–Zrg17 complex, 142
MTF1, 150
MthK, 154

MTP1, 143
muconate, 174
muconate lactonizing enzyme (MLE), 173
mugeneic acid, 123
multi-copper oxidases, 247
multiple anomalous diffraction (MAD), 116
multiple isomorphous replacement (MIR), 116
MurrI, 149
mutations, 154, 300
Mycobacterium, 132
myelination, 307
myoglobin, 53, 200, 217
myohaemerythrin, 53–54

N

NAADP, 190
nacre, 330
NADH, 35
NADH-coenzyme Q (CoQ)
    oxidoreductase, 99
NADPH, 77
naphthalene dioxygenases, 233
Nautilus repertus, 331
necrosis, 350
nedaplatin, 342
Neisseria gonorrhoeae, 118
Neisseria meningitidis, 118
neon, 5
neurodegenerative diseases, 297–319
    calcium 297–299
neuroferritinopathy, 344
neurofibrillary tangles, NFTs, 316
neuromelanin, 306, 311
neuromelanin–iron complex, 312
neuronal survival, 298
neuronal synapses, 298
neurones, 298
neurotoxicity, 299
neurotoxin, 5
neurotransmission, 152
neurotransmitters, 297, 299
    synthesis, 301
nuclear factor of activated T-cells (NFAT), 298
neurofibrillary tangles (NFT), 313
nickel Ni
    enzymes, 257
    Ni–Fe–S proteins, 259

nicotianamine (NA), 123, 137
nicotinamide, 86, 269
nitrogen N, 2–3
    metabolism, 258
    nitric oxide, 3, 308
    nitrile hydratase, 268–269
    nitrite reductase, 243, 251
    nitrite, 251
    nitrogen cycle, 39, 252
    nitrous oxide, 39, 251
    oxide, 6
nitrogenase, 32, 36, 286
    bacterial, 280
nitrous oxide reductase, 39, 252
NML45, 149
noble gas configuration, 13
non-corrin cobalt, 268
    enzymes, 257, 268
norepinephrine (noradrenaline), 307
nuclease, 146, 175, 206
nucleation, 327
nucleic acid, 5, 43, 55
    metabolism, 166, 175
nucleoside triphosphate (NTP), 68
nucleoside, 55
nucleotide, 56
nucleotide polymerization, 175
nucleus, 13

O

Okazaki fragment, 69, 175
oleic acid, 64
oligodendrocytes, 306
oligosaccharides, 65
Oligotropha carboxidovorans, 282
orbital, 13–14, 19, 24–25
    atomic, 23
    d-, 13
    degenerate, 20
    e_g^*, 24
    e_g, 24
    f-, 13
    hybridization, 19
    p- , 13
    molecular orbital, 23
    t_2, 25
    s orbitals, 13

osmium, 10
osmotic balance, 152
osteoblasts, 334
osteocalcin, 334
osteonectin, 334
osteons, 333
ovalbumin, 70
oxaloacetate, 93–94
oxidases, 244
oxidation state, 15
oxidative stress, 305
oxidoreductases, 197
oxygen, 2–3
    activation, 231
    activators, 220–222
    generation, 277
    paradox, 8, 213
oxygenases, 244
oxygenation, 218, 222
oxyhaemoglobin, 219
oxymyoglobin, 220

P

P_{450}, 114
P680, 276
pain signalling, 298
palmitate, 97
palmitoyl, 97
pantothenate, 86
paramagnetism, 10
parenteral nutrition, 351
parkin, 311
Parkinson's disease (PD), 307
passive transport, 151
Pauling, Linus, 47
penicillin, 198
pentose phosphate pathway, 92
pepsin, 115
peptide, 43
peptidyl, 73
peptidyl transferase, 75
peptidylglycine a-hydroxylating
    monooxygenase, 244
Periodic Table, 4
periplasmic transporter FhuB, 120
periplasmic-binding proteins 119, 121
peroxidases, 220

peroxidative degradation, 308
peroxide scavenger, 236
peroxodiferric intermediate, 326
peroxynitrite, 308
phenylalanine hydroxylase, 233–234
phosphatases, 170, 207
phosphate, 5, 29
phosphatidyl ethanolamine, 65
phosphatidylcholine, 206
phosphatidylserine, 345
phosphine, 285
phosphoenolpyruvate, 81, 91, 93
phosphoenolpyruvate carboxykinase
    (PEPCK), 93
phosphofructokinase, 88, 167
phosphoglucomutases, 171
phosphoglycerate (PGA), 77, 90, 173
phosphoglycerate kinase, 167
phosphoinositide cascade, 191
phosphoinositol cascade, 340
phosphoinositol metabolism, 341
phospholambin, 188
phospholipase C, 192, 206, 349
phosphorus, 2, 6
phosphoryl, 5, 170
    transfer, 88, 90, 97, 167, 176
phosphorylation, 87–88, 97, 294
phosphoserine phosphatase, 172–173
photons, 178
photoreception, 178
photoreceptors, 165
photoreduction, 137
photosynthesis, 5, 178, 276
photosystem II (PSII), 1, 276
phthalocyanin, 116
phycobilin, 180
phytoferritin, 137, 139
phytosiderophores (PS), 123
plasma membrane, 65
plastids, 137
plastocyanin, 1, 243
platinum, 1, 10, 341–343
polyamines, 274
polyglutamine (polyQ), 317
    polyQ diseases, 317
polymerases, 175
    PNA αvδ σ factors, 69
polymers, 59
polypeptide, 43

polypeptide fold, 229
polysaccharides, 59, 62
pore loop, 154
porins, 118
porphobilinogen, 126
porphobilinogen synthase, 11, 204
porphyrin, 17, 93, 112, 179
    saddled structure, 30
post-transcriptional processing, 69
post-transcriptional regulation, 146
post-translational modification, 46
Potassium, 2, 151
preventive medicine, 339
primase, 68
prismatic layer, 332
proclavaminate (PCV), 233
procollagen, 334
profibrils, 311
prolidase, 268–269
proline, 43, 308
proline hydroxylase, 231
promoters, 69
propionylCoA, 265
proteins 43, 242
    acyl carrier protein (ACP), 94
    active transport, 153
    blue copper, 242
    chaperone proteins, 139
    elongation factor G, 74
    exit tunnel, 75
    ferric-uptake regulator (Fur), 132
    folding, 350
    membrane, 151
    motifs (β-α-β and α-α), 51
    prepro proteins, 147
    α/β proteins, 53
    structure, 47, 51
    synthesis, 74
    tau, 313
proteolysis, 9–10
proteome, 75
proton, 10
proton gradient, 3, 86
proton transfer, 44, 221
protoporphyrin IX, 179
proximal histidine, 218
Pseudomonas aeruginosa, 118, 133
Pseudomonas putida, 173

pseudouridine, 69
pumps, 151
    ion, 153
    proton, 1, 220
    SERCA, 188
pyridoxal phosphate, 86
pyridoxine, 86
Pyrococcus furiosus, 269
pyrophosphate, 68
pyruvate, 86
pyruvate carboxylase, 197
pyruvate dehydrogenase, 92, 191
pyruvate formate lyase, 215
pyruvate kinase, 161, 167
Pyura pachydermatina, 332

Q

quadrupole splitting, 110
quinol-fumarate reductase, 228–229
quinone, 181, 228

R

Ralstonia metallidurans, 135
Ramachandran plot, 48
reactions,
    elimination, 83
    group-transfer, 82
    hexokinase, 88
    nucleophilic displacements, 82
    redox, 78, 82
reaction centre antenna, 180
reactive nitrogen species (RNS), 308
reactive oxygen species (ROS), 308, 347
redox potential, 19
relative relaxation times, 344
renal disease, 351
replication, 66
replication fork, 68
resting state, 246
restriction endonucleases, 175
rhamnulose kinase, 168
rheumatoid arthritis, 11
Rhizobium, 286
Rhodobacter sphaeroides, 221

Rhodopseudomonas viridis, 180
Rhodospirillum rubrum, 261
riboflavin, 79, 86
ribonucleotide, 176, 216
ribonucleotide reductase (RNR), 214, 237–238
ribose, 55, 59
    cyclic ADP ribose (cADPr), 190
ribosome, 73, 161
ribosomal sites, 73
    aminoacyl, 73
    exit, 74
ribozyme, 5, 75, 176
Rieske dioxygenases, 231, 233
RNA, 43, 57, 59, 175
    RyhB, 133
    maturation, 69
    polymerase, 68, 197
    primer RNA, 69
    ribosomal RNA, 75
    splicing, 176
    5S rRNA, 208
rubredoxins (Rd), 32, 227
rubrerythrins, 236
rusticyanin, 243
ryanodine receptor, 189, 349

S

Saccharomyces cerevisiae, 121
S-adenosylmethionine (AdoMet), 266
salt bridge, 219
sarcoplasmic reticulum, 349
saturnism, 11, 204
sclerocytes, 332
scrapie, 303
secondary active transporters, 159
secondary structures, 51
secretases, 314–315
selectivity filter, 154, 156
selenium, 9
selenomethionine, 116
self-splicing, 176
sequence specificity, 175
serine, 167
serine proteases, 44
serotonin, 307

Serratia marcescens, 118
Shaker mutation, 153
β-sheet, 47, 53, 316
    parallel, 49
    anti-parallel, 49, 51
siderophore, 27, 39–40, 118
    and hydroxamate, 40
signal transduction, 152, 297
    cascades, 299
signalling, 167
'smart' sensor probes, 345
Smf1 permeases, 125
Sodium, 2, 5, 151
    in ionic gradients, 5
    in osmotic regulation, 5
sodium-potassium ATPase, 5, 157
    channels, 153
    gradients, 158, 175
sodium-calcium exchanger, 151, 187–188
sodium-hydrogen exchanger, 151, 160
solubility, 6
Soret band, 113
solute carrier (SLC), 125, 159–160, 187
spectrochemical series, 22
spectroscopy, 105, 111
    ABS, 106
    CD, 106
    correlation (COSY), 111
    ENDOR, 106, 289
    EXAFS, 106
    MCD, 106
    Mössbauer spectroscopy, 105
    multi-dimensional NMR, 111
    nuclear Overhauser effect (NOESY), 111
    Resonance Raman, 106
    Vibrational IR, 106
    Vibrational Raman, 106
    X-ray diffraction, 105
spicules, 331
spin, 13, 22
spinal muscular atrophy (SMA), 300
spliceosome, 69
standard free-energy change $\Delta G°$, 98
standard redox potential, 198
staphyloferrin A, 40
starch, 62
stearic acid, 64

1-stearoyl-2-oleoyl-3-phosphatidylcholine, 65
stellacyanin, 243
β-strands, 48, 49
    parallel, 53
Streptomyces, 132
Strongylocentrotus purpuratus, 333
succinate dehydrogenase, 92, 99, 133
succinylbenzoate synthase, 174
sugar dehydratase, 174
sugar isomerization, 88
sugar-phosphate backbone, 58
sulfite oxidase, 281
sulfur, 2, 6
superconducting quantum interference device
    (SQUID), 107
superoxide, 36, 213, 250
superoxide dismutases (SOD), 3, 35,
    copper zinc superoxide dismutase SOD1,
        140, 250, 272
    iron superoxide dismutase, 133, 272
survival motor neuron (SMN), 300
synaptic dysfunction, 314
synchrotron, 115
synthases
    F_1-ATP, 100
    isopenicillin, 231
    β–ketoacyl-ACP, 97
    methionine, 204
    reductase, 266
synthetase, 72
α-synuclein, 311

T

tantalum, 10
telomeres, 161
tetrahydrobiopterin (BH4), 231, 233
tetrahydrofolate, 86, 266
tetrapyrrole metallation, 30
thalassaemia, 17
thapsigargin, 188
thermolysin, 197, 200
thiamine pyrophosphate, 86
thioester bond, 97
thiohemiacetal, 90
thiolate, 228, 231
threonine, 308

threonyl-tRNA synthetase, 72
thylakoid membrane, 276
thymidylate synthase, 5
thymine, 55, 69
thyroglobin, 10
thyroid, 10
thyroxine (T_4), 9, 10
tin, 9
titanium, 8
toluene monooxygenase
    hydroxylase(ToMOH), 238
toxicity, 5, 11
trace element, 2, 6
transcription, 66, 133
transcription factors, 208, 349
transfer reactions, 5
transferases, 197
transferrin (Tf), 29, 118, 126, 131,
    144–145, 294
    iron-binding site, 145
    to cell cycle, 128, 144
    transferrrin–aluminium complex, 351
transferrin receptor (TFR), 128,
    146–147, 345
    gene, *TFR1*, 147
    transferrin-transferrin-receptor
        pathway, 351
trans-Golgi network (TGN), 301
transient receptor potential (TRP), 185
translation, 66, 112, 133
transition state, 243
translocation step, 74
transmembrane FepA protein, 119
transmembrane organization, 157, 160
    of SLC9AT, 160
    of sodium channel, 157
transport number, 165
transport proteins, 53
transporters, 122
    Arn1–Arn4, 138
    DCT1 (or Nramp2), 122
    Smf1, 122
    yellow stripe 1(YS1), 123
    yellow stripe-like (YSL1), 123
tricarboxylic acid cycle, 77, 86
tricatecholate enterobactin, 119
triglycerides, 65
triiodothyronine (T_3), 9

trinucleotide expansion, 319
triose phosphate isomerase (TIM), 52–53,
    88, 173
    TIM barrel, 53
trioses, 59
tryosine kinase, 192
tryptophan hydroxylase, 307
Tungsten, 10, 279
Tunicates (ascidians or sea-squirts), 293
tyrosinase, 244–245
tyrosines, 30
    tyrosine hydroxylase, 307
    tyrosine kinase, 294

U

ubiquitination, 311
ubiquinol, 224
ubiquinone, 79
ubiquitin carboxy-terminal hydrolase
    (UCH-L1), 311
ubiquitination, 139
ubiquitin-proteasome pathway, 350
uracil, 56, 214
urea, 86, 258
Urease, 258
uridines, 69
uroporphyrinogen, 30

V

valency, 13
valency orbitals, 14
vanadium, 8, 279
    $VOSO_4$, 293
Vibrio cholera, 119
virus coat proteins, 53
vitamins
    B_1, 86
    B_{12}, 17, 86, 118, 257
    B_2, 86
    B_6, 86
    C, 59, 86
    K, 28

W

Water Splitting, 271
Watson-Crick base-pairing, 59
Wilson's disease, 121, 149, 301
Wolinella succinogenes, 229
Wood–Ljungdahl pathway, 262

X

xanthine oxidase, 281
Xenopus, 209
XIP, 187

Y

Yersinia entercolytica, 119
Yersinia pestis, 119
Yersiniabactin, 40

Z

Zap1, 143
zinc, 2, 120, 124, 197–198, 200, 205
    absorption, 128
    binding sites in enzymes, 198
    deficiency, 128, 142, 300
    gene regulator, 197
    hydrolytic reactions, 200
    homeostasis, 125, 143, 148–149
    Lewis Acid, 197
    mononuclear enzymes, 198
    multinuclear and cocatalytic enzymes, 205
    pathway for uptake and efflux of zinc, 136
    storage, 135
    toxicity, 150
    transport and storage, 136, 142, 148
    transporter ZnT3, 313
    uptake, 133, 142
    ZIP family, 121, 123, 125, 128, 149
zinc fingers, 208
zincosomes, 142
Zur proteins, 136